Multiscale Turbulent Transport

Multiscale Turbulent Transport

Special Issue Editors

Marco Martins Afonso
Sílvio Gama

MDPI • Basel • Beijing • Wuhan • Barcelona • Belgrade

Special Issue Editors

Marco Martins Afonso
Centro de Matemática da Universidade do Porto
Portugal

Sílvio Gama
Centro de Matemática da Universidade do Porto
Portugal

Editorial Office
MDPI
St. Alban-Anlage 66
4052 Basel, Switzerland

This is a reprint of articles from the Special Issue published online in the open access journal *Fluids* (ISSN 2311-5521) from 2018 to 2019 (available at: https://www.mdpi.com/journal/fluids/special_issues/turbulent_transports).

For citation purposes, cite each article independently as indicated on the article page online and as indicated below:

LastName, A.A.; LastName, B.B.; LastName, C.C. Article Title. *Journal Name* **Year**, *Article Number*, Page Range.

ISBN 978-3-03928-212-8 (Pbk)
ISBN 978-3-03928-213-5 (PDF)

Contents

About the Special Issue Editors

Marco Martins Afonso obtained his Ph.D. in Physics at the University of Genova (Italy), and has held postdoctoral positions at the Weizmann Institute of Science (Israel), Johns Hopkins University (USA), IMFT (Toulouse, France), I3M (Montpellier, France), M2P2 (Marseille, France), and CMUP (Porto, Portugal), where he is currently an invited scientist. He was also a visiting scholar at the University of Helsinki (Finland) and at LJAD (Nice, France), and a plenary speaker at three international conferences. His interests cover several aspects of fluid mechanics and turbulence, and statistical, nonlinear, and theoretical physics, complex and dynamical systems, applied mathematics, and mechanical engineering.

Sílvio Gama is currently Associate Professor at the Department of Mathematics in the Faculty of Sciences of the University of Porto (FCUP), and a member of CMUP (Centro de Matemática—Universidade do Porto). His background is in fluid mechanics, numerical analysis, and dynamical systems. His research includes fluid turbulence, econophysics, optimization, and optimal control.

 *fluids*

Editorial

Editorial for Special Issue "Multiscale Turbulent Transport"

Marco Martins Afonso * and Sílvio M.A. Gama

Centro de Matemática (Faculdade de Ciências) da Universidade do Porto, Rua do Campo Alegre 687, 4169-007 Porto, Portugal; smgama@fc.up.pt

*   Correspondence: marcomartinsafonso@hotmail.it; Tel.: +351-220402259

Received: 11 October 2019; Accepted: 14 October 2019; Published: 18 October 2019

Turbulent transport is currently a great subject of ongoing investigation at the interface of methodologies running from theory to numerical simulations and experiments, and covering several spatio-temporal scales. Mathematical analysis, physical modelling and engineering applications represent different facets of a classical, long-standing problem still far from achieving a thorough comprehension. The goal of this Special Issue was to offer recent advances covering subjects such as multiscale analysis in turbulent transport processes, Lagrangian and Eulerian descriptions of turbulence, advection of particles and fields in turbulent flows, ideal or non-ideal turbulence (unstationary/inhomogeneous/anisotropic/compressible), turbulent flows in bio-fluid mechanics and magnetohydrodynamics, control and optimization of turbulent transport. The Special Issue was open to regular articles, review papers focused on the state-of-the-art and the progress made over the last few years, as well as new research trends.

Yamanaka et al. (2018) [1] focused on weak defect turbulence in the electroconvection of nematic liquid crystals, from the experimental point of view. Increasing the coarse-graining time, the authors showed that the type of diffusion changes from superdiffusion first to subdiffusion and then to the standard diffusive behaviour, a result reflecting the coexistence of local order and global disorder in the analysis of convective rolls.

Martins Afonso et al. (2019) [2], in the background of linear-stability analysis, passive optimal control and structural sensitivity, investigated the incompressible fluid flow past a backward-slanted step inclined at 25 degrees, which represents a portion of the well-known Ahmed's body mimicking a simplified ground-vehicle geometry of the bluff-body type (such as the rear window of a car). Since this section is responsible for the majority of the aerodynamic drag—due to the unfavourable pressure gradient caused by the recirculation bubble before the reattachment—they studied how to reduce or delay the flow separation, exploiting adjoint-state method, variational formalism, penalization scheme, orthogonalization algorithm and functional derivatives.

Locke et al. (2019) [3] theoretically discussed the B-model for turbulence intermittency, first by pointing out a couple of mistakes in the original derivation and analyzing the corrected results in terms of a set of coefficients, and then by generalizing it and comparing such an extension with experimental findings from a unidirectional oval flume and with other discrete/continuous breakage/cascade models.

Rahman and San (2019) [4] concentrated their attention on the performance of a relaxation filtering approach for Euler turbulence using a central seven-point stencil reconstruction scheme. Two-dimensional inviscid Rayleigh–Taylor instability provided the benchmark for the assessment of high-resolution numerical simulations, for both the time evolution of the flow field and its spectral resolution up to three decades of inertial range.

Araya (2019) [5] explored the performance of several turbulence models for the Reynolds-Averaged Numerical Simulation of steady compressible turbulent flows on complex geometries and unstructured grids. Validation against high-resolution Direct Numerical Simulations and previous experimental data was followed by an application to a few aerodynamic profiles, the main discriminant for the accuracy being the attached or separated character of the velocity field.

Lees and Aluie (2019) [6] underlined the role of baroclinicity (misalignment of pressure and density gradients) in the kinetic-energy budget, by means of coarse graining and scale decomposition. The role of baropycnal work was stressed by successfully comparing an analytical model developed here with on-purpose numerical simulations of compressible turbulence, highlighting the importance of this contributions in the generation of vorticity and strain as well.

Gama et al. (2019) [7] dealt with Padé approximations of the magnetohydrodynamic kinematic dynamo α-effect and of the eddy-viscosity or diffusivity tensors in different flows, employing expressions derived in the framework of the multiple-scale formalism. They performed computations in Fortran in the standard "double" (real*8) and extended "quadruple" (real*16) precision, as well as symbolic calculations in Mathematica.

Bhowmick and Iovieno (2019) [8] simulated the transient evolution of a turbulent cloud top interface, using a pseudo-spectral Direct Numerical Simulation along with Lagrangian droplet equations including collision and coalescence. Clear-air entrainment/detrainment and turbulent mixing are known to affect the droplet spectrum through supersaturation fluctuations, and the effects of collisions—driven by gravitational sedimentation—and condensation/evaporation on the radius distribution were estimated.

López Castaño et al. (2019) [9] tested two algorithms for the temporal integration of the Navier—Stokes equations in the Boussinesq approximation for turbulent flows, within the Large-Eddy Simulation methodology and the OpenFOAM implementation with hexahedral meshes. They proved the validity of an anisotropic filter function, and demonstrated the superiority of a specific method via studies of a wall-bounded channel flow and of Rayleigh—Bénard convection.

Conflicts of Interest: The authors declare no conflict of interest.

References

1. Yamanaka, K.; Narumi, T.; Hashiguchi, M.; Okabe, H.; Hara, K.; Hidaka, Y. Time-Dependent Diffusion Coefficients for Chaotic Advection due to Fluctuations of Convective Rolls. *Fluids* **2018**, *3*, 99. [CrossRef]
2. Martins Afonso, M.; Meliga, P.; Serre, E. Optimal Transient Growth in an Incompressible Flow past a Backward-Slanted Step. *Fluids* **2019**, *4*, 33. [CrossRef]
3. Locke, C.; Seuront, L.; Yamazaki, H. A Correction and Discussion on Log-Normal Intermittency *B*-Model. *Fluids* **2019**, *4*, 35. [CrossRef]
4. Rahman, S.M.; San, O. A Relaxation Filtering Approach for Two-Dimensional Rayleigh–Taylor Instability-Induced Flows. *Fluids* **2019**, *4*, 78. [CrossRef]
5. Araya, G. Turbulence Model Assessment in Compressible Flows around Complex Geometries with Unstructured Grids. *Fluids* **2019**, *4*, 81. [CrossRef]
6. Lees, A.; Aluie, H. Baropycnal Work: A Mechanism for Energy Transfer across Scales. *Fluids* **2019**, *4*, 92. [CrossRef]
7. Gama, S.M.A.; Chertovskih, R.; Zheligovsky, V. Computation of Kinematic and Magnetic α-Effect and Eddy Diffusivity Tensors by Padé Approximation. *Fluids* **2019**, *4*, 110. [CrossRef]
8. Bhowmick, T.; Iovieno, M. Direct Numerical Simulation of a Warm Cloud Top Model Interface: Impact of the Transient Mixing on Different Droplet Population. *Fluids* **2019**, *4*, 144. [CrossRef]
9. López Castaño, S.; Petronio, A.; Petris, G.; Armenio, V. Assessment of Solution Algorithms for LES of Turbulent Flows Using OpenFOAM. *Fluids* **2019**, *4*, 171. [CrossRef]

Article

Time-Dependent Diffusion Coefficients for Chaotic Advection due to Fluctuations of Convective Rolls

Kazuma Yamanaka [1], Takayuki Narumi [2], Megumi Hashiguchi [1], Hirotaka Okabe [1], Kazuhiro Hara [1] and Yoshiki Hidaka [1,*]

[1] Department of Applied Quantum Physics and Nuclear Engineering, Faculty of Engineering, Kyushu University, Fukuoka 819-0395, Japan; yamanaka@athena.ap.kyushu-u.ac.jp (K.Y.); okabe@ap.kyushu-u.ac.jp (H.O.); hara.kazuhiro.590@m.kyushu-u.ac.jp (K.H.)

[2] Graduate School of Sciences and Technology for Innovation, Yamaguchi University, Ube 755-8611, Japan; tnarumi@yamaguchi-u.ac.jp

* Correspondence: hidaka@ap.kyushu-u.ac.jp; Tel.: +81-92-802-3529

Received: 16 October 2018; Accepted: 22 November 2018; Published: 27 November 2018

Abstract: The properties of chaotic advection arising from defect turbulence, that is, weak turbulence in the electroconvection of nematic liquid crystals, were experimentally investigated. Defect turbulence is a phenomenon in which fluctuations of convective rolls arise and are globally disturbed while maintaining convective rolls locally. The time-dependent diffusion coefficient, as measured from the motion of a tagged particle driven by the turbulence, was used to clarify the dependence of the type of diffusion on coarse-graining time. The results showed that, as coarse-graining time increases, the type of diffusion changes from superdiffusion $\rightarrow$ subdiffusion $\rightarrow$ normal diffusion. The change in diffusive properties over the observed timescale reflects the coexistence of local order and global disorder in the defect turbulence.

Keywords: weak turbulence; chaotic advection; Lagrangian chaos; time-dependent diffusion coefficient; nematic electroconvection; defect turbulence

---

1. Introduction

Turbulence occurs by the cascade mechanism (Richardson cascade) because of strong nonlinearities; in ideal cases, the Kolmogorov law (K41) holds [1]. Nevertheless, disorder in fluids arises even in a weakly nonlinear regime through the interaction of a few modes with long timescales based on the slaving principle [2]. For example, in a convective system in which mean flow is hardly relaxed, the convective structure is destabilized or weakly disturbed by the interaction between the convective flow and the mean flow even in the vicinity of the convective threshold [3,4]. A similar phenomenon is observed in an electroconvective system of nematic liquid crystals under an ac electric field [5,6]. For electroconvection in nematics, even in a system in which the mean flow relaxes immediately, the orientation of the rod-shaped molecules of the nematics plays a role corresponding to the mean flow. Because the disturbance in such cases is weak, the convective structure is globally disordered while maintaining local convective rolls. In other words, the correlation length of the disorder is sufficiently longer than the size of the convection roll [7,8]. A disordered state with this property is called *weak turbulence* (To make a distinction, the K41-type turbulence is called developed turbulence). Weak turbulence that occurs in a weakly nonlinear regime near a convection threshold has been investigated as one of the experimental subjects of *spatiotemporal chaos* in spatially extended nonlinear systems [8,9]. However, weak turbulence has seldom been investigated from the viewpoint of turbulent transport. Because order and disorder of different scales coexist, clarifying the differences in properties with coarse-graining scales is required in characterizing it.

Turbulent diffusion that is the transport of mass arising from turbulence is one of the most important phenomena in turbulent transport. Much research on turbulent diffusion caused by developed turbulence have been conducted. Nonetheless, we remark that in time-varying flow, diffusion through mixing occurs even if there is no strong irregularity in the flow. Such phenomenon is called chaotic advection or chaotic mixing. In the context of chaos, it is also called Lagrangian chaos as the chaotic behavior appears within the Lagrangian viewpoint. Chaotic advection arising from weak turbulence is also expected to exhibit properties different from typical turbulent diffusion because of the developed turbulence. Much of the research on weak turbulence as spatiotemporal chaos has been pursued from the Eulerian viewpoint [7,9,10]. In contrast, the chaotic motion of particles driven by weak turbulence is related to the Lagrangian viewpoint and provides other information concerning spatiotemporal chaos.

Electroconvection in nematic liquid crystals under AC electric fields is suitable for investigation of weak turbulence because several advantages exist compared with other systems such as thermal (Rayleigh–Bénard) convection of normal fluids. The timescale of weak turbulence is much shorter than that in Rayleigh-Bénard convection. The refractive index is spatially modulated with the distortion of the nematic director due to convective shear flow. Therefore, the patterns can be optically observed by using the shadowgraph method [11]. An applied voltage used as the control parameter is easily adjustable.

In the electroconvection of nematics, the orientation of the convective roll is determined by the nematic director in the observation plane [12], and the convective flow exerts viscous torque on the director because a nematic substance exhibits viscous anisotropy [13]. These effects act to frustrate the orientation of the convective roll and the nematic director, making the convective roll structure continuously unstable. Thus, weak turbulence occurs. In the electroconvection of nematics, there are two system types of differing symmetry. One is the homeotropic system in which the director can freely rotate as a Nambu-Goldstone mode, and the other is the planar system in which the rotation of the director is limited near the initial orientation of the director. Typical weak turbulence in the homeotropic system is soft-mode turbulence in which the convective roll can also rotate freely [6]. Research aimed at elucidating the transport phenomena arising from soft-mode turbulence has been performed from both the Euler [14] and Lagrange viewpoints. In particular, for the Lagrangian viewpoint, a method called *nonthermal Brownian motion* has been used that measures the motion of isolated particles and analyses the motion from the analogy with Brownian motion [15–21]. In the planar system, *defect turbulence* (see Figure 1) occurs in which defects embedded in the convective stripe pattern nucleate by fluctuations of the convective rolls [22]. Early studies of the defect turbulence focused on the dynamics of defects embedded in the rolls pattern [22]. However, for transport phenomena arising from defect turbulence, fluctuations of the direction of convective rolls (i.e., torsion of the rolls [9]) because of the above-mentioned nonlinear interaction with the director is essential [20,23]. Defect turbulence is similar to the oscillating Rayleigh-Bénard convection by which Lagrangian chaos was studied for the first time [24,25], because it can be regarded as a chaotic oscillation of the convective rolls in a plane normal to the roll axis.

The most important property of weak turbulence is the coexistence of local order and global disorder. Here, this coexistence means there is a hierarchy in the system. This hierarchy was predicted to cause the dependence on observation scale also in the diffusion by chaotic advection. The time-dependent diffusion coefficient is suitable for extracting properties of the diffusion in systems containing such a hierarchy. The time-dependent diffusion coefficient had been already applied to nonthermal Brownian motion in soft-mode turbulence [16,18]. Aside from weak turbulence, biological systems are examples of systems with a hierarchy. The time-dependent diffusion coefficient was also used in extracting properties of biological systems [26,27]. In one study [26], thermal Brownian motion in a system consisting of compartments was simulated to reveal properties of diffusion of a protein molecule in cell membranes. Over very short time scales, a constant time-dependent diffusion coefficient means that the motion within one compartment corresponding to a cell is conventional Brownian motion (normal diffusion). A decreasing time-dependent diffusion coefficient over intermediate time scales means that confinement effect in the compartment by the boundary causes subdiffusion. Over long time scales, the time-dependent diffusion coefficient becomes constant again indicating that hopping diffusion between compartments becomes dominant and consequently normal diffusion occurs. Thus, by increasing the timescale, the type of diffusion in this system changes from normal diffusion $\rightarrow$ subdiffusion $\rightarrow$ normal diffusion.

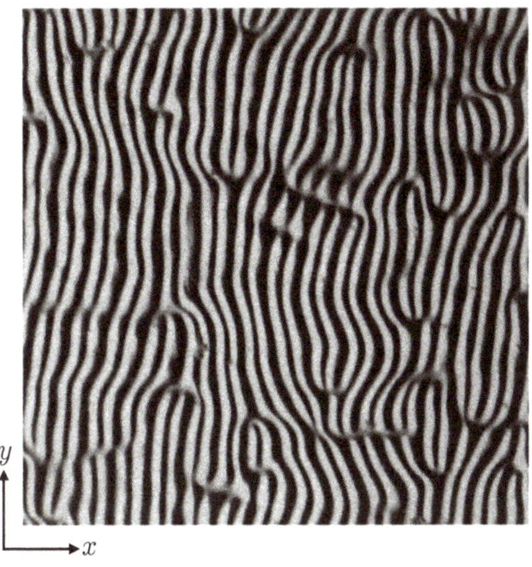

Figure 1. Snapshot of defect turbulence. White lines correspond to convective rolls.

Research on transport phenomena arising from defect turbulence also has been performed from both the Euler [23] and Lagrange [20] viewpoints. The study of chaotic advection in defect turbulence from the Lagrangian viewpoint revealed that hopping is important [20]. That is, a tagged particle is trapped in one convection roll and occasionally hops to an adjacent roll because of fluctuations of the rolls. The activation energy for this hopping was obtained by measuring hopping rates under changes in value of the control parameter. However, detailed analyses using a time-dependent diffusion coefficient have not been conducted. The objective of this paper is to clarify the properties of chaotic advection in the defect turbulence using the time-dependent diffusion coefficients.

2. Results

For applied voltages V beyond a threshold V_c, electroconvection occurs and a convective stripe pattern, referred to as normal rolls, appears in the planar nematic system. The wavevector of the stripe pattern is parallel to the x-direction because the initial nematic director is parallel to the x-direction [12]. Hereafter, we use the normalized voltage $\varepsilon = (V^2 - V_c^2)/V_c^2$ as the control parameter corresponding to the energy injected into the system. To see the difference resulting from changes in ε, measurements were taken for $\varepsilon = 0.5$ and 1.0.

Positions $\mathbf{R}(t) = (X(t), Y(t))$ of a particle that was driven by defect turbulence were measured. Because the motion in the x-direction characterizes the fluctuations of defect turbulence, only $X(t)$ was analyzed [20]. Temporal changes in the particle position $X(t)$ are shown in Figure 2. For $\varepsilon = 0.5$, the particle was trapped in a convective roll, and sometimes hopped to an adjacent roll. As fluctuations of convective rolls became intense with increases in the energy injected, the range of motion under $\varepsilon = 1.0$ becomes much larger than that under $\varepsilon = 0.5$. Furthermore, for $\varepsilon = 1.0$, the trapping time becomes much shorter, and hopping across several rolls is observed.

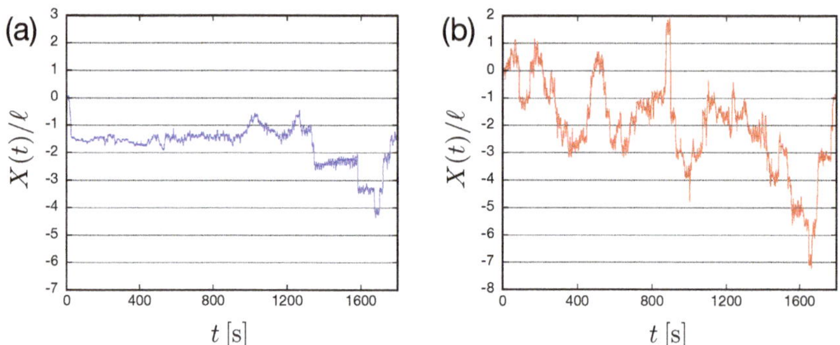

Figure 2. Particle position $X(t)$ for (**a**) $\varepsilon = 0.5$ and (**b**) $\varepsilon = 1.0$. The vertical axes are normalized using the averaged width ℓ of a convective roll, specifically, $\ell = 32.4$ m for $\varepsilon = 0.5$, and 31.5 μm for $\varepsilon = 1.0$.

To clarify the properties of the random motion of the particle, the type of diffusion should be determined. Therefore, the time-dependent diffusion coefficients, defined by

$$D(\tau) = \frac{\langle |X(t+\tau) - X(t)|^2 \rangle_t}{2\tau},$$

(1)

where $\langle \cdots \rangle_t$, denoting the average over t, were obtained from the time series of $X(t)$. The coarse-graining time τ indicates how much time interval the observation is made. $D(\tau)$ in short τ reflects the diffusion in fine structures, and $D(\tau)$ in long τ reflects the diffusion in global structures. The types of diffusion are classified by the index γ of $D(\tau) \propto \tau^\gamma$. In the conventional Brownian motion, also called normal diffusion, $\gamma = 0$; that is, $D(\tau)$ is constant. In contrast, in systems with some complexity, the diffusion behavior becomes anomalous (*anomalous diffusion*), in which $\gamma \neq 0$. Furthermore, the anomalous diffusion is subclassified as $\gamma > 0$ or $\gamma < 0$. For $\gamma > 0$, $D(\tau)$ is an increasing function and corresponds to *superdiffusion*. In this instance, ballistic motion is present [28–30]. For $\gamma < 0$, $D(\tau)$ is a decreasing function and corresponds to *subdiffusion*. In this instance, confinement effects play an important role [31–33].

Figure 3 shows the experimental results for $D(\tau)$. Since order and disorder with different scales coexist in the weak turbulence as mentioned in Introduction, $D(\tau)$ is not monotonous for τ. The first increase in $D(\tau)$ means that superdiffusion occurs at short τ (The conventional (thermal) Brownian motion also exhibits "superdiffusion" at timescales shorter than the viscous relaxation time of the fluid system. However, the behaviors over these timescales are unobservable because the viscous relaxation time corresponding to the mean free time is very short. In contrast, the present superdiffusion occurs over an observable timescale). The following decrease in $D(\tau)$ means that subdiffusion occurs over intermediate τ. Finally, $D(\tau)$ becomes almost constant implying that normal diffusion occurs over longer τ. These behaviors are seen in the two instances $\varepsilon = 0.5$ and 1.0.

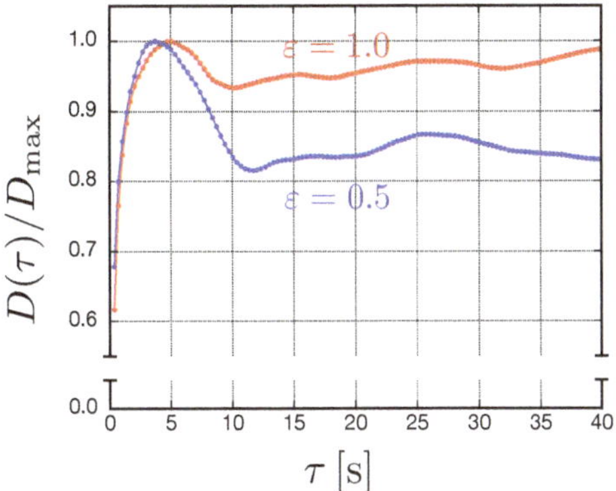

Figure 3. $D(\tau)$ obtained from the data shown in Figure 2. Blue and red lines distinguish $D(tau)$ for *varepsilon* $= 0.5$ and 1.0, respectively. The vertical axis is normalized using the values $D_{\max}$ at the peak of the curves. $D_{\max} = 3.3\ \mu m^2/s$ for $\varepsilon = 0.5$, and 18.8 $\mu m^2/s$ for $\varepsilon = 1.0$.

3. Discussion

The result shown in Figure 3 regarding the changes in diffusive properties of the chaotic advection in defect turbulence with the observed timescales may be interpreted as follows. Superdiffusion over short τ indicates that the particle is moving on the convective flow in a roll (see Figure 4a). Subdiffusion over intermediate τ is caused by the confinement effect for the particle by being trapped in a convective roll (see Figure 4b) [32]. Normal diffusion over longer τ arises because the hopping diffusion becomes dominant. Therefore, it is argued that diffusion arising from defect turbulence changes the properties depending on coarse-graining time, reflecting that the local order still exists.

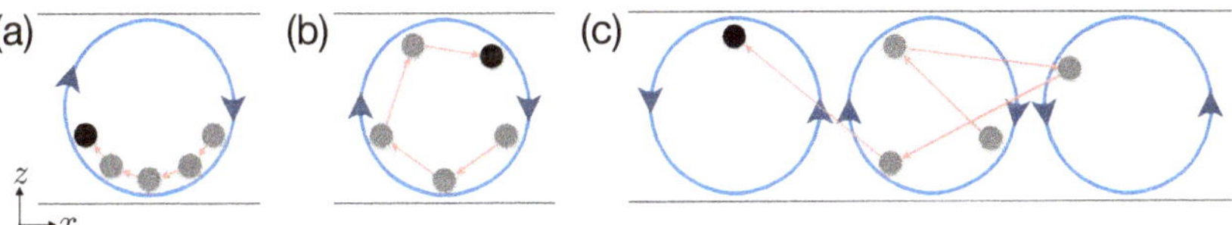

Figure 4. Schematic illustration of the motion of a tagged particle for different coarse-graining times τ. (a) Shorter τ. (b) Intermediate τ. (c) Longer τ.

The behavior of $D(\tau)$ for $\varepsilon = 1.0$ is qualitatively the same as that for $\varepsilon = 0.5$. However, the range of subdiffusion for $\varepsilon = 1.0$ becomes smaller than that for $\varepsilon = 0.5$. The conjecture is that the confinement effect by the rolls weakens as the hopping rate increases with ε [20]. Furthermore, $D(\tau)$ for $\varepsilon = 1.0$ increases slightly in the final range of normal diffusion. This behavior is thought to result from the ballistic motion caused by unidirectional hopping across several convective rolls through intense fluctuations of those rolls with larger amplitudes.

Although this dependence of the diffusive properties on the observation time scale appears also in soft-mode turbulence [16], no subdiffusion was seen. The difference in the homeotropic system in which the soft-mode turbulence appears from the present planar system in which the defect turbulence appears is that the director in the observation plane behaves as a Nambu-Goldstone mode, as mentioned in Introduction [6]. To clarify the reason subdiffusion does not occur in the soft-mode turbulence despite the existence of local convective rolls may lead to elucidating the role of the Nambu-Goldstone mode for chaotic advection.

In conclusion, the properties of chaotic advection arising from defect turbulence were investigated using the time-dependent diffusion coefficients. We found a hierarchy in the diffusive properties that depended on the observation time scale. Specifically, the type of diffusion changes from superdiffusion $\rightarrow$ subdiffusion $\rightarrow$ normal diffusion as the coarse-graining time increases. This result reflects the coexistence of the local order and the global disorder in the defect turbulence.

4. Materials and Methods

The sample cell used in the present study was prepared employing the standard method for the planar alignment of the nematic liquid crystal *p*-methoxybenzylidene-*p'*-*n*-butylaniline (MBBA) [20]. The area of the square electrodes was 1.0×1.0 cm^2, and polymer films of 50 μm thickness were used to maintain the cell gap. To obtain uniformly planar alignment of the MBBA molecules in the *x*-*y* plane parallel to the electrodes, the surfaces of the glass plates with the electrodes were rubbed by a velvet cloth in one direction (defined as the *x*-direction) after the spin-coating of a surfactant (polyvinyl alcohol). The initial planar director thus obtained was parallel to the *x*-direction. An ac voltage $V_{\mathrm{ac}}(t) = \sqrt{2}V\cos(2\pi f t)$ was applied to the sample in the *z*-direction. The present study was performed for $f = 1000$ Hz to achieve normal defect turbulence, and specifically to avoid oblique rolls or defect turbulence with abnormal rolls instability [34]. In all measurements, the temperature was kept at 30.00 ± 0.05 °C.

To observe the nonthermal Brownian motion in the *x*-*y* plane, small particles (Micropearl KBS-5065, Sekisui Chemical, Osaka, Japan) were introduced to the sample cell. The particles had a diameter spread of 6.48 ± 0.17 μm and their density of particles was 1.22×10^3 kg/m^3, which was a little higher than that of the MBBA (1.02×10^3 kg/m^3). A CMOS camera (NY-X7i Super System, Canon, Tokyo, Japan) mounted on a stereomicroscope (SMZ1000, Nikon, Tokyo, Japan) was used to capture images including the particles in a period of 1800 s at time intervals of 0.033 s. The image size was 640 pixels $\times$ 480 pixels, and the resolution was 2.83 μm/pixel. A time series of a particle's position $\mathbf{R}(t) = (X(t), Y(t))$ was obtained from the images using motion analysis software (VW-H1MA, Keyence, Osaka, Japan). Since the measuring period (1800 s) was considerably longer than the timescale of the phenomena of interest, the statistical accuracy in the analysis was sufficient. To reduce noise due to thermal Brownian motion and motion blur, coarse-graining by taking averages for every ten intervals of $\mathbf{R}(t)$ was performed. As a result, the time interval of the time series of $\mathbf{R}(t)$ is 0.33 s.

Author Contributions: Conceptualization, Y.H.; Methodology, T.N., M.H. and Y.H.; Formal Analysis, T.N.; Investigation, K.Y.; Data Curation, K.Y.; Writing—Original Draft Preparation, K.Y. and Y.H.; Writing—Review & Editing, T.N., H.O. and K.H.; Supervision, H.O. and K.H.; Funding Acquisition, T.N. and Y.H.

Funding: This work was supported by JSPS KAKENHI (Grant Nos., JP15K05799, JP25103006, and JP17K14593).

Acknowledgments: We thank Richard Haase, from Edanz Group (www.edanzediting.com/ac) for editing a draft of this manuscript.

Conflicts of Interest: The authors declare no conflict of interest.

References

1. Kolmogorov, A.N. The local structure of turbulence in incompressible viscous fluid for very large Reynolds numbers. *Dokl. Akad. Nauk SSSR* **1941** *30*, 301–305; reprinted in *Proc. R. Soc. Lond. A* **1991**, *434*, 9–13. [CrossRef]

2. Haken, H. Self-Organization. In *Synergetics*, 3rd ed.; Springer: Berlin/Heidelberg, Germany, 1983; pp. 191–227, ISBN 978-3-642-88340-8.

3. Siggia, E.D.; Zippelius, A. Pattern Selection in Rayleigh-Bénard Convection near Threshold. *Phys. Rev. Lett.* **1981**, *47*, 835–838. [CrossRef]

4. Xi, H.-W.; Li, X.-J.; Gunton, J.D. Direct Transition to Spatiotemporal Chaos in Low Prandtl Number Fluids. *Phys. Rev. Lett.* **1997**, *78*, 1046–1049. [CrossRef]

5. Kai, S.; Hayashi, K.-I.; Hidaka, Y. Pattern Forming Instability in Homeotropically Aligned Liquid Crystals. *J. Phys. Chem.* **1996**, *100*, 19007–19016. [CrossRef]

6. Hidaka, Y.; Tamura, K.; Kai, S. Soft-Mode Turbulence in Electroconvection of Nematics. *Prog. Theor. Phys. Suppl.* **2006**, *161*, 1–11. [CrossRef]

7. Cross, M.C.; Hohenberg, P.C. Pattern formation outside of equilibrium. *Rev. Mod. Phys.* **1993**, *65*, 851–1112. [CrossRef]

8. Hidaka, Y.; Oikawa, N. Chaos and Spatiotemporal Chaos in Convective Systems. *Forma* **2014**, *29*, 29–32. [CrossRef]

9. Manneville, P. *Dissipative Structures and Weak Turbulence*; Academic Press: San Diego, CA, USA, 1990; ISBN 978-0124692602.

10. Dennin, M.; Ahlers, G.; Cannell, D.S. Spatiotemporal Chaos in Electroconvection. *Science* **1996**, *272*, 388–390. [CrossRef]

11. Rasenat, S.; Hartung, G.; Winkler, B.L.; Rehberg, I. The shadowgraph method in convection experiments. *Exp. Fluids* **1989**, *7*, 412–420. [CrossRef]

12. Helfrich, W. Conduction-Induced Alignment of Nematic Liquid Crystals: Basic Model and Stability Considerations. *J. Chem. Phys.* **1969**, *51*, 4092–4105. [CrossRef]

13. Plaut, E.; Pesch, W. Extended weakly nonlinear theory of planar nematic convection. *Phys. Rev. E* **1999**, *59*, 1747–1769. [CrossRef]

14. Narumi, T.; Yoshitani, J.; Suzuki, M.; Hidaka, Y.; Nugroho, F.; Nagaya, T.; Kai, S. Memory function of turbulent fluctuations in soft-mode turbulence. *Phys. Rev. E* **2013**, *87*, 012505. [CrossRef] [PubMed]

15. Tamura, K.; Yusuf, Y.; Hidaka, Y.; Kai, S. Nonlinear Transport and Anomalous Brownian Motion in Soft-Mode Turbulence. *J. Phys. Soc. Jpn.* **2001**, *70*, 2805–2808. [CrossRef]

16. Tamura, K.; Hidaka, Y.; Yusuf, Y.; Kai, S. Anomalous diffusion and Lévy distribution of particle velocity in soft-mode turbulence in electroconvection. *Phys. A* **2002**, *306*, 157–168. [CrossRef]

17. Hidaka, Y.; Hosokawa, Y.; Oikawa, N.; Tamura, K.; Anugraha, R.; Kai, S. A nonequilibrium temperature and fluctuation theorem for soft-mode turbulence. *Phys. D* **2010**, *239*, 735–738. [CrossRef]

18. Suzuki, M.; Sueto, H.; Hosokawa, Y.; Muramoto, N.; Narumi, T.; Hidaka, Y.; Kai, S. Duality of diffusion dynamics in particle motion in soft-mode turbulence. *Phys. Rev. E* **2013**, *88*, 042147. [CrossRef] [PubMed]

19. Takahashi, K.; Kimura, Y. Dynamics of colloidal particles in electrohydrodynamic convection of nematic liquid crystal. *Phys. Rev. E* **2014**, *90*, 012502. [CrossRef] [PubMed]

20. Hidaka, Y.; Hashiguchi, M.; Oikawa, N.; Kai, S. Lagrangian chaos and particle diffusion in electroconvection of planar nematic liquid crystals. *Phys. Rev. E* **2015**, *92*, 032909. [CrossRef] [PubMed]

21. Maeda, K.; Narumi, T.; Anugraha, R.; Okabe, H.; Hara, K.; Hidaka, Y. Sub-Diffusion in Electroconvective Turbulence of Homeotropic Nematic Liquid Crystals. *J. Phys. Soc. Jpn.* **2018**, *87*, 014401. [CrossRef]

22. Kai, S.; Chizumi, N.; Kohno, M. Pattern Formation, Defect Motions and Onset of Defect Chaos in the Electrohydrodynamic Instability of Nematic Liquid Crystals. *J. Phys. Soc. Jpn.* **1989**, *58*, 3541–3554. [CrossRef]

23. Narumi, T.; Mikami, Y.; Nagaya, T.; Okabe, H.; Hara, K.; Hidaka, Y. Relaxation with long-period oscillation in defect turbulence of planar nematic liquid crystals. *Phys. Rev. E* **2016**, *94*, 042701. [CrossRef] [PubMed]

24. Solomon, T.H.; Gollub, J.P. Chaotic particle transport in time-dependent Rayleigh-Bénard convection. *Phys. Rev. A* **1988**, *38*, 6280–6286. [CrossRef]

25. Ouchi, K.; Mori, H. Anomalous Diffusion and Mixing in an Oscillating Rayleigh-Bénard Flow. *Prog. Theor. Phys.* **1992**, *88*, 467–484. [CrossRef]

26. Ritchie, K.; Shan, X.-Y.; Kondo, J.; Iwasawa, K.; Fujiwara, T.; Kusumi, A. Detection of Non-Brownian Diffusion in the Cell Membrane in Single Molecule Tracking. *Biophys. J.* **2005**, *88*, 2266–2277. [CrossRef] [PubMed]

27. Masuda, A.; Ushida, K.; Okamoto, T. Direct observation of spatiotemporal dependence of anomalous diffusion in inhomogeneous fluid by sampling-volume-controlled fluorescence correlation spectroscopy. *Phys. Rev. E* **2005**, *72*, 060101. [CrossRef] [PubMed]

28. Feng, Y.; Goree, J.; Liu, B.; Intrator, T.P.; Murillo, M.S. Superdiffusion of two-dimensional Yukawa liquids due to a perpendicular magnetic field. *Phys. Rev. E* **2014**, *90*, 013105. [CrossRef] [PubMed]

29. Budini, A.A. Memory-induced diffusive-superdiffusive transition: Ensemble and time-averaged observables. *Phys. Rev. E* **2017**, *95*, 052110. [CrossRef] [PubMed]

30. Nizkaya, T.V.; Asmolov, E.S.; Vinogradova, O.I. Advective superdiffusion in superhydrophobic microchannels. *Phys. Rev. E* **2017**, *96*, 033109. [CrossRef] [PubMed]

31. Doliwa, B.; Heuer, A. Cage Effect, Local Anisotropies, and Dynamic Heterogeneities at the Glass Transition: A Computer Study of Hard Spheres. *Phys. Rev. Lett.* **1998**, *80*, 4915–4918. [CrossRef]
32. Falkovich, G.; Gawędzki, K.; Vergassola, M. Particles and fields in fluid turbulence. *Rev. Mod. Phys.* **2001**, *73*, 913–975. [CrossRef]
33. Weeks, E.R.; Weitz, D.A. Subdiffusion and the cage effect studied near the colloidal glass transition. *Chem. Phys.* **2002**, *284*, 361–367. [CrossRef]
34. Oikawa, N.; Hidaka, Y.; Kai, S. Formation of a defect lattice in electroconvection of nematics. *Phys. Rev. E* **2004**, *70*, 066204. [CrossRef] [PubMed]

Article

Optimal Transient Growth in an Incompressible Flow past a Backward-Slanted Step

Marco Martins Afonso [1,2,*], **Philippe Meliga** [2] **and Eric Serre** [2]

[1]   Centro de Matemática (Faculdade de Ciências) da Universidade do Porto, Rua do Campo Alegre 687, 4169-007 Porto, Portugal
[2]   Aix Marseille Université, CNRS, Centrale Marseille, M2P2 UMR 7340, 13451 Marseille, France; philippe.meliga@univ-amu.fr (P.M.); eric.serre@univ-amu.fr (E.S.)
*   Correspondence: marcomartinsafonso@hotmail.it; Tel.: +351-220402259

Received: 6 February 2019; Accepted: 14 February 2019; Published: 20 February 2019

Abstract: With the aim of providing a first step in the quest for a reduction of the aerodynamic drag on the rear-end of a car, we study the phenomena of separation and reattachment of an incompressible flow by focusing on a specific aerodynamic geometry, namely a backward-slanted step at 25° of inclination. The ensuing recirculation bubble provides the basis for an analytical and numerical investigation of streamwise-streak generation, lift-up effect, and turbulent-wake and Kelvin–Helmholtz instabilities. A linear stability analysis is performed, and an optimal control problem with a steady volumic forcing is tackled by means of a variational formulation, adjoint methods, penalization schemes, and an orthogonalization algorithm. Dealing with the transient growth of spanwise-periodic perturbations, and inspired by the need of physically-realizable disturbances, we finally provide a procedure attaining a kinetic-energy maximal gain on the order of 10^6, with respect to the power introduced by the external forcing.

Keywords: linear stability analysis; separation and reattachment; optimal control; streak lift-up; turbulent-wake and Kelvin–Helmholtz instabilities; incompressibility; 3D perturbations of 2D steady base flow; structural sensitivity; recirculation bubble; 25° backward-slanted step

---

1. Introduction

The research field of hydrodynamic stability has the objective of elucidating how the structures of some specific temporal frequency and spatial scale are selected and emerge, owing to the amplification of small-magnitude perturbations. The comprehension of these effects is of huge relevance, since many flows of practical interest are dominated by genuine instability mechanisms that can be either enhanced or alleviated to improve performances. Typical expected benefits consist of the reduction of the operational cost of vehicles by decreasing skin friction or aerodynamic drag, or the extension of the operating conditions of turbomachinery by increasing the surface heat flux. The investigation is based on structural sensitivity [1,2], a theoretical concept stemming from the framework of stability analysis in laminar flows. This allows one to identify, beforehand, which regions of a given flow are most sensitive to a prescribed actuation, without the need for calculating the actual controlled flow and of resorting systematically to a trial-and-error procedure, which would represent an insurmountable bottleneck. Here, we apply this concept to determine where and how to control efficiently the turbulent-flow separation occurring at the rear-end of a ground vehicle. Such an approach can, thus, be used to obtain valuable information about the most sensitive regions for open-loop control, based on the underlying physics.

Separated flows often arise in industrial applications, resulting from an adverse pressure gradient stemming from either operating conditions or geometrical constraints (airfoil at high angles of attack, rear-end of a blunt body). They are usually associated with a loss of performance. For a ground vehicle, the flow separation taking place at its rear-end contributes to a huge increase in the drag force and, thus, also in the fuel consumption and pollutant emissions. For instance, a drag increase by 10% is expected to augment the fuel consumption by 5% at highway speeds. Moreover, flow separation causes low-frequency instabilities, which can trigger the excitation of aeroacoustic noise (sunroof cavities, side mirrors). The implementation of efficient control strategies, aimed at preventing separation itself or—when this is prohibitively costly or inevitable—at alleviating its detrimental consequences, is, therefore, a great environmental and economical issue. The dynamics are triggered by complex interactions between small-scale structures inside the shear layer, huge flow separations, and trailing vortices expanding far in the wake.

Many complex phenomena are investigated by means of linear perturbation dynamics, aimed at describing the fate of infinitesimal disturbances superimposed on a steady basic flow, and providing a rigorous mathematical foundation to investigate the control of fluid systems. Various perspectives have emerged, depending on whether the disturbance growth is characterized over large or short time intervals: on one hand, the archetype of disturbance energy amplified over asymptotically large times is the occurrence of vortex-shedding in wake flows, a behaviour called a modal instability [1]; on the other hand, the transient amplification over finite times is typically observed in channel flows, and is referred to as a non-modal instability [3–5].

For boundary-layer-like flows exhibiting a marginal separation, as occurs at the rear-end of a vehicle with small slant angle, a non-modal theoretical analysis can identify flow regions where the transient amplification of streamwise streaks (by the lift-up effect) is most sensitive to steady spanwise periodic disturbances [6]. In the experiments, such disturbances can result from either steady jets or roughness elements positioned upstream of the separation location, reproducing a parietal or a volumic forcing, respectively. Indeed, it is very well known that the global dynamics of complex flows can be modified by imposing local disturbances. Typical examples are the use of surface rugosities to delay the transition to turbulence in boundary layers, or the injection of fluid into the wake of a bluff body to alleviate unsteadiness [7]. We remind that the lift-up effect is related to the vertical mixing of large-speed fluid from higher layers to lower ones, and vice versa for small-speed fluid. A streamwise vorticity perturbation arises, and evolves into a set of streamwise stripes characterized by relevant variations of the streamwise velocity, possibly with a periodic structure in the spanwise direction: the streaks.

In practice, one can use the adjoint-state method to calculate the gradient of some objective function (energy gain over a specified time horizon, growth rate of unstable disturbances) with respect to each actuation parameter, thus making it possible to cover large parameter spaces with a limited number of computations. This capability is useful as an aid to guide the design of efficient, tractable control strategies. In the past, it has mainly been applied to related problems of vortex shedding in compressible or incompressible laminar wakes, and the agreement between the experimental results and the theoretical predictions is excellent, near the threshold of instability. Its application to flow separation for ground vehicles of practical importance constitutes a major issue, since substantial developments are necessary in order to encompass the complexity of turbulent-flow regimes, where large intervals of temporal and spatial scales strongly interact.

The paper is organized as follows. In Section 2, we describe our numerical approach and its validation. In Section 3, we specify the geometry under consideration and the main equations in play. In Section 4, we introduce the base flow we have adopted. In Section 5, we perform the linear-stability analysis and focus on the direct and adjoint perturbations. In Section 6, we analyze the control mechanisms and the associated kinetic-energy gain. Conclusions and perspectives follow, in Section 7. The Appendix A is devoted to showing some further details about boundary conditions and adjoint equations.

2. Description and Validation of Numerical Tools

We have made use of the FreeFEM++ software [8] to build a Finite-Element Method code. This tool solves the continuity and Navier–Stokes equations in their variational formulation, with prescribed boundary conditions (Dirichlet, Neumann, or mixed). We have implemented a P1 scheme for the pressure field, and a P1b scheme for the velocity field (for which we have also tested a P2 scheme, without any appreciable change in the results) [9].

We have performed two main validation tests. In both cases, the quantitative validation makes use of a software which, by scanning printed figures from scientific articles, gives the numerical values of plotted points or lines with sufficient precision.

First, we have considered the 2D open cavity from [10]. We have implemented the same exact geometry and mesh from their Figure 7, consisting of a long flat floor interrupted by a unit square excavated below it. We have performed a qualitative validation against their Figures 8c,d, and 10, for the direct and adjoint perturbations and the eigenvalues. More importantly, we have focused on their Figure 9b, reporting the generalized displacement thickness (to be defined more precisely by Equation (4)), and have found a good quantitative agreement, as shown in our Figure 1. (Note that the prefactor 2.71 reported in the caption of their Figure 9b in [10] was wrong, the correct coefficient from the theory and in the plot is actually 1.72 [11].)

Second, we have considered the 3D backward-facing step (inclined at 90°) from [12]. We have implemented the same exact geometry from their Figure 1, and (because of the different numerical scheme) an approximate mesh from their Figure 2. We have performed a qualitative validation against their Figures 3, 4, and 7 for the base flow, the skin friction, and the eigenvalues. More importantly, we have focused on their Figure 5, reporting the separation/reattachment points, and have found a good quantitative agreement, as shown in our Figure 2.

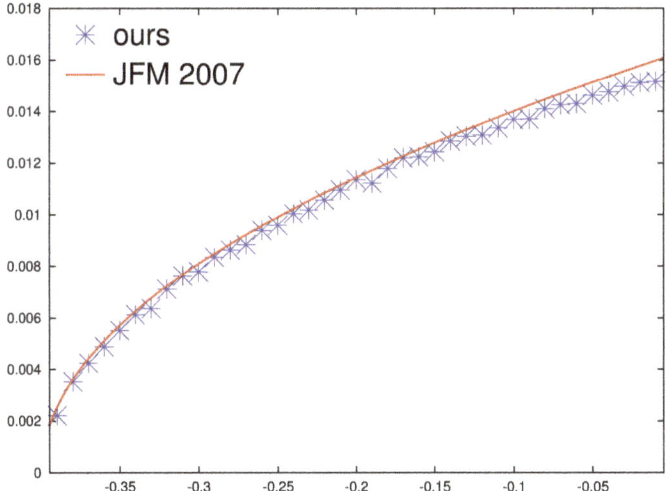

Figure 1. Validation against Figure 9b of reference [10]. Plotted in ordinate is the generalized displacement thickness (Equation (4)) versus the streamwise coordinate (in abscissa).

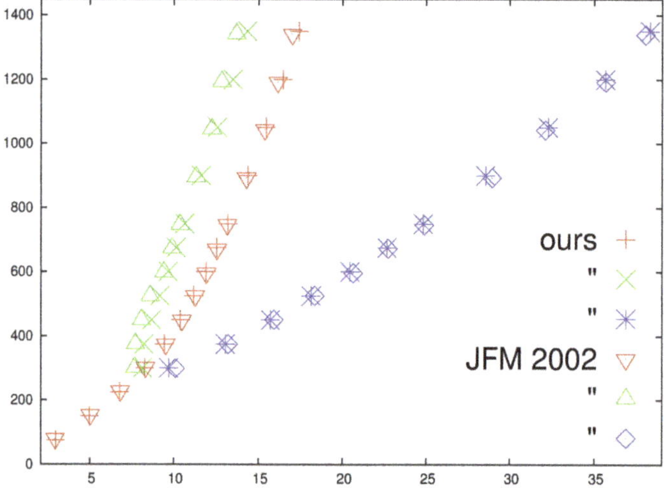

Figure 2. Validation against Figure 5 of reference [12]. As functions of the Reynolds number (in ordinates), plotted are the abscissae of the reattachment point on the lower wall for all Re (red plus signs and inverted triangles), and—only for Re $\geq$ 300—of the separation (green times signs and standard triangles) and reattachment (blue stars and diamonds) points on the upper wall.

3. Geometry and Equations

We have focused on a geometry issued from a standard ERCOFTAC benchmark, namely a backward-slanted step with a slope of 25° with respect to the horizontal surface. This configuration, plotted in Figure 3, represents a simplification of the rear end of a car and of a portion of Ahmed's body [13]. The x and y axes correspond to the streamwise and wall-normal components, respectively, with the origin placed at the leftmost/uppermost point of the sloping zone. This picture is assumed to be invariant in the spanwise z direction, which implies that the base flow is assumed to be two-dimensional, while the perturbations can present a three-dimensional character. (Different geometries investigated through this scheme can be found in, e.g., [14–16]).

The linear density of meshing points for the automatic triangulation process has been assumed to be 4 on segments UD, DC, and CV; 14 on segments WU, UV, and VZ; and 24 on segments EW, WX, XY, YZ, ZB, BA, AO, OI, and IE—a finer grid is obviously required close to the lower physical boundary. The resulting number of triangular elements employed in the numerical simulations is about 5×10^5.

To non-dimensionalize, we have assumed as reference units the vertical projection of the step and the uniform inlet speed. As the two quantities have unitary values, the non-dimensionalized kinetic viscosity ν equals the inverse of the Reynolds number, based on the step height.

We have imposed standard inlet and outlet conditions on the left and right boundaries, respectively, and a free-slip condition on the upper boundary. At the lower boundary (a physical wall) we have imposed the no-slip condition, except for the beginning part, EI, where a free-slip condition has been used, in order to allow for the evolution of a boundary-layer profile [17]. (See also Appendix A).

Tests have also been made, in which we have varied the streamwise length of the domain (both upstream and downstream), its normal height, and the length of the segment EI for the imposition of the boundary condition. The chosen reference geometry falls in a range where convergence has already taken place. The case in which the upper boundary is a physical wall has also been briefly investigated, both in the case of the standard segment DC, and in a modified domain where the latter has a curved S-shaped profile to simulate a streamline and to study the influence of confinement [18,19]. The effect of the resolution has been tested as well, by implementing a discretization of up to almost 7×10^5 triangles, without appreciable changes.

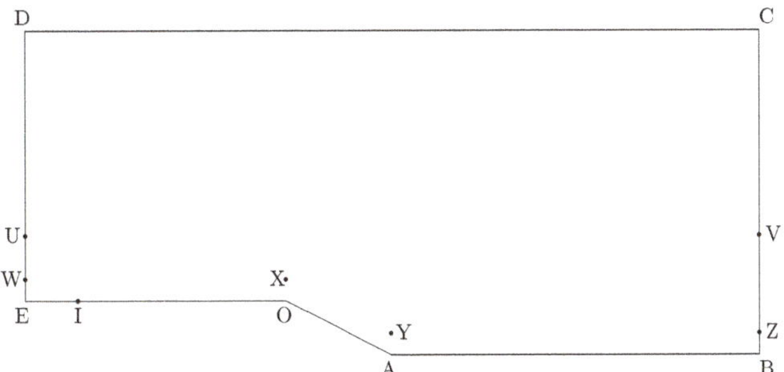

Figure 3. Our reference geometry, with the following point coordinates (note that the figure is not to scale). Physical points: O = (0,0), A = (2.1445, −1), B = (100, −1), C = (100, 30), D = (−25, 30), E = (−25, 0). Only for boundary conditions: I = (−20, 0). Only for meshing: U = (−25, 0.5), V = (100, 0.5), W = (−25, 0.1), X = (0, 0.1), Y = (2.1445, −0.9), Z = (100, −0.9). The x axis points to the right and the y axis to the top, with invariance with respect to the z axis. The length of the sloping portion is $\overline{OA} = 2.3662$.

The full incompressible flow $\begin{pmatrix} u \\ p \end{pmatrix}(x, y, z, t)$, comprising both the velocity and the pressure fields, satisfies the Navier–Stokes and continuity equations,

$$\begin{cases} \partial_t u + u \cdot \nabla u = -\nabla p + \nu \nabla^2 u \, , \\ \qquad \nabla \cdot u = 0 \, . \end{cases} \tag{1}$$

(For an interesting discussion of the role of compressibility—not considered here—see, e.g., [20,21].) In what follows, we decompose the flow into a 2D steady solution plus a 3D small perturbation:

$$\begin{pmatrix} u \\ p \end{pmatrix} = \begin{pmatrix} U \\ P \end{pmatrix} + \begin{pmatrix} u' \\ p' \end{pmatrix} , \text{ with } U = \begin{pmatrix} U \\ V \\ 0 \end{pmatrix} (x, y) \, , \text{ and } u' = \begin{pmatrix} u' \\ v' \\ w' \end{pmatrix} (x, y, z, t) \, , \tag{2}$$

for $|u'| \ll |U|$ and $|p'| \ll |P|$.

4. Base Flow

We have assumed as our base flow, $\begin{pmatrix} U \\ P \end{pmatrix}$, a steady solution of the Navier–Stokes and continuity equations,

$$\begin{cases} U \cdot \nabla U = -\nabla P + \nu \nabla^2 U \, , \\ \qquad \nabla \cdot U = 0 \, , \end{cases} \tag{3}$$

satisfying the same boundary conditions as the full flow. We have obtained this flow, numerically, by means of Newton's iterative method [22]. (The relevance of small modifications in the base flow was studied in, e.g., [23,24].)

Notice that, because of mass conservation, this type of base flow presents a speed overshoot (i.e., for some range of y the horizontal velocity exceeds unity). The vertical profile is not monotonic, as in a standard boundary layer, so it is not appropriate to define a typical width as the height at which the velocity reaches a definite percentage of the far-field value. It is, therefore, more convenient to quantify the boundary layer by means of the so-called "generalized displacement thickness" [10]:

$$\delta_1(x) \equiv \frac{\int dy\, y\, \omega(x,y)}{\int dy\, \omega(x,y)}\,, \tag{4}$$

where $\omega \equiv \partial_x V - \partial_y U$ is the vorticity of the base flow (a scalar quantity, i.e., the z component—the only non-zero—of the vector given by the curl of the base velocity). When this profile reaches the step, we find $\delta_1(x = 0) \in [0.08, 0.18]$, depending on the Reynolds number.

We have taken into consideration Reynolds numbers ranging from 500 to 3000, with increments of 500. The flow separates from the bottom boundary at the step and, with growing Reynolds number, a larger and larger recirculation bubble develops in the wake, until reattachment takes place. Figure 4 displays the dependence of the reattachment point (i.e., the abscissa after the beginning of the step, at which the vertical derivative of the horizontal velocity at the lower wall turns from negative to positive) as a function of Re. A sketch of the base flow for Re $= 1000$ is shown in Figure 5.

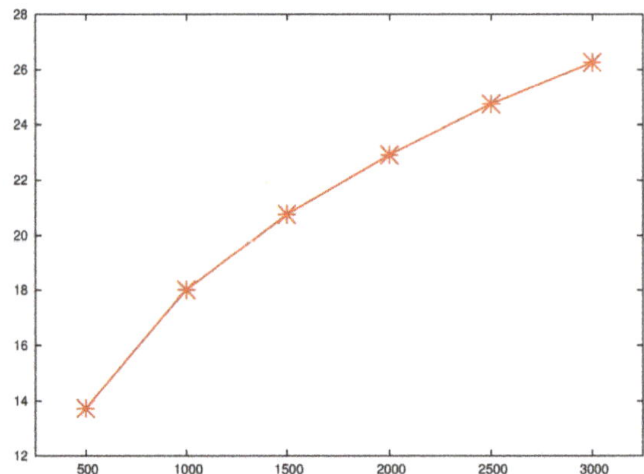

Figure 4. Streamwise coordinate (after the beginning of the step) of the reattachment point for the base flow (in ordinate), as a function of the Reynolds number (in abscissa).

Figure 5. Horizontal component U of the base-flow velocity at Re $= 1000$.

5. Linear Stability Analysis

Upon fixing our base flow, we have performed a linear stability analysis (see e.g., [25,26]). Owing to the steadiness of the base flow and to its invariance in the spanwise direction, we consider the perturbation introduced in Equation (2) in the form of a Fourier mode in z, exponentially evolving in time:

$$\begin{pmatrix} u' \\ p' \end{pmatrix}(x,t) = \begin{pmatrix} \mathcal{U} \\ \mathcal{P} \end{pmatrix}(x)\, e^{\sigma t} + \text{c.c.} = \begin{pmatrix} u'' \\ p'' \end{pmatrix}(x,y)\, e^{i\beta z + \sigma t} + \text{c.c.} \qquad \beta \in \mathbb{R},\ \sigma \in \mathbb{C}. \tag{5}$$

The resulting linearized equation for the (direct) perturbation is:

$$\begin{cases} \sigma \mathcal{U} + \mathcal{U} \cdot \nabla U + U \cdot \nabla \mathcal{U} = -\nabla \mathcal{P} + \nu \nabla^2 \mathcal{U}, \\ \nabla \cdot \mathcal{U} = 0. \end{cases} \tag{6}$$

We then seek those complex values of σ such that Equation (6) has nontrivial solutions $\begin{pmatrix} \mathcal{U} \\ \mathcal{P} \end{pmatrix}$, which can accordingly be defined as (direct) eigenfunctions. The baseline case, Re $= 1000$ and $\beta = 0$ (i.e., no spanwise dependence) is stable, as all the eigenvalues have negative real part, as plotted in Figure 6. Instability can be reached in two ways: either by modifying the spanwise wavenumber (e.g., $\beta = 1$ in Figure 7), or by augmenting the Reynolds number (e.g., Re $= 3000$ in Figure 8). It is worth noticing that the former operation induces stationary instabilities—as the imaginary part of the rightmost eigenvalue still vanishes—while the latter introduces unstationary ones ($\Im(\sigma) \neq 0$ for those points where $\Re(\sigma) > 0$, and the picture is, of course, symmetric with respect to the horizontal axis). The critical Reynolds number, for which the flow develops its first linear instability at some value of β, is approximately 750.

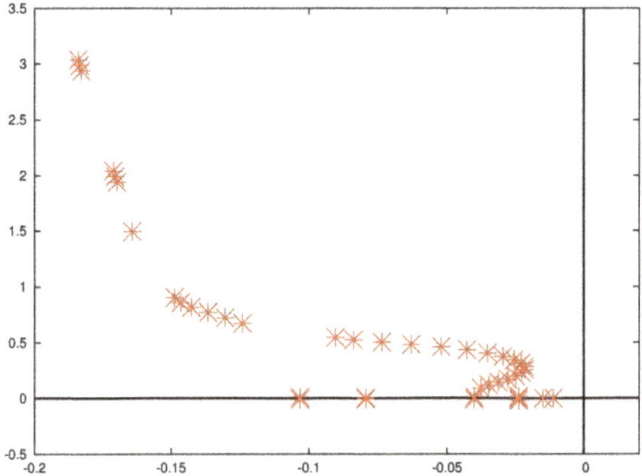

Figure 6. Complex spectrum of the operator described by Equation (6) for the perturbation field, at Re $=$ 1000 and $\beta = 0$. The real and imaginary parts of σ are plotted in abscissa and ordinate, respectively.

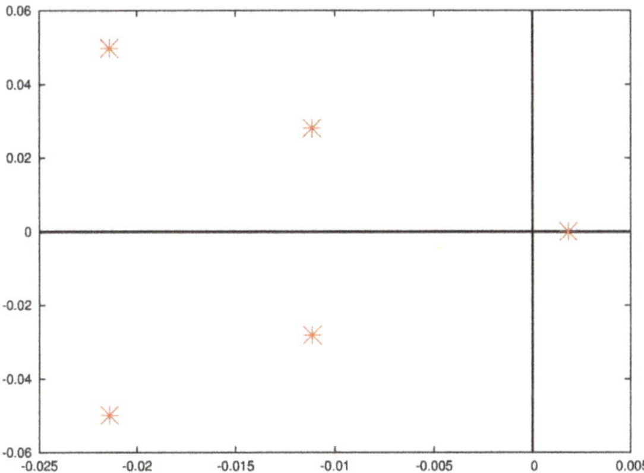

Figure 7. Same as in Figure 6, but for $\beta = 1$.

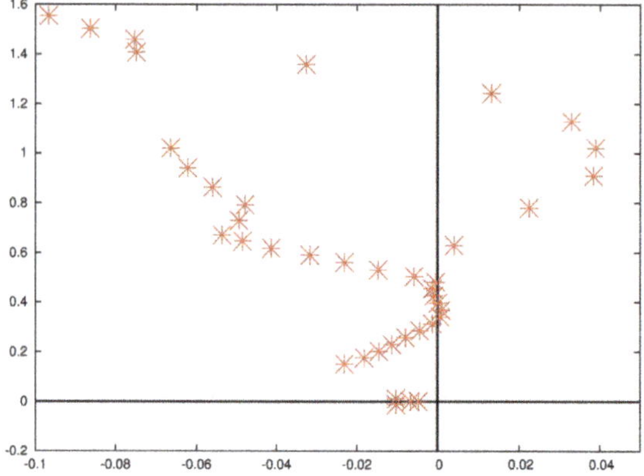

Figure 8. Same as in Figure 6, but for Re = 3000.

A sketch of the perturbation (for the largest-real-part eigenvalue, depicted in Figure 7), at Re = 1000 and $\beta = 1$, is presented in Figure 9. It is evident that this eigenvector is a physically meaningful one, because it is concentrated in the recirculation bubble, which is the zone where instability develops.

Figure 9. Horizontal component u'' of the dominant perturbation field at Re = 1000 and $\beta = 1$.

6. Control and Gain

In this section, we follow [27,28] by introducing a forcing on the right-hand side of the Navier–Stokes equation, which we assume as steady and volumic: $\mathcal{F}(x,t) = f(x,y)e^{i\beta z}$. A key point is that this allows us not only to leave the boundary conditions unchanged with respect to Equation (6), but, more importantly, to confine ourselves to steady solutions $\binom{\mathcal{U}}{\mathcal{P}}$. Indeed, we have already analyzed and found the temporal evolution of the general unforced solutions (eigenvalues and eigenfunctions) in the previous section, and what we are looking for here is just a particular solution to a steady forcing, which can, thus, be assumed as time-independent.

Therefore, we focus on the equations:

$$\begin{cases} \mathcal{U} \cdot \nabla U + U \cdot \nabla \mathcal{U} = -\nabla \mathcal{P} + \nu \nabla^2 \mathcal{U} + \mathcal{F}, \\ \nabla \cdot \mathcal{U} = 0. \end{cases} \tag{7}$$

We define as gain the quantity

$$g \equiv \frac{E_u}{E_f} = \frac{\int dx \int dy\, |u''|^2}{\int dx \int dy\, |f|^2}, \tag{8}$$

and as optimal gain:

$$G(\beta, \text{Re}) \equiv \max_{\mathcal{F}} g. \tag{9}$$

The procedure to find the optimal forcing consists in an iterative algorithm making use of the adjoint variables $\begin{pmatrix} \boldsymbol{u}^\dagger \\ \mathcal{P}^\dagger \end{pmatrix}$ satisfying

$$\begin{cases} (\boldsymbol{\nabla u}) \cdot \boldsymbol{u}^\dagger - \boldsymbol{u} \cdot \boldsymbol{\nabla u}^\dagger = \boldsymbol{\nabla}\mathcal{P}^\dagger + \nu \nabla^2 \boldsymbol{u}^\dagger + \dfrac{\boldsymbol{u}}{2E_f}\,, \\ \boldsymbol{\nabla} \cdot \boldsymbol{u}^\dagger = 0\,, \end{cases} \tag{10}$$

coupled with the suitable outlet boundary conditions (see, e.g., [29]) specified in Appendix A.

6.1. Standard (Non-Penalized) Case

In our reference case (standard geometry with Re $= 1000$ and $\beta = 1$), the magnitudes of what we have obtained numerically as optimal response and corresponding optimal forcing are sketched in Figure 10.

A clear problem arises here: even if, on one hand, the optimal response develops streamwise for the whole length, on the other hand, the optimal forcing is localized in the vicinity of the step and of the sloping portion of the wall. This is not what one would expect for physical realizability, as, on the contrary, a localization on the horizontal upstream part would be suitable. Indeed, our aim is to take advantage of the formation of counter-rotating longitudinal vortices (i.e., the streak lift-up; see, e.g., [6,30,31]), and to let these interact with the recirculation bubble, in an interesting example of interaction between the Kelvin–Helmholtz and the wake instabilities. In [7,32,33], the formation of streaks was implemented experimentally by placing a series of small cylinders, acting as rugosity elements. (The modified flow that one would obtain after physically placing the roughness elements is clearly not the same as the one in their absence. What is meant here is that this discrepancy must be small for our theory to work—which clearly poses severe restrictions on the applicable elements—so that the modified flow should be obtainable as the sum of the original basic flow plus the weak perturbations currently analyzed.) Notice that the spanwise periodicity of this array of cylinders can be described effectively through our periodic expansion in z (i.e., by means of the wavenumber β which should equal 2π divided by the array spacing). In principle, one could introduce step functions in the integrals defining the gain, and we have briefly explored this option preliminarily. However, in the next subsection, we are going to study this problem by means of a penalization method (see, e.g., [34,35]).

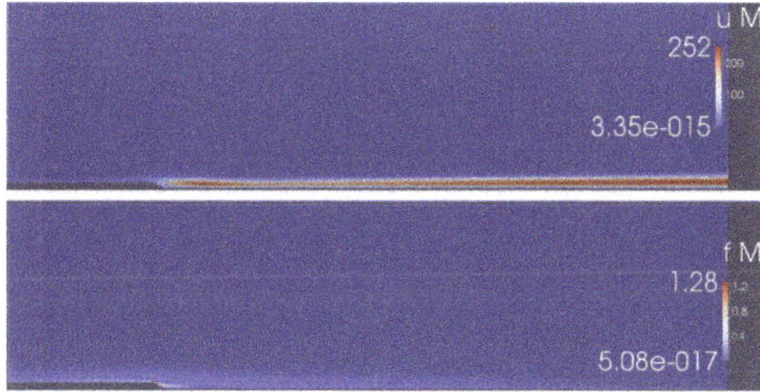

Figure 10. Magnitude of the optimal response and of the corresponding optimal forcing fields, $|\boldsymbol{u}''|$ and $|\boldsymbol{f}|$, in the upper and lower panels, respectively, at Re $= 1000$ and $\beta = 1$. The gain is maximized according to (7).

6.2. Penalized Case

For the present section, let us introduce an effective viscosity $\nu_{\text{eff}}(x)$, defined to be equal to ν upstream and until the beginning of the step, and to a value some orders of magnitude larger for abscissae downstream of it. We, then, focus on:

$$
\begin{cases}
\mathcal{U} \cdot \nabla U + U \cdot \nabla \mathcal{U} = -\nabla \mathcal{P} + \nu_{\text{eff}}(x)\nabla^2 \mathcal{U} + \mathcal{F}, \\
\qquad\qquad \nabla \cdot \mathcal{U} = 0.
\end{cases}
\tag{11}
$$

In this way, we obtain the optimal response and forcing sketched in Figure 11, which should more precisely be defined as sub-optimal, because of the penalization scheme. We expect the forcing with such a shape to be physically realizable, due to its localization on the upstream portion of the wall, but the same cannot be said about the velocity response, due to its concentrated character; very different in look from the envisaged streaks appearing in the previous subsection.

In Figure 12, we plot the optimal gain G as a function of β at different Re. The maximum of the curve not only obviously grows at larger and larger with Re, but also shifts to the right. As, on the contrary, the boundary-layer thickness shrinks when increasing the Reynolds number, we focus on the product between the generalized displacement computed at the step, $\delta_1|_{x=0}$, and the optimal wavenumber β_{opt}. This is shown in Figure 13, and proves that the ratio between the thickness and the optimal spanwise wavelength is almost independent of Re.

Figure 11. Magnitude of the sub-optimal (penalized) response and forcing fields, $|u''|$ and $|f|$, in the upper and lower panels, respectively, at Re $= 1000$ and $\beta = 1$. The gain is maximized following (11).

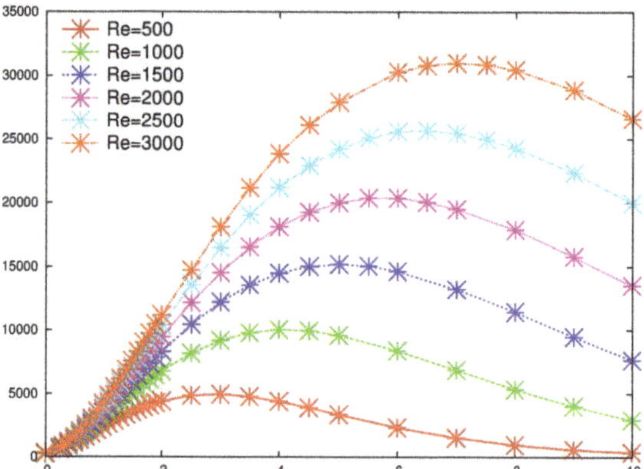

Figure 12. Optimal gain (in ordinates) versus spanwise wavenumber (in abscissa) at different Reynolds numbers. The gain is maximized according to (11).

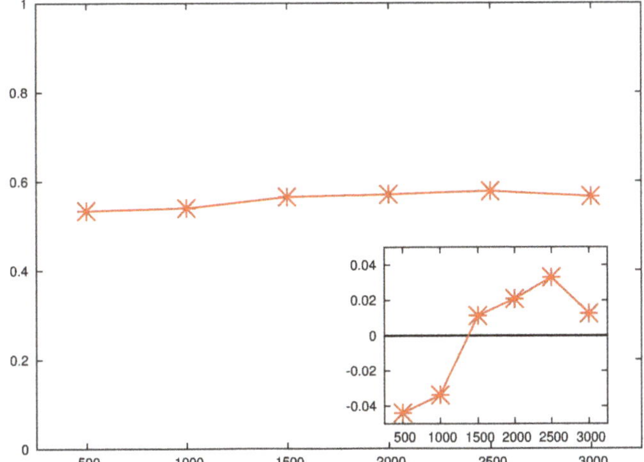

Figure 13. Product between optimal spanwise wavenumber—maxima of Figure 12—and generalized displacement thickness at the step $\delta_1|_{x=0}$ (in ordinate), versus Reynolds number (in abscissa). The inset shows that, within a relative maximum error of less than 5%, this product is independent of Re.

6.3. Penalized Control with Non-Penalized Response

The way to circumvent the paradox, presented in the previous subsection, is very simple. One can indeed find the optimal control through the penalized scheme, but, of course, once this forcing has been found, its real action on the physical velocity must be computed with the actual (space-independent) viscosity ν. We then implement what one could call a "non-penalized response to penalized-optimal control":

$$
\text{LAST}\left\{\begin{array}{l} \mathcal{U}\cdot\nabla u + u\cdot\nabla\mathcal{U} = -\nabla\mathcal{P} + \nu_{\text{eff}}(x)\nabla^2\mathcal{U} + \mathcal{F}, \\ \nabla\cdot\mathcal{U} = 0, \\ \mathcal{U}\cdot\nabla u + u\cdot\nabla\mathcal{U} = -\nabla\mathcal{P} + \nu\nabla^2\mathcal{U} + \mathcal{F}_{\text{opt}}. \end{array}\right. \left.\begin{array}{l} \text{LOOP TO} \\ \text{FIND } \mathcal{F}_{\text{opt}} \end{array}\right\} \tag{12}
$$

This way we obtain the response sketched in Figure 14, together with the aforementioned penalized-optimal forcing field. The fact that streaks are actually generated is confirmed by Figure 15, which represents vertical cuts of the domain $y \in [0,1] \times z \in [0,2\pi)$ at eight different streamwise locations.

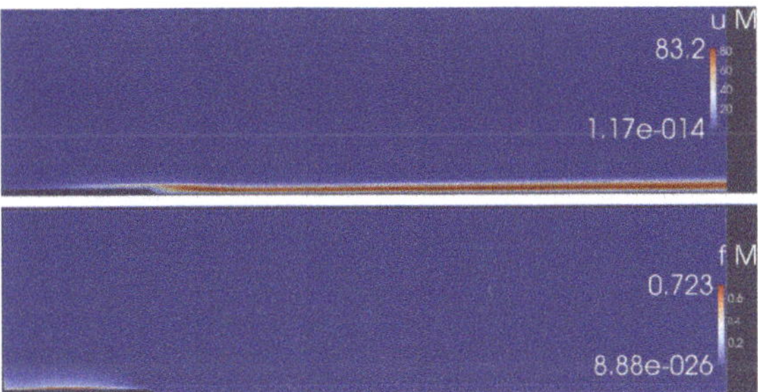

Figure 14. Magnitude of the non-penalized response to penalized-optimal control, $|u''|$, in the upper panel, according to the scheme (12), at Re $= 1000$ and $\beta = 1$. The gain is maximized according to (11), and the magnitude of the corresponding penalized-optimal forcing field $|f|$ (the same as in Figure 11) is reported, again, in the lower panel for the sake of simplicity.

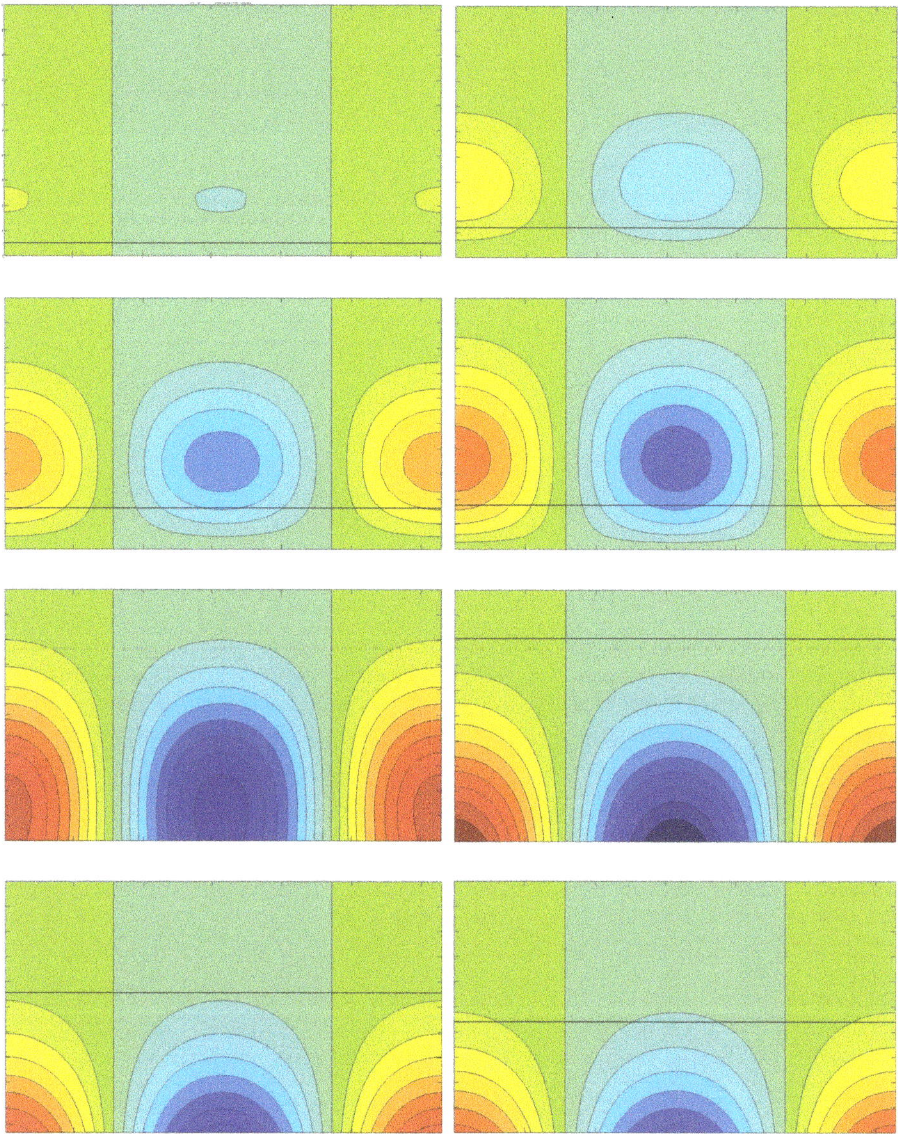

Figure 15. Streamwise component u'' of the optimal response (velocity field, depicted in the upper panel of Figure 14), with positive values in red and negative ones in blue, in vertical cuts at eight different streamwise coordinates: $x = -15$ and -10 (top row), -5 and 0, 5 and 10, and 15 and 20 (bottom row). The horizontal axis is $z \in [0, 2\pi)$, and the vertical one is $y \in [0, 1]$; notice that, for the four latter plots, the physical domain extends below the bottom border of the figure, namely at a depth $-1 \leq y \leq 0$ which exactly equals the height shown. The black horizontal lines represent the height of the generalized displacement thickness $\delta_1(x)$ at each location; the line is not shown in the fifth panel because happening to be placed above the top border (i.e., $\delta_1|_{x=5} > 1$).

In Figure 16, we plot a comparison for the optimal G as a function of β at $\text{Re} = 500$, according to the three aforementioned schemes. In this completely-stable situation, one can see that—moving from the initial non-penalized scheme (black) to the final scheme, proposed in this subsection (blue)—the loss in the optimal gain is on less than one order of magnitude, but with the advantage of delivering an optimal forcing definitely feasible, in terms of physical realizability.

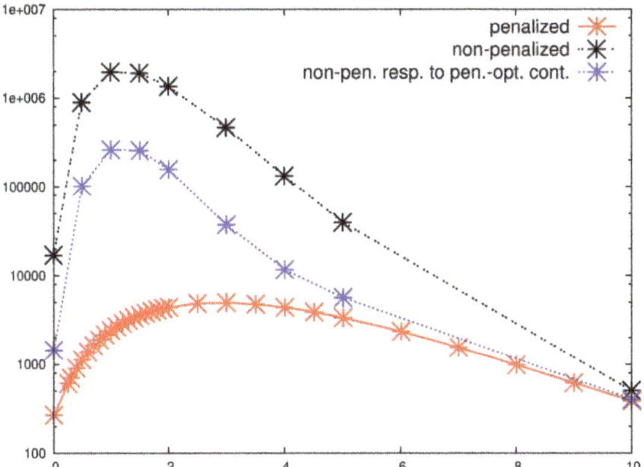

Figure 16. Optimal gain (in ordinates) versus spanwise wavenumber (in abscissa) at Re $=$ 500, according to the three maximization schemes (7) (black), (11) (red), and (12) (blue).

The comparison between (11) and (12) is also plotted in Figure 17 at Re $=$ 3000. Notice that (7) cannot be enforced here, because the Reynolds number is larger than the critical value. Moreover, since this configuration is unstable for some perturbations, we also plot (in black) the gain corresponding to a non-modal response field, which is computed through an orthogonalization procedure in order to exclude spurious peaks related to the modal amplification (which we are not interested in).

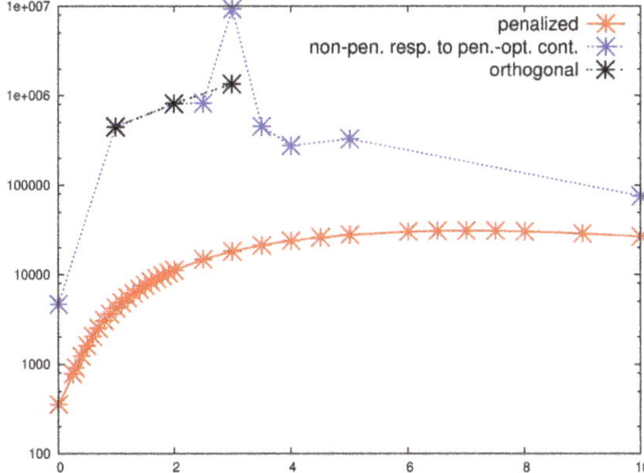

Figure 17. Optimal gain (in ordinates) versus spanwise wavenumber (in abscissa) at Re $=$ 3000, according to the two maximization schemes (11) (red) and (12) (blue). The black points are the result of a process of orthogonalization aimed at excluding spurious amplifications, as happens at the blue peak with $\beta = 3$.

7. Conclusions and Perspectives

We have studied the transient growth of perturbations in a separated boundary layer, namely in the wake of a backward-slanted step at 25°. We have shown that, by means of a suitable penalization method, unstable cases are also tractable in our formalism. We have been able to find situations where the optimal control is spanwise-periodic and localized on the horizontal upstream portion of the wall (which can be experimentally reproduced by an array of rugosity elements), and the corresponding response is represented by streaks.

Among future perspectives, the following questions are of interest.

First, one could change the nature of the forcing term, from volumic to parietal. This would represent an external blowing or suction on the lower wall upstream of the step, implemented by imposing, on this piece of boundary, a condition on the velocity, which should keep the zero tangential component but have a prescribed nonzero normal component (a function of x). We have already made a preliminary test for this situation, but a more profound investigation is definitely required.

Second, it would be interesting to relax the assumption of steady forcing and to investigate the problem also in the temporal domain. One should, then, fix a finite time horizon for the optimization, and perform back-and-forth temporal loops until convergence [36]. This is due to the fact that, if one keeps the time dependence in the equations, the evolution of the adjoint field corresponds to a well-known backward-in-time integration, with "final" conditions imposed on the final time horizon. Of course, the evolution of the direct field is forward-in-time, and one has to perform an optimization on the initial conditions.

Also, the incompressibility of the flow is a key ingredient for the results shown here. It might be worth investigating how they change if a compressible flow is considered, instead. We expect the theoretical analysis to be much more difficult, in view of the necessity of introducing a state equation.

Moreover, the present study is a linear one—rigorously speaking, valid only for infinitesimal perturbations. If the perturbations are small but finite, we expect our framework to be still in excellent agreement with the real picture. However, it is evident that this check can be done only numerically, by performing simulations of the full problem, in order to understand whether the nonlinear coupling in the Navier–Stokes advection term induces significant modifications [37]. This issue could be investigated, for example, by means of appropriate Large-Eddy Simulations and provide the basis for a more direct comparison with experiments [38].

Last, but definitely not least, a relevant question arises about the stability of the considered steady flows to perturbations involving large spatial scales. Further weakly-nonlinear analysis (as in [39] for hydrodynamic flows, and in [40] for MHD flows) may reveal the complex dynamics of large-scale perturbations affecting the performance of the vehicle.

Author Contributions: The authors contributed equally to this work, except for writing—original draft preparation, review, and editing—performed by M.M.A.

Funding: This research was supported by: LabEx "Mécanique et Complexité" (AMU, France); CMUP (UID/MAT/00144/2019), funded by FCT with national (MCTES) and European structural funds through the programs FEDER under the partnership agreement PT2020; Project STRIDE-NORTE-01-0145-FEDER-000033, funded by ERDF-NORTE 2020; and Project MAGIC-POCI-01-0145-FEDER-032485, funded by FEDER via COMPETE 2020-POCI and by FCT/MCTES via PIDDAC.

Acknowledgments: We thank Michel Pognant and Dominique Fougère for technical assistance.

Appendix A. Detailed Description of Boundary Conditions and Adjoint Equations

The boundary conditions for the full velocity field $\boldsymbol{u} = \begin{pmatrix} u \\ v \\ w \end{pmatrix}$ are as follows:

- Inlet on segment ED: $u = 1$, $v = w = 0$;
- Outlet on segment BC: $p\mathrm{I} - \nu\boldsymbol{\nabla}\boldsymbol{u} = 0$;
- Free slip on segments EI and DC: $v = 0$, $\nabla_y u = \nabla_y w = 0$;
- No slip on segments IO, OA and AB: $u = v = w = 0$.

The base flow $\boldsymbol{U}$ inherits the same exact conditions. On the contrary, the perturbation must satisfy fully-homogeneous boundary conditions, so that all the formulae above hold also for the quantities with a prime, except for the very first one, which becomes (at inlet ED) $u' = 0$.

Equation (6) can be rewritten, in terms of the field $\begin{pmatrix} u'' \\ p'' \end{pmatrix}(x,y)$, as:

$$
\begin{cases}
\sigma u'' + (u''\nabla_x + v''\nabla_y)U + (U\nabla_x + V\nabla_y)u'' = -\nabla_x p'' + \nu(\nabla_x^2 + \nabla_y^2 - \beta^2)u'' , \\
\sigma v'' + (u''\nabla_x + v''\nabla_y)V + (U\nabla_x + V\nabla_y)v'' = -\nabla_y p'' + \nu(\nabla_x^2 + \nabla_y^2 - \beta^2)v'' , \\
\sigma w'' + (U\nabla_x + V\nabla_y)w'' = -i\beta p'' + \nu(\nabla_x^2 + \nabla_y^2 - \beta^2)w'' , \\
\nabla_x u'' + \nabla_y v'' + i\beta w'' = 0
\end{cases}
\tag{A1}
$$

(where $\boldsymbol{u}'' = \begin{pmatrix} u'' \\ v'' \\ w'' \end{pmatrix}$).

The derivation of the adjoint equations involves calculating the scalar product of Equations (6) (regarded as a 4D vector) with the adjoint field $\begin{pmatrix} \boldsymbol{u}^{\dagger} \\ \mathcal{P}^{\dagger} \end{pmatrix}$, and in integrating by parts on the whole domain, benefitting from our boundary conditions. In particular, the ones for the adjoint variables are the same as for the direct counterpart, except for the outlet condition which is (on segment BC): $\mathcal{P}^{\dagger}\mathbf{I} + \nu\boldsymbol{\nabla}\boldsymbol{u}^{\dagger} + \boldsymbol{U}\otimes\boldsymbol{u}^{\dagger} = 0$.

However, when a forcing is also present in the direct equations (as in Equation (7)), the procedure is more complex, as it involves the whole formalism of Lagrange multipliers and functional derivatives [27,28]. We do not report it here, and we simply remind the reader of the final result (10).

References

1. Giannetti, F.; Luchini, P. Structural sensitivity of the first instability of the cylinder wake. *J. Fluid Mech.* **2007**, *581*, 167–197. [CrossRef]
2. Marquet, O.; Sipp, D.; Jacquin, L. Sensitivity analysis and passive control of cylinder flow. *J. Fluid Mech.* **2008**, *615*, 221–252. [CrossRef]
3. Schmid, P.J. Nonmodal stability theory. *Ann. Rev. Fluid Mech.* **2007**, *39*, 129–162. [CrossRef]
4. Åkervik, E.; Ehrenstein, U.; Gallaire, F.; Henningson, D.S. Global two-dimensional stability measures of the flat plate boundary-layer flow. *Eur. J. Mech. B Fluids* **2008**, *27*, 501–513. [CrossRef]
5. Sipp, D.; Marquet, O.; Meliga, P.; Barbagallo, A. Dynamics and control of global instabilities in open flows: A linearized approach. *Appl. Mech. Rev.* **2010**, *63*, 030801. [CrossRef]
6. Cossu, C.; Pujals, G.; Depardon, S. Optimal transient growth and very large-scale structures in turbulent boundary layers. *J. Fluid Mech.* **2009**, *619*, 79–94. [CrossRef]
7. Pujals, G.; Cossu, C.; Depardon, S. Forcing large-scale coherent streaks in a zero-pressure-gradient turbulent boundary layer. *J. Turb.* **2010**, *11*, 25. [CrossRef]
8. Hecht, F. New development in FreeFem++. *J. Numer. Math.* **2012**, *20*, 251–265. [CrossRef]
9. Meliga, P.; Gallaire, F. Control of axisymmetric vortex breakdown in a constricted pipe: Nonlinear steady states and weakly nonlinear asymptotic expansions. *Phys. Fluids* **2011**, *23*, 084102. [CrossRef]
10. Sipp, D.; Lebedev, A. Global stability of base and mean flows: A general approach and its applications to cylinder and open cavity flows. *J. Fluid Mech.* **2007**, *593*, 333–358. [CrossRef]
11. Ehrenstein, U.; Gallaire, F. On two-dimensional temporal modes in spatially evolving open flows: The flat-plate boundary layer. *J. Fluid Mech.* **2005**, *536*, 209–218. [CrossRef]
12. Barkley, D.; Gomes, M.G.M.; Henderson, R.D. Three-dimensional instability in flow over a backward-facing step. *J. Fluid Mech.* **2002**, *473*, 167–190. [CrossRef]
13. Ahmed, S.; Ramm, G.; Faltin, G. Some salient features of the time-averaged ground vehicle wake. *SAE Transactions* **1984**, 473–503. [CrossRef]
14. Sipp, D.; Jacquin, L. Three-dimensional centrifugal-type instabilities of two-dimensional flows in rotating systems. *Phys. Fluids* **2000**, *12*, 1740–1748. [CrossRef]
15. Gallaire, F.; Marquillie, M.; Ehrenstein, U. Three-dimensional transverse instabilities in detached boundary layers. *J. Fluid Mech.* **2007**, *571*, 221–233. [CrossRef]
16. Meliga, P.; Chomaz, J.-M. Global modes in a confined impinging jet: Application to heat transfer and control. *Theor. Comput. Fluid Dyn.* **2011**, *25*, 179–193. [CrossRef]

17. Biancofiore, L.; Gallaire, F.; Pasquetti, R. Influence of confinement on a two-dimensional wake. *J. Fluid Mech.* **2007**, *688*, 297–320. [CrossRef]

18. Marquet, O.; Sipp, D.; Chomaz, J.-M.; Jacquin, L. Amplifier and resonator dynamics of a low-Reynolds-number recirculation bubble in a global framework. *J. Fluid Mech.* **2008**, *605*, 429–443. [CrossRef]

19. Marquet, O.; Lombardi, M.; Chomaz, J.-M.; Sipp, D.; Jacquin, L. Direct and adjoint global modes of a recirculation bubble: Lift-up and convective non-normalities. *J. Fluid Mech.* **2009**, *622*, 1–21. [CrossRef]

20. Meliga, P.; Sipp, D.; Chomaz, J.-M. Effect of compressibility on the global stability of axisymmetric wake flows. *J. Fluid Mech.* **2010**, *660*, 499–526. [CrossRef]

21. Meliga, P.; Sipp, D.; Chomaz, J.-M. Open-loop control of compressible afterbody flows using adjoint methods. *Phys. Fluids* **2010**, *22*, 054109:1–054109:18. [CrossRef]

22. Fornberg, B. Steady viscous flow past a circular cylinder up to Reynolds number 600. *J. Comput. Phys.* **1985**, *61*, 297–320. [CrossRef]

23. Bottaro, A.; Corbett, P.; Luchini, P. The effect of base flow variation on flow stability. *J. Fluid Mech.* **2003**, *476*, 293–302. [CrossRef]

24. Brandt, L.; Sipp, D.; Pralits, J.O.; Marquet, O. Effect of base-flow variation in noise amplifiers: The flat-plate boundary layer. *J. Fluid Mech.* **2011**, *687*, 503–528. [CrossRef]

25. Meliga, P.; Pujals, G.; Serre, E. Sensitivity of 2-D turbulent flow past a D-shaped cylinder using global stability. *Phys. Fluids* **2012**, *24*, 061701. [CrossRef]

26. Vozella, L. Transizione alla turbolenza per moti in condotti. Ph.D. Thesis, Università di Genova, Genova, Italy, 2007.

27. Pujals, G. Perturbations optimales dans les écoulements de paroi turbulents et application au contrôle de décollement. Ph.D. Thesis, LadHyx, Palaiseau, France, 2009.

28. Marquet, O. Stabilité globale et contrôle d'écoulements de recirculation. Ph.D. Thesis, Université de Poitiers, Poitiers, France, 2007.

29. Meliga, P.; Boujo, E.; Pujals, G.; Gallaire, F. Sensitivity of aerodynamic forces in laminar and turbulent flow past a square cylinder. *Phys. Fluids* **2014**, *26*, 104101. [CrossRef]

30. Del Guercio, G.; Cossu, C.; Pujals, G. Stabilizing effect of optimally amplified streaks in parallel wakes. *J. Fluid Mech.* **2014**, *739*, 37–56. [CrossRef]

31. Del Guercio, G.; Cossu, C.; Pujals, G. Optimal streaks in the circular cylinder wake and suppression of the global instability. *J. Fluid Mech.* **2014**, *752*, 572–588. [CrossRef]

32. Pujals, G.; Depardon, S.; Cossu, C. Drag reduction of a 3D bluff body using coherent streamwise streaks. *Exp. Fluids* **2010**, *49*, 1085–1094. [CrossRef]

33. Pujals, G.; Depardon, S.; Cossu, C. Transient growth of coherent streaks for control of turbulent flow separation. *Int. J. Aerodyn.* **2011**, *1*, 318–336. [CrossRef]

34. Minguez, M.; Pasquetti, R.; Serre, E. High-order large-eddy simulation of flow over the Ahmed body car model. *Phys. Fluids* **2008**, *20*, 095101:1–095101:17. [CrossRef]

35. Minguez, M.; Brun, C.; Pasquetti, R.; Serre, E. Experimental and high-order LES analysis of the flow in near-wall region of a square cylinder. *Int. J. Heat Fluid Flow* **2011**, *32*, 558–566. [CrossRef]

36. Turek, S.; Rivkind, L.; Hron, J.; Glowinski, R. Numerical study of a modified time-stepping θ-scheme for incompressible flow simulations. *J. Sci. Comput.* **2006**, *28*, 533–547. [CrossRef]

37. Del Guercio, G.; Cossu, C.; Pujals, G. Optimal perturbations of non-parallel wakes and their stabilizing effect on the global instability. *Phys. Fluids* **2014**, *26*, 024110. [CrossRef]

38. Meliga, P.; Sipp, D.; Chomaz, J.-M. Elephant modes and low frequency unsteadiness in a high Reynolds number, transonic afterbody wake. *Phys. Fluids* **2009**, *21*, 054105. [CrossRef]

39. Gama, S.; Vergassola, M.; Frisch, U. Negative eddy viscosity in isotropically forced two-dimensional flow: Linear and nonlinear dynamics. *J. Fluid Mech.* **1994**, *260*, 95–126. [CrossRef]

40. Chertovskih, R.; Zheligovsky, V. Large-scale weakly nonlinear perturbations of convective magnetic dynamos in a rotating layer. *Physica D* **2015**, *313*, 99–116. [CrossRef]

 *fluids*

Article

A Correction and Discussion on Log-Normal Intermittency *B*-Model

Christopher Locke [1],*, Laurent Seuront [2] and Hidekatsu Yamazaki [1]

[1]   Department of Ocean Sciences, Tokyo University of Marine Science and Technology, 4-5-7 Konan, Minato-ku, Tokyo 108-8477, Japan; hide@kaiyodai.ac.jp

[2]   CNRS, Univ. Lille, Univ. Littoral Côte d'Opale, UMR 8187, LOG, Laboratoire d'Océanologie et de Géosciences, F 62930 Wimereux, France; laurent.seuront@cnrs.fr

*   Correspondence: chris.locke@outlook.com

Received: 2 December 2018; Accepted: 13 February 2019; Published: 21 February 2019

Abstract: This paper discusses a turbulent intermittency model introduced in 1990, the *B*-model. It was found that the original manuscript which introduced the *B*-model contained a couple arithmetic errors in the equations. This work goes over corrections to the original equations, and explains where problems arose in the derivations. These corrections cause the results to differ from those in the original manuscript, and these differences are discussed. A generalization of this *B*-model is then introduced to explore the range of behaviors this style of model provides. To distinguish between the different intermittency models discussed in this paper requires structure function power exponents of order greater than 12. As a source of comparison, data from a flume experiment is introduced, and, with the corrections introduced, this data seems to imply that an intermittency coefficient between 0.17 and 0.2 gives good agreement. Higher quality future measurements of high order moments could help with distinguishing the different intermittency models.

Keywords: turbulence theory; intermittency; breakage model

---

1. Introduction

Fluid turbulence is a phenomenon characterized by the presence of small scale fluctuations in the velocity and pressure fields, along with an increased rate of mixing of mass and momentum [1]. Turbulent flows also exhibit characteristic phenomena like coherent structures in the flow and intermittency. As the analytic and numerical solution of such flows is expensive and susceptible to chaos, one has to rely on models to simulate and simplify their dynamics. Such turbulence models include two-equation models (like the k-ϵ model and k-ω model [2]), Reynolds stress models (like the Speziale–Sarkar–Gatski model [3] and the Mishra–Girimaji model [4]), along with models in Large Eddy Simulations [5]. The *B*-model is an improvement on the log-normal model for the pdf of the dissipation rate in turbulent flows.

The first aim of this paper is to discuss the *B*-model introduced in the paper "Breakage models: lognormality and intermittency" [6] (hereinafter referred to as Y90) which introduces a model of turbulent intermittency, and to identify some errors in the original manuscript. We go over these corrections to the original equations and in Section 2.3 explain where the errors occurred. Originally, the Y90 model only considered the beta function for the distribution of the intermittency coefficient. A generalization of this model to any probability density function (pdf) is introduced in Section 2.4 to probe the range of behaviours this style of model provides. The model corrections cause the results to differ from those in the original manuscript. These differences are discussed in Section 4. This section also gives a brief discussion of the differences existing between the discrete *B*-model and the phenomenologically more realistic family of continuous cascade models.

2. Breakage Models

Turbulence obeys the Navier–Stokes equation. When fluid flow is at a low Reynolds number, the Navier–Stokes equation can predict the flow since nonlinear terms are not dominant. As the Reynolds number increases, the nonlinear terms in the Navier–Stokes equation create chaotic flow. However, the flow exhibits coherent structures at fully developed turbulence, which can only be dealt with in the statistical sense. As stressed by Frisch [7] in his seminal work on the stochastic nature of turbulence, "the understanding of chaos in deterministic systems gives us confidence that a probabilistic description of turbulence is justified". As a matter of fact, Kolmogorov [8] is the most successful statistical description of fully developed turbulence and is the root of intermittency models. Turbulent intermittency is the phenomenon where the instantaneous dissipation rate intermittently reaches very high values, and more often as the Reynolds number increases. An introduction to turbulent intermittency can be found in Frisch [9].

2.1. The Gurvich–Yaglom Model

Gurvich & Yaglom [10] (hereinafter referred to as GY) derived a theory of log-normality for the dissipation rate ϵ (the rate at which turbulent kinetic energy dissipates), extending the model of Kolmogorov [8]. Here, we explain parts of the GY model that are relevant to the Y90 paper and our discussion.

The GY model applied a breakage cascade to study how the statistics of turbulent fluctuations vary with the length scale under consideration. In this model, the average dissipation rate over some spatial domain Q is

$$\langle \epsilon \rangle = \frac{1}{Q} \int_Q \mathrm{d}\mathbf{x}\, \epsilon(\mathbf{x})\,. \tag{1}$$

In this equation, the instantaneous local dissipation rate $\epsilon(\mathbf{x})$ is given by

$$\epsilon(\mathbf{x}) = 2\nu \sum_{i,j} (s_{ij}(\mathbf{x}))^2\,, \tag{2}$$

where ν is viscosity and s_{ij} are the strain rates, defined from the fluid velocity u as

$$s_{ij}(\mathbf{x}) = \tfrac{1}{2}\left(\frac{\partial u_i(\mathbf{x})}{\partial x_j} + \frac{\partial u_j(\mathbf{x})}{\partial x_i} \right) \quad \text{for } i,j = 1,2,3\,. \tag{3}$$

Here, both the turbulent velocities u_i and spatial coordinates x_i are components of 3D vectors, with $\mathbf{x}$ being shorthand for the whole vector itself.

The original spatial domain Q is then divided into successively smaller subdomains Q_i with average dissipation rate in this subdomain given by a random variable ϵ_i,

$$\epsilon_i = \frac{1}{Q_i} \int_{Q_i} \mathrm{d}\mathbf{x}\, \epsilon(\mathbf{x})\,. \tag{4}$$

The breakage coefficient is then defined as the ratio of two successive dissipation rates ϵ_i,

$$\alpha_i = \epsilon_i/\epsilon_{i-1} \quad \text{for } i = 1,\ldots,N_b\,, \tag{5}$$

where N_b is the number of breakage processes. Given this relation, conservation of energy demands that the expected ratio of two successive dissipation rates must be 1,

$$E[\alpha] = 1\,. \tag{6}$$

The length scale of a given subdomain Q_i is represented by l_i. In the GY model, the ratio of successive length scales is constant,

$$\lambda \equiv l_{i-1}/l_i .\tag{7}$$

This breakage process is taken to continue until the length scale l_{N_b} where fluctuations of $\epsilon(\mathbf{x})$ can be neglected. We also define the largest length scale as L, and the smallest scale at which fluctuations can be neglected as η.

Using the above notation, the dissipation rates can be written as (noting that $\epsilon_0 = \langle \epsilon \rangle$, the expectation value of the dissipation rate over the whole spatial domain)

$$\epsilon_n = \alpha_n \alpha_{n-1} \cdots \alpha_2 \alpha_1 \langle \epsilon \rangle ,\tag{8}$$

$$\log \epsilon_n = \log \langle \epsilon \rangle + \sum_{i=1}^{n} \log \alpha_i ,\tag{9}$$

$$\log \epsilon = \log \langle \epsilon \rangle + \sum_{i=1}^{N_b} \log \alpha_i ,\tag{10}$$

where ϵ is the value of $\epsilon(\mathbf{x})$ in the domain Q_{N_b}. Using the central limit theorem, Gurvich & Yaglom [10] then argue that both $\log \epsilon_n$ and $\log \alpha_i$ are normally distributed, and hence ϵ_n and α_i are *log-normally distributed*. The means and variances are denoted

$$m_n = \overline{\log \epsilon_n} , \quad m = \overline{\log \epsilon} ,\tag{11}$$

$$\sigma_n^2 = \overline{(\log \epsilon_n - m_n)^2} , \quad \sigma^2 = \overline{(\log \epsilon - m)^2} .\tag{12}$$

Next, because the summands $\log \alpha_i$, beyond the first ones, are identically distributed, the above variables may be represented as

$$m_n = \log \langle \epsilon \rangle + A_1(\mathbf{x}) + n\xi , \quad m = \log \langle \epsilon \rangle + A_1(\mathbf{x}) + N_b \xi ,\tag{13}$$
$$\sigma_n^2 = A(\mathbf{x}) + n\mu , \quad \sigma^2 = A(\mathbf{x}) + N_b \mu ,\tag{14}$$

where

$$\xi = \overline{\log \alpha_i} , \quad \mu = \overline{(\log \alpha_i - \xi)^2} .\tag{15}$$

The terms A and A_1 come from the non-universality of the first summands $\log \alpha_i$ that depend on behaviour of the flow on the order of L. By noting that $n = \log_\lambda(L/l_n)$ and $N_b = \log_\lambda(L/\eta)$, the expressions for m and σ can then be written as

$$m_r = \log \langle \epsilon \rangle + A_1(\mathbf{x}) + \xi_1 \log(L/r) , \quad m = \log \langle \epsilon \rangle + A_1(\mathbf{x}) + \xi_1 \log(L/\eta) ,\tag{16}$$
$$\sigma_r^2 = A(\mathbf{x}) + \mu_1 \log(L/r) , \quad \sigma^2 = A(\mathbf{x}) + \mu_1 \log(L/\eta) ,\tag{17}$$

where the new variables ξ_1 and μ_1 are given by

$$\xi_1 = \xi / \log \lambda , \quad \mu_1 = \mu / \log \lambda .\tag{18}$$

2.2. B-Model

The *B*-model follows the discussion laid out above in Section 2.1. Here, we will explain how the *B*-model builds upon and differs from the GY model. Y90 first ignores the spatially dependent terms

$A_1(\mathbf{x})$ and $A(\mathbf{x})$. The resulting equations for the mean and variance of $\log \epsilon_r$, where now r represents both the length scale and breakage step, are

$$m_r = \log\langle\epsilon\rangle + \xi\log_\lambda(L/r),\tag{19}$$

$$\sigma_r^2 = \mu\log_\lambda(L/r).\tag{20}$$

This agrees with Equations (13) and (14) after dropping the spatially dependent terms.

Next, Y90 makes the observation that, in going from one breakage step to the next, the maximum value of α should occur when all the dissipation takes place in just one of the subdomains. In this case, the maximum value is $\alpha_{max} = \lambda^3$. This differs from GY, which assumed a log-normal distribution for α with no upper bound. To enforce this upper bound on α, the B-model used a beta distribution for the pdf of α,

$$f_\alpha(\alpha; a, b, \alpha_{max}) = \frac{\alpha^{a-1}(\alpha_{max} - \alpha)^{b-1}}{B(a,b)\,\alpha_{max}^{a+b-1}} \quad \text{in } (0, \alpha_{max}),\tag{21}$$

with a and b being positive parameters and $B(a,b)$ being the beta function. Y90 then uses this pdf to calculate ξ and μ.

From Monin & Ozmidov [11], the structure function $R_\epsilon(r) = E[\epsilon_r(x)\epsilon_r(x+r)]$ can be found using the second order moment

$$M_2 = \frac{2}{r^2}\int_0^r\int_0^{r'} R_\epsilon(r'')\mathrm{d}r''\mathrm{d}r',\tag{22}$$

$$\begin{aligned}R_\epsilon(r) &= \frac{1}{2}\frac{\mathrm{d}^2}{\mathrm{d}r^2}(r^2 M_2)\\ &= \langle\epsilon\rangle^2(L/r)^{2\xi+2\mu}\\ &= \langle\epsilon\rangle^2(L/r)^\theta,\end{aligned}\tag{23}$$

where the intermittency coefficient is defined to be $\theta = 2\xi + 2\mu$.

The next equations from Y90 we just quote here for completeness. The dissipation spectrum $S_\epsilon(k)$ has the behaviour

$$S_\epsilon(k) \sim k^{-1+\theta}.\tag{24}$$

The $-\frac{5}{3}$ law of [12] becomes

$$E(k) \sim \langle\epsilon\rangle^{2/3}k^{-\frac{5}{3}+\chi},\tag{25}$$

$$\chi = \tfrac{2}{3}\xi + \tfrac{2}{9}\mu.\tag{26}$$

Finally, the slope of the nth order structure function of velocity is

$$R_n(\mathbf{r}) \sim \langle\epsilon\rangle^{n/3}r^{\zeta_n},\tag{27}$$

$$\zeta_n = \tfrac{n}{3} - \tfrac{n}{3}\xi - \tfrac{n^2}{18}\mu.\tag{28}$$

2.3. B-Model Corrections

We have identified and will describe two mistakes in the original Y90 paper. The first mistake is related to the calculation of the expectation value ξ and variance μ of $\log\alpha$ for the beta distribution. Using the beta pdf in Equation (21), the results were

$$\xi = \log\alpha_{max} + \Psi(a) - \Psi\left((\alpha_{max} - 1)a\right),\tag{29}$$

$$\mu = \{\Psi(a) - \Psi(\alpha_{max}a)\}^2 + \Psi'(a) - \Psi'(\alpha_{max}a) - \{\Psi(a) - \Psi\left((\alpha_{max} - 1)a\right)\}^2,\tag{30}$$

where Ψ is the digamma function

$$\Psi(x) \equiv \frac{\mathrm{d}}{\mathrm{d}x} \log\left(\Gamma(x)\right). \tag{31}$$

The correct formula and derivation of ξ is

$$\xi \equiv E\left[\log \alpha\right] = \frac{1}{B(a,b)\,\alpha_{max}^{a+b-1}} \int_0^{\alpha_{max}} \log \alpha \; \alpha^{a-1}(\alpha_{max} - \alpha)^{b-1}\mathrm{d}\alpha \tag{32}$$

$$= \log \alpha_{max} + \frac{1}{B(a,b)} \int_0^1 \log x \; x^{a-1}(1-x)^{b-1}\mathrm{d}x$$

$$= \log \alpha_{max} + \Psi(a) - \Psi(a+b)$$

$$= \log \alpha_{max} + \Psi(a) - \Psi(\alpha_{max}a). \tag{33}$$

The mistake from Y90 occurs in the evaluation of the integral on the third to last line. It should evaluate to $\Psi(a) - \Psi(a+b)$ [13], but was mistakenly taken to be $\Psi(a) - \Psi(b)$. Using this corrected result, the formula for μ greatly simplifes:

$$\mu \equiv E\left[(\log \alpha - \xi)^2\right] = E\left[\log^2 \alpha\right] - \xi^2 \tag{34}$$

$$= \frac{1}{B(a,b)} \int_0^1 (\log x + \log \alpha_{max})^2 \; x^{a-1}(1-x)^{b-1}\mathrm{d}x \; - \; \xi^2$$

$$= \frac{1}{B(a,b)} \int_0^1 \log^2 x \; x^{a-1}(1-x)^{b-1}\mathrm{d}x \; - \; (\Psi(a) - \Psi(\alpha_{max}a))^2$$

$$= \Psi'(a) - \Psi'(\alpha_{max}a). \tag{35}$$

The second mistake involves a simple misreading of the GY paper. The Y90 paper says "*... we assume the breakage process is applicable in the inertial subrange of velocity power spectrum, so we adopt the same approximation of* $\log_\lambda (L/r) \approx \log_e (L/r)$". However, this approximation is only valid when λ is close to e. The confusion potentially arose in going from Equations (13) and (14) to (16) and (17), when the $\log \lambda$ factor was absorbed into the definitions ξ_1 and μ_1. The effect of reintroducing the $1/\log \lambda$ factor is that each of ξ and μ must be divided by $\log \lambda$. The new equations for θ, χ and ζ_n therefore become

$$\theta = (2\xi + 2\mu) / \log \lambda, \tag{36}$$

$$\chi = \left(\tfrac{2}{3}\xi + \tfrac{2}{9}\mu\right) / \log \lambda, \tag{37}$$

$$\zeta_n = \tfrac{n}{3} - \tfrac{1}{3}\left(n\xi + \tfrac{1}{6}n^2\mu\right) / \log \lambda. \tag{38}$$

2.4. Model Extension

To further investigate the foundations on which the B-model is based, and the sensitivity to the choice of pdf, we propose the following extension of this model. We keep the assumption that the breakage coefficient α is restricted to the range $0 < \alpha < \alpha_{max}$, but instead of choosing a specific pdf like the beta distribution, we generalize to any pdf

$$f_\alpha(\alpha; a, b, \alpha_{max}, C) \text{ in the range } (0, \alpha_{max}), \tag{39}$$

where a and b represent two parameters that characterize the pdf, and C is the normalization constant. Keeping in mind that α_{max} is identified with λ^3, the number of subdivisions of a given cell at each breakage step, the following system of equations determines the values of a, b, and C:

$$\int_0^{\alpha_{max}} f(\alpha) \, \mathrm{d}\alpha = 1 \quad \text{(normalization)}, \tag{40}$$

$$E\left[\alpha\right] = 1 \quad \text{(breakage constraint)}, \tag{41}$$

$$\frac{2\xi + 2\mu}{\log \lambda} = \theta \quad \text{(intermittency coefficient)}. \tag{42}$$

The first equation gives normalization of the pdf, the second comes from Equation (6), and the last is the definition of the intermittency coefficient in Equation (36). $E[\alpha]$ is calculated through the integral $\int_0^{\alpha_{max}} \alpha f(\alpha)\, d\alpha$, while ξ and μ are also calculated through similar integrals using their definitions (32) and (34). Since there are three equations and three unknowns, once a value of θ and a pdf are prescribed, the model parameters are completely determined.

In addition to the beta distribution defined in (21), the following two trial pdfs are also considered (normalization constant left out for clarity)

$$f_\alpha(\alpha; a, b, \alpha_{max}) = \begin{cases} 1 & \text{for } a < \alpha < b \\ 0 & \text{otherwise} \end{cases} \quad \text{in } (0, \alpha_{max}), \tag{43}$$

$$f_\alpha(\alpha; a, b, \alpha_{max}) = \sin^a \left[\pi \left(\frac{\alpha}{\alpha_{max}} \right)^b \right] \quad \text{in } (0, \alpha_{max}). \tag{44}$$

The first is just a straightforward uniform pdf, while the second is a trigonometric function that was chosen because it is zero on the end-points and the parameters a and b are such that they provide fairly independent variation of the mean and standard deviation through variation of a and b. The uniform pdf (43) was chosen as a simple yet extreme pdf, while the trigonometric pdf (44) was chosen as a different smooth pdf that could give similar distribution to the beta function or log-normal case. For completeness, the log-normal model of GY is also included,

$$f_\alpha(\alpha; a, b, \alpha_{max}) = \frac{1}{\alpha b \sqrt{2\pi}} e^{-(\log \alpha - a)^2 / 2b^2}, \tag{45}$$

where a and b are identified with the mean and standard deviation, respectively.

Naturally, just like in GY, the breakage constraint (41) for the log-normal case forces $\xi = -\frac{1}{2}\mu$, and therefore there is a very simple algebraic formula for θ,

$$\theta = \mu / \log \lambda. \tag{46}$$

Since the computer code we wrote for this work was written to solve the equations for any input pdf numerically, Equation (46) was not explicitly used. Nevertheless, for a log-normal distribution, the values obtained agree with the analytical values, as shown in Table 1. The exponent $K(q)$ is given by $\langle \epsilon_\lambda^q \rangle = \lambda^{K(q)}$ and is related to the power exponents ζ by

$$K(q/3) = q/3 - \zeta_q. \tag{47}$$

Since turbulent energy is conserved over the inertial subrange, $\langle \epsilon_\lambda \rangle$ equals $\langle \epsilon_1 \rangle$, and so $K(1) = 0$ is expected. While the models proposed in this paper do not explicitly enforce the condition $K(1) = 0$, the values reported in Table 1 cannot be statistically differentiated from this theoretical expectation.

Note that Kolmogorov's 4/5 law [14] implies the third order moment is proportional to $\langle \epsilon \rangle$. In this case $\zeta_3 = 1$ and the result forces $\xi = -\mu/2$, the condition for the lognormal model. The B-model focuses on keeping α_{max} finite at the expense of a small deviation from Kolmogorov's 4/5 law (see Table 1). On the other hand, the lognormal model strictly follows Kolmogorov's 4/5 law while α is allowed to be arbitrarily large, which is physically unrealistic.

Table 1. Comparison of ζ, μ, and $K(1)$ for the four choices of pdf.

Model		$\lambda = 2$		$\lambda = 5$	
		$\theta = 0.2$	$\theta = 0.25$	$\theta = 0.2$	$\theta = 0.25$
B-model	ζ	−0.06250	−0.07643	−0.1352	−0.1634
	μ	0.1318	0.1631	0.2961	0.3645
	$K(1)$	0.0049	0.0074	0.0080	0.012
Uniform pdf	ζ	−0.06124	−0.07446	−0.1243	−0.1475
	μ	0.1306	0.1611	0.2853	0.3488
	$K(1)$	0.0059	0.0088	0.011	0.017
Trigonometric pdf	ζ	−0.06389	−0.07848	−0.1485	−0.1824
	μ	0.1332	0.1651	0.3094	0.3836
	$K(1)$	0.0039	0.0059	0.0039	0.0058
Log-normal (GY)	ζ	−0.06930	−0.08663	−0.1609	−0.2012
	μ	0.1386	0.1733	0.3219	0.4024
	$K(1)$	0.000007	0.000007	0.000003	0.000002

3. Flume Experiment

In order to compare the results of the breakage models introduced here, we will make use of both the historical power exponent data of Anselmet et al. [15], as well as data that was illustrated in Seuront et al. [16] and collected in April 1998. We elaborate on this experiment in this section.

Turbulence was generated by means of fixed PVC grids illustrated in Figure 1 (grid diameter 2 mm, mesh size 1 cm) located every 50 cm in a unidirectional oval flume (2 meters long and 1 meter wide, with a working channel section 30 cm wide and 30 cm deep), where a mean flow was generated by the friction of 10 vertical parallel PVC disks (5 mm thick and 0.6 m in diameter) on the surface of the water. Instantaneous horizontal turbulent velocity was measured by high frequency (100 Hz) hot-film velocimetry (DANTEC Serial #9055R0111), and the turbulent energy dissipation rate ϵ (m^2 s^{-3}) was derived following Tennekes & Lumley [17] from the turbulence spectrum obtained from Fourier analysis of time series data recorded by the hot-film probe as

$$\epsilon = 15\nu \int_0^\infty k^2 E(k)\,\mathrm{d}k, \tag{48}$$

where ν is the kinematic viscosity (m^2 s^{-1}), k the wavenumber and $E(k)$ the turbulence spectrum (m s^{-2}).

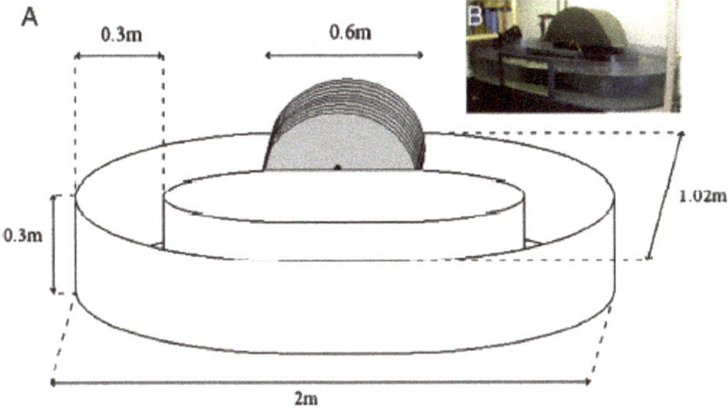

Figure 1. Schematic illustration of the flume experiment set up (see Section 3).

The data used here was specifically derived from an experiment with a mean flow velocity $v = 50$ cm s^{-1} (Reynolds number 1.6×10^5), resulting in velocity fluctuations exhibiting a $-5/3$ power law behavior (i.e., $E(k) \propto k^{-5/3}$) over two decades, and a dissipation rate $\epsilon = 10^{-6}$ m^2 s^{-3} (see

Figure 2). The flume data collected from this experiment (100 Hz sampling rate) consists of 25 sets of 158,720,000 data points.

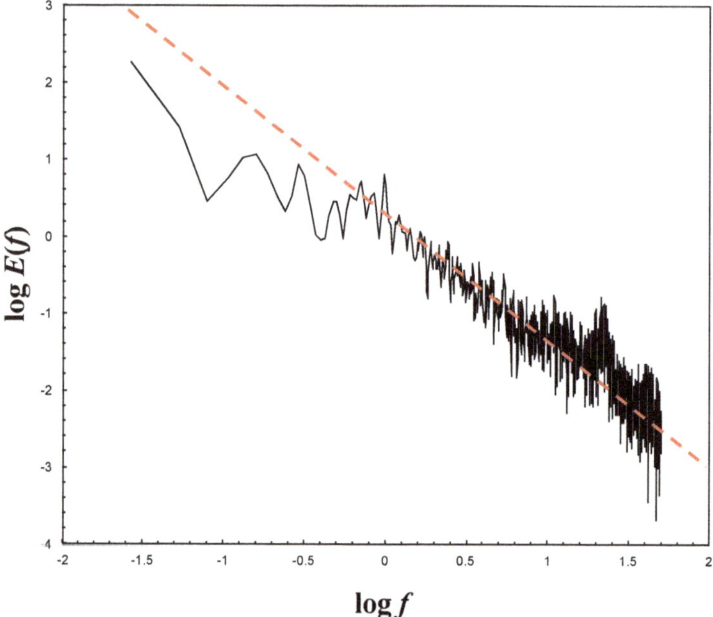

Figure 2. Power spectrum as a function of the wavenumber of data generated from flume experiment (normalized to be unitless, and displayed in log-scale). Dotted line is the $-5/3$ slope.

The role of the sample size in estimating the structure function exponents ζ_n is illustrated in Figure 3. The sample size N is the number of distinct sections of 1024 data points of flume data recorded at 100 Hz. The exponents ζ_n are obtained from the relation $\langle (\Delta v_l)^n \rangle \sim l^{\zeta_n}$. Taking an ensemble average of all the values of the moments n, $(\Delta v_l)^n$, up to a sample size of $N = 155{,}000$ (158,720,000 data points) gives our estimate of the exponent ζ_n. Figure 3 shows how the estimate for the exponents ζ_{12} and ζ_{18} changes as the sample size is increased. ζ_{12} and ζ_{18} converge at different rates, illustrating the dependence of ζ_n on sample size, but it can be seen in this figure that both estimates converge to constant values when $N > 3 \times 10^4$. This convergence is indicative of exponent convergence.

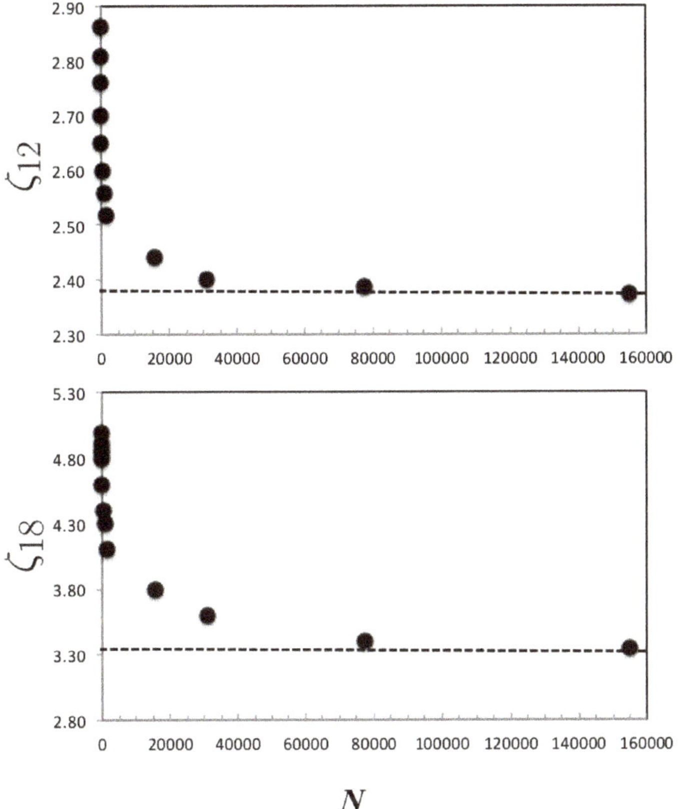

Figure 3. Illustration of the role of sample size in the convergence of the functions ζ_{12} (**top panel**) and ζ_{18} (**bottom panel**) as sample size N increases, for a sample size of $N > 3 \times 10^4$.

4. Discussion

In Yamazaki [6], it was found that, for values of λ less than 5, it was impossible to solve the equations with the condition $\theta = 0.2$. However, the corrected equations for the B-model permit solutions for λ all the way down to a value of 2. For this analysis, we consider two possible values of λ, 2 and 5, keeping in mind comparisons to the old B-model are only meaningful for $\lambda = 5$.

4.1. Breakage Coefficient: α

This work considers four pdfs: the three defined in Equations (21), (43) and (44), and the log-normal pdf of GY in Equation (45). First, we investigate how the distributions of these various pdfs compare with one another once the system of Equations (40) to (42) is solved to determine the pdf parameters. The distributions for the case where $\theta = 0.2$ are plotted in Figure 4. In Y90 (see Figure 1 of the Y90 manuscript [6]), the corresponding log-normal and beta distributions were markedly different; the beta distribution was about half the width of the log-normal case. The reason for this was that in Y90 the plots were done for $\lambda = 5$, and so the parameters were off by a factor of $\log 5$. A consequence of the above corrections is that the difference virtually disappears. The log-normal distribution is slightly shifted to the left of the beta-distribution, with a longer logarithmically falling off tail. However, the values of ζ and μ from Table 1 show that they are very close to each other statistically. In fact, the trigonometric pdf also agrees very well with the two. Naturally, given its very different character, the uniform distribution does not line up as nicely as the others do; however, the parameters in Table 1 derived from solving the system of Equations (40) to (42) give very similar values of ξ and μ for the uniform case also.

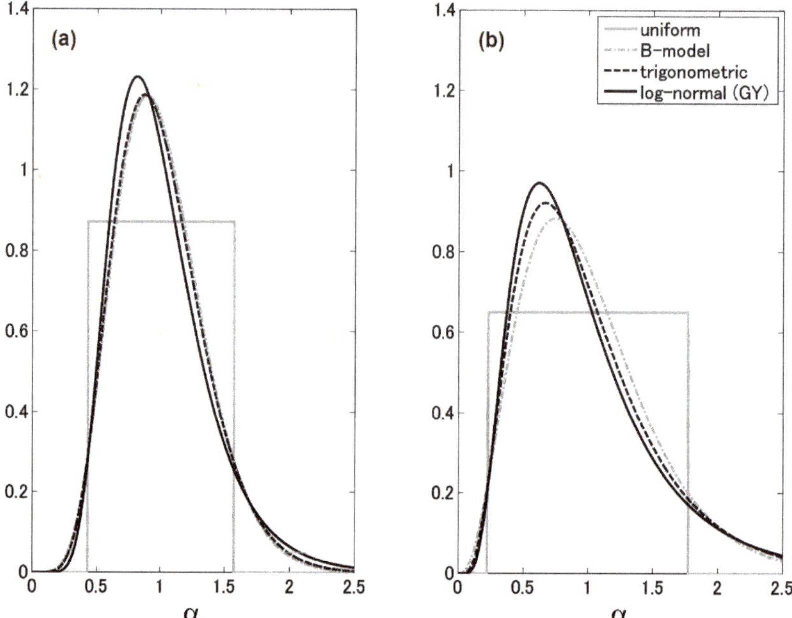

Figure 4. Pdf distributions for (**a**) $\lambda = 2$ and (**b**) $\lambda = 5$ corresponding to the case where $\theta = 0.2$.

4.2. Correction Factor: χ

The results of Y90 show a positive value of χ, in contrast to the negative correction factors of the GY model and β-model. However, the corrected values for χ are negative, as seen in Figure 5. When λ is set to 2, all choices of pdfs give similar negative curves, but when λ is 5, the differences between the curves become more pronounced. In this case, the uniform pdf increases with θ, becoming slightly positive out past $\theta = 0.5$. In the range of θ considered though, χ remains distinctly negative. It is also worth mentioning here, that unlike in Section 4.1, when calculating χ for GY, the $\log \lambda$ factors cancel out in such a way that the χ curve is correct in both papers.

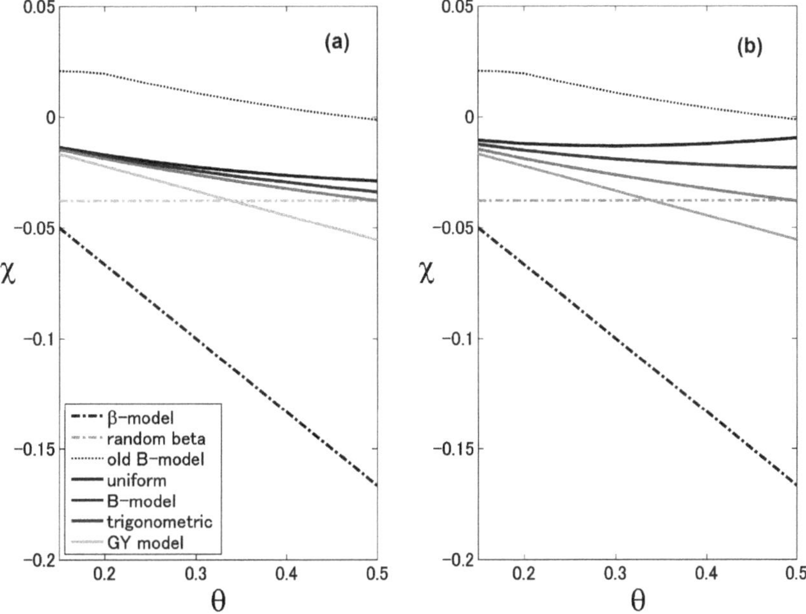

Figure 5. Correction factor χ for the universal spectrum slope with (**a**) $\lambda = 2$, and (**b**) $\lambda = 5$. The dotted line of the old B-model is done for $\lambda = 5$ in both cases. Gray-scale solid lines are listed in the legend in the order they appear in the figure from top to bottom.

4.3. Power Law Coefficient: ζ_n

It was noted in Y90 that ζ_n for the GY model and B-model are convex functions of n, so there is the pathological characteristic that ζ_n eventually becomes negative at high orders, but that the B-model exhibits this tendency less than GY. When we extrapolated out the plots of ζ_n in Figure 6, it can be seen that, while the B-model is less convex than GY, the difference between the two is much less pronounced than in Y90.

Similarly to the case of χ in Figure 5, when $\lambda = 2$, the different pdfs tend to give similar ζ_n curves, but when λ is increased to 5 the differences are enhanced. This dependence of the B-model on λ, the ratio of length scales, stresses a possible limitation of the model. In all cases, the curves are much lower and closer to GY than the B-model curves of Y90.

4.4. Range of λ and θ

One interesting question is the sensitivity of the model to the scale ratio λ. Y90 only considered $\lambda = 5$ because, as defined, the incorrect B-model equations were only solvable for $\lambda \geq 5$. In this study, solutions exist all the way down to $\lambda = 2$, and when λ is greater than 5 the changes are negligible. When ζ_n was plotted for $\lambda = 9$ and compared with those of $\lambda = 5$, the two cases were almost indistinguishable from each other.

The B-model of Y90 considered two possibilities for θ: 0.2 and 0.25. With the above changes, these two values of θ give less agreement with the empirical data of Anselmet et al. [15]. In order for the power law coefficient ζ_n to agree with data, the intermittency coefficient could be as low as 0.17. Such lower θ curves are included in Figure 7 and are discussed in more detail next.

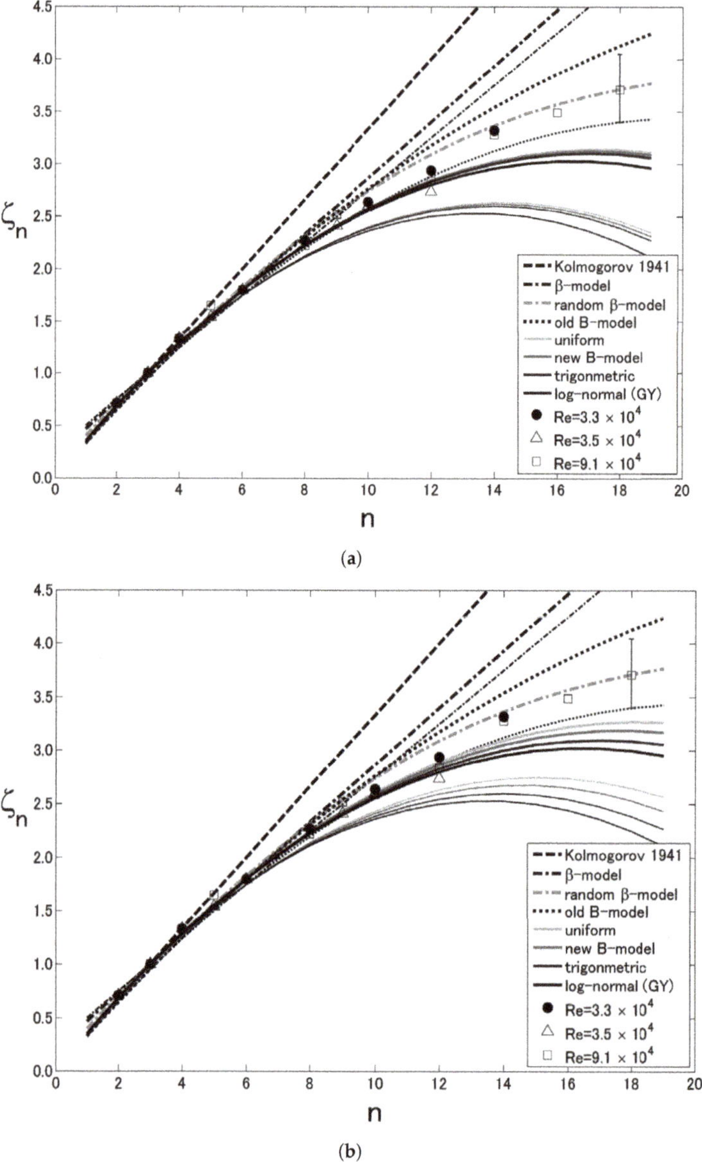

Figure 6. Power exponents of nth order structure function for (**a**) $\lambda = 2$ and (**b**) $\lambda = 5$. In the cases where there are two lines of a given model (e.g., old B-model), the highest curve (thicker line) corresponds to $\theta = 0.2$; the lowest (thinner line) to $\theta = 0.25$. The old B-model curves are done for $\lambda = 5$ in both plots. Grey-scale solid lines are listed in the legend in the order they appear in the figure from top to bottom. Experimental data points are compiled from Anselmet et al. [15] with the corresponding Reynolds number in the legend.

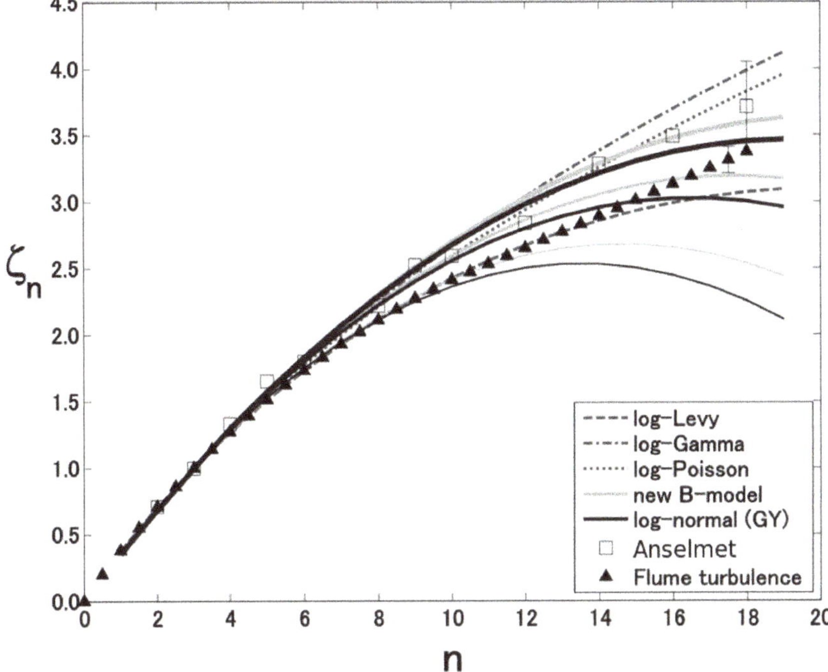

Figure 7. Power exponents of nth order structure function for $\lambda = 5$, including continuous cascade models. In the case of the B-model, and GY model where there are three lines each, the highest curve (thickest line) corresponds to $\theta = 0.17$; the middle (moderate thickness), to $\theta = 0.2$; the lowest (thinnest line), to $\theta = 0.25$. Parameters for the log-Lévy model are $C_1 = 0.16$ and $\alpha = 1.55$ (Seuront & Yamazaki, unpublished data) and are taken from Saito [18] for the log-Gamma, and She & Leveque [19] for the log-Poisson model. Experimental data points are from Anselmet et al. [15] for white boxes (Reynolds number of 9.1×10^4), while the black triangles correspond to velocity fluctuations recorded in a circular flume using hot film velocimetry (Re $= 1.6 \times 10^5$) (see [16] and Section 3). Error bars (only last shown for clarity) are standard deviations of the independent estimates ζ_n from the flume experiment of Section 3.

4.5. Model Comparisons

The original and revised versions of the B-model [6] belong to a family of discrete cascade models where $\log \epsilon_\lambda = \log\langle \epsilon \rangle + \sum \log X_i$, that includes the log-normal model [10], the β-model [20], the random β-model [21], and the α-model and the p-model [22–24] (see also [16] for a review). However, discrete models are less realistic in a phenomenological sense, as they imply an elementary scale ratio λ (e.g., $\lambda = 5$ for the B-model) corresponding to discrete scales, whereas in turbulence intermittent fluctuations intrinsically exist at all scales. As such, continuous cascade models may provide a better alternative to study the stochastic properties of turbulent intermittent fluctuations. To our knowledge, three such models have been described in the literature: the log-Lévy model of Schertzer & Lovejoy [25], the log-Gamma model of Saito [18], and the log-Poisson model of Dubrulle [26], She & Leveque [19], and She & Waymire [27]. The corresponding formulations for the structure function exponents ζ_q are given below:

1. Log-Lévy model:

$$\zeta_q = \frac{q}{3} - \frac{C_1}{\alpha - 1}\left(\left(\frac{q}{3}\right)^\alpha - \frac{q}{3}\right). \tag{49}$$

C_1 is the codimension of the mean events ($0 \leq C_1 \leq d$, where d is the dimension of the observation space, i.e., $d = 1$ for a time-series), and α is the Lévy index, bounded between 0 and 2. Specifically, the log-Lévy model is equivalent to the β-model when $\alpha = 0$ and to the log-normal model when $\alpha = 2$. The log-Lévy models can then be regarded as a family of models bounded between the β-model and the log-normal model.

2. Log-Gamma model:

$$\zeta_q = \frac{q}{3} - \frac{q}{3}d + \frac{\log(\phi/(\phi + q/3))}{\log(\phi/(\phi + 1))}d \,. \tag{50}$$

$\phi > 0$ is a characteristic parameter of the gamma distribution (see e.g., [28]), and d is the dimension of the observation space.

3. Log-Poisson model:

$$\zeta_q = \frac{q}{3} - c\left((1 - \gamma)\frac{q}{3} - 1 + \gamma^{q/3}\right) \,. \tag{51}$$

$c > 0$ is the codimension, which plays a similar role to C_1 of the log-Lévy model, but this parameter characterizes the extreme events. Another parameter, $0 < \gamma < 1$, is linked to the maximum singularity (i.e., the most extreme event) reachable from a finite sample. She & Leveque [19] proposed the general relation with $c = 2$ and $\gamma = 2/3$,

$$\zeta_q = \frac{q}{9} + 2 - 2(2/3)^{q/3} \,, \tag{52}$$

which is in remarkable agreement with the experimental results of Benzi et al. [29].

The log-Lévy, log-Gamma, and log-Poisson models all provide very good fits to empirical data in both atmospheric and oceanic turbulence [16,30]. It is nevertheless stressed that, in contrast with the log-Lévy model, the log-Gamma and log-Poisson models have some limitations. First, the Gamma and Poisson distributions are not stable. In other words, a linear combination of Poisson variables does not follow a Poisson distribution [28], and hence the limitation of the multiplicative approach is clear from $\log \epsilon_\lambda = \log\langle \epsilon \rangle + \sum \log X_i$. Secondly, in the log-Poisson model, the values of the parameters c and γ are linked to the sample size of a given data set. The parameter γ is associated with the maximum singularity: the longer the available data set, the higher the probability of encountering new rare events, and thus, the higher the value of γ. Note that this directly affects the formulation of ζ_q in Equation (51), and so implicitly limits its generality and fitting power. The same limitations apply to c insofar as this parameter describes the absolute distribution of rare events γ.

A comparison of these continuous cascade models to the B-model and the GY model are given in Figure 7. The values of the parameters used in Figure 7 for the continuous cascade models are $C_1 = 0.16$ and $\alpha = 1.55$ for log-Lévy (Seuront & Yamazaki, unpublished data), and taken from Saito [18] for log-Gamma and She & Leveque [19] for log-Poisson. An intermittency coefficient value of $\theta = 0.17$ was also included in this figure, which shows good agreement with the data of [15] for very large values of n. This may suggest that the true θ is less than 0.2. However, significant differences between the fitting power of these models only start appearing for power exponents above 12th order. Hence, it is stressed that an unambiguous quantitative assessment of these models relies on large (i.e., typically with more than 10^7 data points) experimental data sets with lower uncertainties and/or high quality.

To look at higher power exponents, we used large high quality data sets which have become available for in situ and ex situ with noticeable improvement in both the resolution and precision of turbulence measuring techniques such as shear sensors, particle imaging velocimetry and hot film and hot wire velocimetry. This has been illustrated using large data sets (ca. 5×10^8 data points) of velocity fluctuations recorded in the laboratory via hot film velocimetry in a turbulent flume as discussed in Section 3. The corresponding exponents ζ_n consistently converge to constant values up to 18th order and are shown in Figure 7. Note that, while ζ_n, the slope of the nth order structure function, converges well, the moments themselves are much more sensitive, as recognized early on by [15]. This work only looks at the slope of the structure function though, so this difficulty is avoided, and the data presented here implies that θ is bounded between 0.17 and 0.2. Nevertheless, future experiments with high quality data of the moments themselves for orders beyond 12 could go a long way in helping distinguish between different competing breakage models, including the B-model.

5. Conclusions

Two mistakes in the Y90 paper were identified and corrected, and the effects on the model of these corrections were discussed. We have tested the range of behaviours that the B-model can produce. Two new pdfs were introduced and an analogous analysis was carried out. It was found that, for $\theta = 0.2$ or 0.25 and $\lambda = 2$ or 5, these new pdfs gave behaviour with minor differences from the B-model and log-normal model. We also discussed the differences existing between the B-model, an example of a family of discrete cascade models, and the continuous cascade models available in the literature.

For the above values of θ the power exponent curves do not match up very well with the historical data of Anselmet et al. [15]. This data is not enough to completely distinguish between different values of θ data going past 12th order in structure function exponents. Beyond these relatively low Reynolds number data (i.e., Re between 3.3×10^4 in a turbulent duct and below 9.1×10^4 in a turbulent jet), structure function analysis has been proved to be a powerful tool to distinguish the fitting power of both discrete and continuous cascade models on highly turbulent atmospheric and oceanic turbulence data [16,30–33] for moments lower than 10th order. When a large data set ($>10^8$) of high quality data is considered (Figure 7 and Section 3), the slope of the structure function ζ_n can be estimated well to higher orders. Good agreement with the model is then found up to 18th order moments for θ between 0.17 and 0.2. Future experiments that can probe not just the slope of the structure function, but the moments themselves too, would enable studies to better distinguish between the various breakage models.

Author Contributions: H.Y. conceived this study; C.L. found mistakes in Y90 manuscript [6] and developed new models; L.S. provided laboratory data set; C.L. prepared manuscript with input from H.Y. and L.S.

Funding: This work was partially supported by a grant (JPMJCR12A6) from Core Research for Evolution Science and Technology (CREST), a Japan Science and Technology Agency.

Conflicts of Interest: The authors declare no conflict of interest.

References

1. Pope, S.B. *Turbulent Flows*; Cambridge University Press: Cambridge, UK, 2000.
2. Menter, F.R. Two-equation eddy-viscosity turbulence models for engineering applications. *AIAA J.* **1994**, *32*, 1598–1605. [CrossRef]
3. Speziale, C.G.; Sarkar, S.; Gatski, T.B. Modelling the pressure—Strain correlation of turbulence: An invariant dynamical systems approach. *J. Fluid Mech.* **1991**, *227*, 245–272. [CrossRef]
4. Mishra, A.A.; Girimaji, S.S. Toward approximating non-local dynamics in single-point pressure–strain correlation closures. *J. Fluid Mech.* **2017**, *811*, 168–188. [CrossRef]
5. Sagaut, P. *Large Eddy Simulation for Incompressible Flows: An Introduction*; Springer Science & Business Media: Berlin, Germany, 2006.
6. Yamazaki, H. Breakage models: Lognormality and intermittency. *J. Fluid Mech.* **1990**, *219*, 181–193.
7. Frisch, U. *The Legacy of A. N. Kolmogorov*; Cambridge University Press: Cambridge, UK, 1995.
8. Kolmogorov, A.N. The local structure of turbulence in incompressible viscous fluid for very large Reynolds numbers. *Doklady Akad. Nauk SSSR* **1941**, *30*, 299–303. [CrossRef]
9. Frisch, U. Fully developed turbulence and intermittency. *Ann. N. Y. Acad. Sci.* **1995**, *357*, 359–367. [CrossRef]
10. Gurvich, A.S.; Yaglom, A.M. Breakdown of eddies and probability distributions for small scale turbulence. *Phys. Fluids* **1967**, *10*, 59–65. [CrossRef]
11. Monin, A.S.; Ozmidov, R.V. *Turbulence in the Ocean*; Springer: Dordrecht, The Netherlands, 1985; p. 247.

12. Kolmogorov, A.N. A refinement of previous hypotheses concerning the local structure of turbulence in a viscous incompressible fluid at high Reynolds number. *J. Fluid Mech.* **1962**, *13*, 82–85. [CrossRef]
13. Gradshteyn, I.S.; Ryzhik, I.M. *Table of Integrals, Series, and Products*; Academic Press: Cambridge, MA, USA, 1980.
14. Batchelor, G.K. *The Theory of Homogeneous Turbulence*; Cambridge University Press: Cambridge, UK, 1953.
15. Anselmet, F.; Gagne, Y.; Hopfinger, E.J.; Antonia, R.A. High-order velocity structure functions in turbulent shear flows. *J. Fluid Mech.* **1984**, *140*, 63–89. [CrossRef]
16. Seuront, L.; Yamazaki, H.; Schmitt, F. Intermittency (Chapter 7). In *Marine Turbulence: Theories, Observations and Models*; Cambridge University Press: Cambridge, UK, 2005; pp. 66–78.
17. Tennekes, H.; Lumley, J.L. *A First Course In Turbulence*; MIT Press: Cambridge, MA, USA, 1972.
18. Saito, Y. Log-gamma distribution model of intermittency in turbulence. *J. Phys. Soc. Jpn.* **1992**, *61*, 403–406.
19. She, Z.-S.; Leveque, E. Universal scaling laws in fully developed turbulence. *Phys. Rev. Lett.* **1994**, *72*, 336–339.
20. Frisch, U.; Sulem, P.-L.; Nelkin, M. A simple dynamical model of intermittent fully developed turbulence. *J. Fluid Mech.* **1978** , *87*, 719–736. [CrossRef]
21. Benzi, R.; Paladin, G.; Parisi, G.; Vulpiani, A. On the multifractal nature of fully developed turbulence and chaotic systems. *J. Phys. A* **1984**, *17*, 3521–3531. [CrossRef]
22. Schertzer, D.; Lovejoy, S. On the dimension of atmospheric motions. In *Turbulence and Chaotic Phenomena in Fluids*; Tatsumi, T., Ed.; Elsevier Science Ltd.: Amsterdam, The Netherlands, 1984; pp. 505–512.
23. Schertzer, D.; Lovejoy, S. *The Dimension and Intermittency of Atmosphereic Dynamics*; Springer: Berlin/Heidelberg, Germany, 1985; pp. 7–33.
24. Meneveau, C.; Sreenivasan, K.R. Simple multifractal cascade model for fully developed turbulence. *Phys. Rev. Lett.* **1987**, *59*, 1424–1427.
25. Schertzer, D.; Lovejoy, S. Physical modeling and analysis of rain and clouds by anisotropic scaling multiplicative processes. *J. Geophys. Res.* **1987**, *92*, 9693–9714. [CrossRef]
26. Dubrulle, B. Intermittency in fully developed turbulence: Log-Poisson statistics and generalized scale covariance. *Phys. Rev. Lett.* **1994**, *73*, 959–962. [CrossRef]
27. She, Z.-S.; Waymire, E.C. Quantized energy cascade and log-Poisson statistics in fully developed turbulence. *Phys. Rev. Lett.* **1995**, *74*, 262–265. [CrossRef]
28. Feller, W. *An Introduction to Probability Theory and Its Applications*; Wiley: New York, NY, USA, 1968; Volume 1.
29. Benzi, R.; Ciliberto, S.; Baudet, C.; Ruiz Chavarria, G.; Tripiccione, R. Extended self-similarity in the dissipation range of fully developed turbulence. *Europhys. Lett.* **1993**, *24*, 275–279. [CrossRef]
30. Yamazaki, H.; Mitchell, J.G.; Seuront, L.; Wolk, F.; Li, H. Phytoplankton microstructure in fully developed oceanic turbulence. *Geophys. Res. Lett.* **2006**, *33*, L01603. [CrossRef]
31. Benzi, R.; Ciliberto, S.; Baudet, C.; Chavarria, G.R. On the scaling of three-dimensional homogeneous and isotropic turbulence. *Phys. D Nonlinear Phenom.* **1995**, *80*, 385–398. [CrossRef]
32. Ruiz-Chavarria, G.; Baudet, C.; Ciliberto, S. Scaling laws and dissipation scale of a passive scalar in fully developed turbulence. *Phys. D Nonlinear Phenom.* **1996**, *99*, 369–380. [CrossRef]
33. Schmitt, F.; Schertzer, D.; Lovejoy, S.; Brunet, Y. Multifractal temperature and flux of temperature variance in fully developed turbulence. *Eur. Phys. Lett.* **1996**, *34*, 195–200. [CrossRef]

 *fluids*

Article

A Relaxation Filtering Approach for Two-Dimensional Rayleigh–Taylor Instability-Induced Flows

Sk. Mashfiqur Rahman and Omer San *

School of Mechanical and Aerospace Engineering, Oklahoma State University, Stillwater, OK 74078, USA; skmashfiqur.rahman@okstate.edu

* Correspondence: osan@okstate.edu; Tel.: +1-405-744-2457; Fax: +1-405-744-7873

Received: 12 February 2019; Accepted: 16 April 2019; Published: 21 April 2019

Abstract: In this paper, we investigate the performance of a relaxation filtering approach for the Euler turbulence using a central seven-point stencil reconstruction scheme. High-resolution numerical experiments are performed for both multi-mode and single-mode inviscid Rayleigh–Taylor instability (RTI) problems in two-dimensional canonical settings. In our numerical assessments, we focus on the computational performance considering both time evolution of the flow field and its spectral resolution up to three decades of inertial range. Our assessments also include an implicit large eddy simulation (ILES) approach that is based on a fifth-order weighted essential non-oscillatory (WENO) with built-in numerical dissipation due to its upwind-based reconstruction architecture. We show that the relaxation filtering approach equipped with a central seven-point stencil, sixth-order accurate discrete filter yields accurate results efficiently, since there is no additional cost associated with the computation of the smoothness indicators and interface Riemann solvers. Our a-posteriori spectral analysis also demonstrates that its resolution capacity is sufficiently high to capture the details of the flow behavior induced by the instability. Furthermore, its resolution capability can be effectively controlled by the filter shape and strength.

Keywords: Rayleigh–Taylor instability; relaxation filtering; implicit LES; WENO schemes; Euler equations

---

1. Introduction

Rayleigh–Taylor instability (RTI) is an interfacial hydrodynamic instability that occurs at the interface separating two fluids of different densities in the presence of relative acceleration [1]. Understanding the behavior of the RTI-induced flows are of great importance because of the prevalence of such instability phenomena in many natural, industrial, and astrophysical systems with unstably stratified interfaces, such as coastal upwelling near the surface of the oceans [2], atmosphere and clouds [3], plasma physics such as magnetic or inertial confinement fusion implosions [4], the ignition of supernova [5,6], air bubble formation in the blood of deep sea divers [7], premixed combustion [8,9] and many more. In general, RTI phenomenon is one of the easiest hydrodynamics instabilities to observe, for example, if we invert a glass filled with water, the RTI occurs which makes the water falling [10,11]. Lord Rayleigh first described theoretically an instability that occurs in Cirrus cloud formation when a dense fluid is supported by a lighter one in a gravitational field [12]. Later Sir G. Taylor demonstrated the same instability experimentally for accelerated fluids [1], and honouring to their contributions, this instability is named after Lord Rayleigh and Sir G. Taylor which is the Rayleigh–Taylor instability. A detailed overview of the application of RTI phenomena with definitions, physical interpretations and terminologies can be found in [13–17]. Although RTI is a part of many diverse areas of scientific research, and there have been a substantial body of works conducted on this instability over the last few decades, still there have been many open questions yet to be answered about the nature of the RTI-induced flows [18–20]. Moreover, the numerical simulation of RTI studies was not popular much before the 1980s due to the limitation of computational resources. Additionally, the simulation of the flows with RTI is comparatively challenging task since the instability grows up from small scales to initiate secondary instabilities. With a goal to enhance our understandings of RTI phenomenon numerically, the main purpose of this paper is to resolve and analyze the flow field of the RTI test problem with two different initial condition setups using the relaxing filtering modeling approach and compare the high- and coarse-resolution simulation results with the traditional ILES-Riemann solver simulation results.

With the advancement of computational resources, numerical study of RTI has become popular to many research groups for last few years. However, many of the earlier numerical studies in this direction show some variations in the results, such as growth rates of mixing or instability, from the experimental results [21] and it was required to find more improvements in model development and in finding the parameter dependence on the problem setup. Nevertheless, for the last few years, there have been a considerable number works conducted on developing our understandings on RTI related problems [20,22–32]. In general, the single-mode RTI is studied most and has also been used as a building block for multi-mode RTI development [20]. Among some recent notable works on RTI study, the late-time growth in single-mode RTI was studied in [20,31]. In [31], the authors used the implicit large eddy simulation (ILES) approach in a three-dimensional computation of RTI whereas in [20], the author illustrated the growth stages in single-mode RTI using direct numerical simulation (DNS) for two-dimensional computational domain. The formation of the Kelvin-Helmholtz (KH) roll-up of the spike for two-dimensional single-mode RTI simulation was shown in [33]. In [25], it was found that the two-dimensional flow simulation results vary substantially from the three-dimensional ones. Also, it was observed in other literature that the two-dimensional single-mode RTI flows grow faster than the three-dimensional ones [26,34,35]. Youngs et al. [36–38] also studied on the comparison between two and three-dimensional RTI to find out the variation in growth rate at different time stage of the simulations. Also, the authors showed the level of mixing was lower for two-dimensional case. For the multi-mode RTI simulations, similar findings have been shown for growth rate of mixing layer of both two and three-dimensional test setup in [36,39,40]. On the other hand, Zhou et al. and Shvarts et al. [41,42] did a scaling analysis of RT turbulence. In his seminal paper, Chertkov [43] proposed a phenomenological theory for the two-dimensional RT system corresponding to so-called Bolgiano scaling. Quantitatively, also explained in [16], the theory leads to the $k^{-11/5}$ scaling for velocity spectrum, and $k^{-7/5}$ scaling for the density/temperature spectrum. Such theoretical predictions have been also confirmed by direct numerical simulations [44]. In this investigation, we consider the two-dimensional computational domain since it is computationally economical, and we can investigate such scaling behaviors in simulating flows with large inertial range. In our two-dimensional setup, we simulate both multi-mode and single-mode RTI cases using relaxation filtering and ILES modeling approach and obtain the density field contour and kinetic energy spectra to analyze the two-dimensional flow behavior. Our primary focus of this work is to compare between the numerical models implemented on this particular test problem through analyzing the density field plots and statistical tools. However, we will also analyze the resolution and scale resolving capacity of both models through high-resolution simulations.

Numerical simulation of the turbulent flows with instabilities has always been a challenging task since the turbulent features and discontinuities coexist in the flow for a very large spatial and temporal scales. Resolving these enormous scales not only require a huge amount of computational resources but also development of a suitable scheme which has the regularization capability to prevent any oscillations near the discontinuities or shocks. A desired scheme should add a sufficient amount of artificial numerical dissipation near the instability to capture the shocks; however, it should not be too dissipative to damp the small-scale structures of the flow. Over the years, a vast number of successful shock-capturing algorithms have been introduced. For any numerical examination or assessment of a turbulent flow governed by Eulerian hyperbolic conservation laws, ILES methodology can be a good choice to consider which is proven to show a good performance on resolving turbulent flows with shock and discontinuities [45–48]. One of the popular ILES framework is to use an upwind scheme (e.g., Weighted Essentially Non-Oscillatory (WENO) scheme) along with a Riemann solver to incorporate the artificial through the use of numerical truncation errors [49]. The upwind-biased and nonlinearly weighted WENO schemes are widely used in resolving highly compressible turbulent flows because of their robustness to capture discontinuities in shock dominated flows and high order of accuracy in preserving turbulence features [50–52]. It should be mentioned that the development of improved WENO scheme is an active research field and still, there are a lot of works going on in this research direction [53–56]. Another candidate modeling approach is the explicit filtering approach using relaxation filtering which add dissipation on the truncated scales in LES through a low-pass filter [57–60]. In this approach, an additional low-pass spatial filter is used to estimate the effect of unresolved scales. The selection of relaxation filter also affects the solution field for explicit filtering approach and there are a significant number of literature available on the formulation of suitable and efficient LES filters [61–63]. In this work, we use the sixth-order symmetric central scheme with a 7-point stencil Simpson's filter (SF7) as a relaxation filter for RTI test case. We also implement ILES scheme combined with Roe and Rusanov Riemann solver to compare the results obtained by our relaxation filtering solver. The main purposes of this paper are to simulate RTI-induced flow (for both single and multi-mode perturbation) using a relaxation filtering approach to observe the resolution capability of this scheme, analyze the flow behavior by observing the density field contours, and compare the results obtained by relaxation filtering scheme and ILES scheme through kinetic energy (with and without density-weighted velocity) and power density spectra plots. The results show that the relaxation filtering scheme captures more scale in inertial subregion whereas the ILES scheme resolves more scales in high wavenumber regime. For both multi-and single-mode RTI, the kinetic energy spectra plots tend to follow the $k^{-11/5}$ scaling law. On the other hand, the power density spectra plots are observed to be aligned to $k^{-7/5}$ at high resolution. More rigorous derivations and mathematical analyses of scaling laws can be found elsewhere [64–69]. Also, the density contour plots for single-mode RTI reveal that the symmetry of the falling spike break at high resolution because of the formation of secondary instabilities from smaller scales resolved at high resolution. However, the lower resolution simulations resolve the symmetry for all schemes since the numerical dissipation surpasses the formation of the secondary instabilities. Also, it has been seen that the filter strength of the relaxation filter allows us to add more or less dissipation to the solver which eventually affects the flow behavior.

The rest of the paper is organized as follows. Section 2 gives a brief description of the governing equations. Section 3 illustrates the numerical methodology implemented in this study. In Section 4, we detail the results obtained by the numerical schemes used in our investigation along with the problem definitions of the RTI test problem. We demonstrate our findings through high- and coarse-resolution density field contour and density-weighted energy spectra plots. Section 5 gives the summary of our findings and conclusions.

2. Governing Equations

In our study, we consider two-dimensional Euler equations in their conservative dimensionless form as underlying governing equation for the Rayleigh–Taylor instability-induced flow evolution and can be expressed as:

$$\frac{\partial q}{\partial t} + \frac{\partial F}{\partial x} + \frac{\partial G}{\partial y} = S, \tag{1}$$

where F, G account for the inviscid flux contributions to the governing equation and S represents the gravitational term acting on the vertically downward direction (i.e., $g = -1$). The quantities included in q, F, G and S are:

$$\left.\begin{aligned}
q &= [\rho,\ \rho u,\ \rho v,\ \rho e]^T, \\
F &= [\rho u,\ \rho u^2 + p,\ \rho u v,\ \rho u h]^T, \\
G &= [\rho v,\ \rho u v,\ \rho v^2 + p,\ \rho v h]^T, \\
S &= [0,\ 0,\ -\rho,\ -\rho v]^T.
\end{aligned}\right\}$$
(2)

Here, ρ, p, e, u, and v are the density, pressure, total energy per unit mass, and the horizontal and vertical velocity components, respectively. The total enthalpy, h and pressure, p can be obtained by:

$$\begin{aligned}
h &= e + p/\rho, \\
p &= \rho(\gamma - 1)\left(e - \frac{1}{2}(u^2 + v^2)\right),
\end{aligned}$$
(3)

where $\gamma = 7/5$ is chosen as the ratio of specific heats in our study. We refer the reader to [50,70] for details on the development of the eigensystem of the equations to devise hyperbolic conservation laws.

3. Numerical Methods

To develop the computational algorithm for our test problem governed by hyperbolic conservation laws, we formulate a finite volume framework by using different numerical strategies and schemes. In this section, we briefly introduce the numerical methods considered in the present study. We use the method of lines to cast our system of partial differential equations given in Equation (1) in the following form of ordinary differential equation through time:

$$\frac{dq_{i,j}}{dt} = \pounds(q_{i,j}),$$
(4)

where $q_{i,j}$ is the cell-averaged vector of dependent variables, and $\pounds(q_{i,j})$ represents the convective flux terms in the governing equation which can be expressed in the following discretized form:

$$\pounds(q_{i,j}) = \frac{1}{\Delta x}\left(F_{i-1/2,j} - F_{i+1/2,j}\right) + \frac{1}{\Delta y}\left(G_{i,j-1/2} - G_{i,j+1/2}\right) + S_{i,j}.$$
(5)

Here, $F_{i\pm1/2,j}$ are the cell face flux reconstructions in x-direction and $G_{i,j\pm1/2}$ are the cell face flux reconstructions in y-direction. We use the optimal third-order accurate total variation diminishing Runge-Kutta (TVDRK3) scheme [71] for the time integration:

$$q_{i,j}^{(1)} = q_{i,j}^n + \Delta t\pounds(q_{i,j}^n),$$
(6)

$$q_{i,j}^{(2)} = \frac{3}{4}q_{i,j}^n + \frac{1}{4}q_{i,j}^{(1)} + \frac{1}{4}\Delta t\pounds(q_{i,j}^{(1)}),$$
(7)

$$q_{i,j}^{n+1} = \frac{1}{3}q_{i,j}^n + \frac{2}{3}q_{i,j}^{(2)} + \frac{2}{3}\Delta t\pounds(q_{i,j}^{(2)}),$$
(8)

where the time step, Δt should be obtained by (satisfying the Courant-Friedrichs-Lewy (CFL) criterion):

$$\Delta t = \min\left(\eta\frac{\Delta x}{\max(|u|,\ |u + a|,\ |u - a|)},\ \eta\frac{\Delta y}{\max(|v|,\ |v + a|,\ |v - a|)}\right),$$
(9)

where a is the speed of the sound that can be computed from the primitive flow variables (i.e., $a = \sqrt{\gamma p/\rho}$). In our current investigation, we use $\eta = 0.5$ for all the simulations ($\eta \leq 1$ for numerical stability). For the cell face flux reconstructions, we have implemented the ILES and relaxation filtering modeling approaches on our test problem which will be discussed briefly in the subsequent sections.

3.1. ILES Approach

To develop our ILES framework, we first use the WENO interpolation scheme to reconstruct the left and right state of the cell boundaries. Later, we calculate the fluxes at cell edges from the reconstructed left and right states using a Riemann solver. The finite volume framework of a system of Euler conservation equations usually requires a Riemann solver to avoid the Riemann problem [72]. The damping characteristics of nonlinear WENO schemes acts as an implicit filter to prevent the energy accumulation near the grid cut-off [73,74]. In this work, we use the 5th order accurate WENO scheme followed by two widely used Riemann solver, Roe and Rusanov Riemann solver, to determine the flux at cell boundaries.

3.1.1. Weno Reconstruction

The WENO scheme is first introduced in [75] for problems with shocks and discontinuity to get an improvement over the essentially non-oscillatory (ENO) method [76,77]. In this work, we use an implementation of the WENO reconstruction using 7-point stencils (i.e., updating any quantity located at index i depends on the information coming from $i-3, i-2, ..., i+3$) which can be written as:

$$q^L_{i+1/2} = w_0\left(\frac{1}{3}q_{i-2} - \frac{7}{6}q_{i-1} + \frac{11}{6}q_i\right) + w_1\left(-\frac{1}{6}q_{i-1} + \frac{5}{6}q_i + \frac{1}{3}q_{i+1}\right) + w_2\left(\frac{1}{3}q_i + \frac{5}{6}q_{i+1} - \frac{1}{6}q_{i+2}\right), \quad (10)$$

$$q^R_{i-1/2} = w_0\left(\frac{1}{3}q_{i+2} - \frac{7}{6}q_{i+1} + \frac{11}{6}q_i\right) + w_1\left(-\frac{1}{6}q_{i+1} + \frac{5}{6}q_i + \frac{1}{3}q_{i-1}\right) + w_2\left(\frac{1}{3}q_i + \frac{5}{6}q_{i-1} - \frac{1}{6}q_{i-2}\right). \quad (11)$$

Here, $q^L_{i+1/2}$ and $q^R_{i-1/2}$ are the left state (positive) and right state (negative) fluxes, respectively, approximated at midpoints between cell nodes. The left (L) and right (R) states correspond to the possibility of advection from both directions. Since the procedures are similar in the y-direction, we shall present stencil expressions only in the x-direction for the rest of this document. w_k are the nonlinear WENO weights of the kth stencil where $k = 0, 1, ..., r$ and r is the number of stencils ($r = 2$ for the WENO5 scheme). The nonlinear weights are proposed by Jiang and Shu [78] in their classical WENO-JS scheme as:

$$w_k = \frac{\alpha_k}{\sum\limits_{k=0}^{2} \alpha_k}, \quad \alpha_k = \frac{d_k}{(\beta_k + \epsilon)^p}, \quad (12)$$

but the nonlinear weights defined by the WENO-JS scheme are found to be more dissipative than many low-dissipation linear schemes in both smooth region and regions around discontinuities or shock waves [79]. In our study, we have used an improved version of WENO approach proposed by [80], often referred to as WENO-Z scheme. One of the main reasons behind selecting WENO-Z can be less dissipative behavior than classical WENO-JS to capture shock waves. Also, there is a smaller loss in accuracy at critical points for improved nonlinear weights. The new nonlinear weights for the WENO-Z scheme are defined by:

$$w_k = \frac{\alpha_k}{\sum\limits_{k=0}^{2} \alpha_k}, \quad \alpha_k = d_k\left(1 + \left(\frac{|\beta_2 - \beta_0|}{\beta_k + \epsilon}\right)^p\right), \quad (13)$$

where β_k and p are the smoothness indicator of the kth stencil and a positive integer, respectively. Here, $\epsilon = 1.0 \times 10^{-20}$, a small constant preventing zero division, and $p = 2$ is set in the present study to get the optimal fifth-order accuracy at critical points. The expressions for β_k in terms of cell values of q are given by:

$$\left.\begin{aligned}
\beta_0 &= \frac{13}{12}(q_{i-2} - 2q_{i-1} + q_i)^2 + \frac{1}{4}(q_{i-2} - 4q_{i-1} + 3q_i)^2, \\
\beta_1 &= \frac{13}{12}(q_{i-1} - 2q_i + q_{i+1})^2 + \frac{1}{4}(q_{i-1} - q_{i+1})^2, \\
\beta_2 &= \frac{13}{12}(q_i - 2q_{i+1} + q_{i+2})^2 + \frac{1}{4}(3q_i - 4q_{i+1} + q_{i+2})^2.
\end{aligned}\right\} \tag{14}$$

d_k are the optimal weights for the linear high-order scheme which are given by:

$$d_0 = \frac{1}{10}, \ d_1 = \frac{3}{5}, \ d_2 = \frac{3}{10}.$$

3.1.2. Roe Riemann Solver

Based on the Godunov theorem [72], Roe developed an approximate Riemann solver, known as the Roe Riemann solver [81]. In our computational algorithm, we use the flux difference splitting (FDS) scheme of Roe [81] where the exact values of the fluxes at the interface can be computed in the x-direction by:

$$F_{i+1/2,j} = \frac{1}{2}(F^R_{i+1/2,j} + F^L_{i+1/2,j}) - \frac{1}{2}\Delta F. \tag{15}$$

Here, ΔF is the flux difference, calculated as:

$$\begin{aligned}
\Delta F =\ &\phi^{(1)}[1,\ \tilde{u} - \tilde{a},\ \tilde{v},\ \tilde{h} - \tilde{u}\tilde{a}]^T + \phi^{(2)}\left[1,\ \tilde{u},\ \tilde{v},\ \frac{1}{2}(\tilde{u}^2 + \tilde{v}^2)\right]^T + \phi^{(3)}[1,\ \tilde{u} + \tilde{a},\ \tilde{v},\ \tilde{h} + \tilde{u}\tilde{a}]^T + \\
&\phi^{(4)}[0,\ 0,\ 1,\ \tilde{v}]^T,
\end{aligned} \tag{16}$$

where

$$\left.\begin{aligned}
\phi^{(1)} &= \frac{1}{2\tilde{a}^2}(\Delta p - \tilde{\rho}\tilde{a}\Delta u)\lambda_3, \\
\phi^{(2)} &= \frac{1}{\tilde{a}^2}(\tilde{a}^2\Delta\rho - \Delta p)\lambda_1, \\
\phi^{(3)} &= \frac{1}{2\tilde{a}^2}(\Delta p + \tilde{\rho}\tilde{a}\Delta u)\lambda_2, \\
\phi^{(4)} &= \tilde{\rho}\Delta v\lambda_1.
\end{aligned}\right\} \tag{17}$$

Here, Δ denotes the difference between right and left state fluxes for the variables ρ, p, u, v (e.g., $\Delta u = u_R - u_L$), and eigenvalues are defined as $\lambda_1 = |\tilde{u}|$, $\lambda_2 = |\tilde{u} + \tilde{a}|$ and $\lambda_3 = |\tilde{u} - \tilde{a}|$, where $\tilde{a}$ is the speed of the sound at averaged state. In the equations, the tilde represents the density-weighted average, or the Roe average, between the left and right states. The Roe average values can be found by:

$$\left.\begin{aligned}
\tilde{\rho} &= \frac{\rho_R\sqrt{\rho_R} + \rho_L\sqrt{\rho_L}}{\sqrt{\rho_R} + \sqrt{\rho_L}}, \\
\tilde{u} &= \frac{u_R\sqrt{\rho_R} + u_L\sqrt{\rho_L}}{\sqrt{\rho_R} + \sqrt{\rho_L}}, \\
\tilde{v} &= \frac{v_R\sqrt{\rho_R} + v_L\sqrt{\rho_L}}{\sqrt{\rho_R} + \sqrt{\rho_L}}, \\
\tilde{h} &= \frac{h_R\sqrt{\rho_R} + h_L\sqrt{\rho_L}}{\sqrt{\rho_R} + \sqrt{\rho_L}}, \\
\tilde{a} &= \sqrt{(\gamma - 1)\left[\tilde{h} - \frac{1}{2}(\tilde{u}^2 + \tilde{v}^2)\right]},
\end{aligned}\right\} \tag{18}$$

where the left and right states of the un-averaged conserved variables are available from the WENO5 reconstruction described earlier. However, it is later realized that the stationary expansion shocks are not dissipated appropriately by this method. To fix the entropy in the expansion shocks, Harten proposed the following approach [82] replacing Roe averaged eigenvalues by:

$$\lambda_i = \frac{\lambda_i^2 + \epsilon^2}{2\epsilon} \quad \text{if } \lambda_i < \epsilon. \tag{19}$$

Here, $\epsilon = 2\kappa\tilde{a}$ where κ is a small positive number, is set 0.1 in our computations. Similarly, in y-direction, $\lambda_1 = |\tilde{v}|$, $\lambda_2 = |\tilde{v} + \tilde{a}|$ and $\lambda_3 = |\tilde{v} - \tilde{a}|$. The interfacial fluxes in y-direction can be estimated by:

$$G_{i,j+1/2} = \frac{1}{2}(G^R_{i,j+1/2} + G^L_{i,j+1/2}) - \frac{1}{2}\Delta G, \tag{20}$$

where

$$\Delta G = \phi^{(1)}[1,\ \tilde{u},\ \tilde{v} - \tilde{a},\ \tilde{h} - \tilde{v}\tilde{a}]^T + \phi^{(2)}[1,\ \tilde{u},\ \tilde{v},\ \frac{1}{2}(\tilde{u}^2 + \tilde{v}^2)]^T + \phi^{(3)}[1,\ \tilde{u},\ \tilde{v} + \tilde{a},\ \tilde{h} + \tilde{v}\tilde{a}]^T +$$

$$\phi^{(4)}[0,\ 1,\ 0,\ \tilde{u}]^T, \tag{21}$$

with

$$\left.\begin{aligned}
\phi^{(1)} &= \frac{1}{2\tilde{a}^2}(\Delta p - \tilde{\rho}\tilde{a}\Delta v)\lambda_3, \\
\phi^{(2)} &= \frac{1}{\tilde{a}^2}(\tilde{a}^2\Delta \rho - \Delta p)\lambda_1, \\
\phi^{(3)} &= \frac{1}{2\tilde{a}^2}(\Delta p + \tilde{\rho}\tilde{a}\Delta v)\lambda_2, \\
\phi^{(4)} &= \tilde{\rho}\Delta u\lambda_1,
\end{aligned}\right\} \tag{22}$$

where $\Delta u = u_R - u_L$, $\Delta v = v_R - v_L$, $\Delta \rho = \rho_R - \rho_L$, and $\Delta p = p_R - p_L$.

3.1.3. Rusanov Riemann Solver

Rusanov proposes a Riemann solver based on the information obtained from maximum local wave propagation speed [83], sometimes referred to as local Lax-Friedrichs flux [84,85]. The expression for Rusanov solver in the x-direction is as follows:

$$F_{i+1/2,j} = \frac{1}{2}(F^R_{i+1/2,j} + F^L_{i+1/2,j}) - c_{i+1/2}\left(q^R_{i+1/2,j} - q^L_{i+1/2,j}\right), \tag{23}$$

where the right constructed state flux component, F^R is $F(q^R_{i+1/2,j})$, the left constructed state flux component, F^L is $F(q^L_{i+1/2,j})$ and the characteristic speed, $c_{i+1/2} = \tilde{a} + |\tilde{u}|$. The density-weighted average of the conserved variables can be calculated by Equation (18). Similarly, the expression for Rusanov solver in y-direction is:

$$G_{i,j+1/2} = \frac{1}{2}(G^R_{i,j+1/2} + G^L_{i,j+1/2}) - c_{j+1/2}\left(q^R_{i,j+1/2} - q^L_{i,j+1/2}\right), \tag{24}$$

where $c_{j+1/2} = \tilde{a} + |\tilde{v}|$.

3.2. Central Scheme with Relaxation Filtering (Cs+Rf) Approach

In our relaxation filtering approach, we consider a symmetric flux reconstruction using a purely central scheme (CS) combined with a low-pass spatial filter, 7-point stencil Simpson's filter (SF7) in our case, as a relaxation filter (RF). We denoted this solver as CS+RF. For the cell interfacial reconstruction of the conserved quantity, the following symmetric non-dissipative scheme is used [86]:

$$q_{i+1/2,j} = a(q_{i+1,j} + q_{i,j}) + b(q_{i+2,j} + q_{i-1,j}) + c(q_{i+3,j} + q_{i-2,j}), \tag{25}$$

for the interpolation in x-direction, and similarly in y-direction, the conservative interpolation formula reads:

$$q_{i,j+1/2} = a(q_{i,j+1} + q_{i,j}) + b(q_{i,j+2} + q_{i,j-1}) + c(q_{i,j+3} + q_{i,j-2}), \tag{26}$$

where the stencil coefficients are given by:

$$a = 37/60; \quad b = -8/60; \quad c = 1/60.$$

Here, $q_{i,j}$ represents the flow variables (at cell centers) given in Equation (4). The calculated fluxes from the relevant face quantities determined from the nodal values can be used in discretized finite volume equation. In this approach, we assume that the explicit filtering removes the frequencies higher than a selected cut-off threshold through the use of the low-pass spatial filter. A low-pass filter is commonly used in explicit filtering approaches which can be considered to be a free modeling parameter with a specific order of accuracy and a fixed filtering strength [87]. The filtering operation is done at the end of every timestep to remove high frequency content from the solution which eventually prevents the oscillations [58,88,89]. A discussion and analysis of the characteristics on different low-pass filter can be found in [90]. In our investigation, the expression for the sixth-order sequential RF for any quantity f is:

$$\bar{f}_{i,j} = f_{i,j}^* - \sigma\left(a_0 f_{i,j}^* + a_1(f_{i+1,j}^* + f_{i-1,j}^*) + a_2(f_{i+2,j}^* + f_{i-2,j}^*) + a_3(f_{i+3,j}^* + f_{i-3,j}^*)\right), \tag{27}$$

where

$$f_{i,j}^* = f_{i,j} - \sigma\left(a_0 f_{i,j} + a_1(f_{i,j+1} + f_{i,j-1}) + a_2(f_{i,j+2} + f_{i,j-2}) + a_3(f_{i,j+3} + f_{i,j-3})\right). \tag{28}$$

Here, the discrete quantity $f_{i,j}$ yields the filtered value $\bar{f}_{i,j}$ and the filtering coefficients are:

$$a_0 = 5/16; \quad a_1 = -15/64; \quad a_2 = 3/32; \quad a_3 = -1/64. \tag{29}$$

and σ is a parameter that controls filter dissipation strength in a range of $[0,1]$ where $\sigma = 0$ indicates no filtering effect at all, i.e., completely non-dissipative and $\sigma = 1$ indicates the highest filtering effect, i.e., most dissipative with a complete attenuation at the grid cut-off wavenumber. The transfer function of the SF7 filter displays a trend of more dissipation with the increase of the parameter σ [49].

4. Results

In this section, we present our numerical assessment of the modeling approaches outlined in the previous section for both multi-mode and single-mode two-dimensional RTI test problem. We first illustrate the problem definitions of our test case which is followed by the results obtained by different numerical solvers. We perform our quantitative comparisons between the ILES and CS+RF models using the density contours, density-weighted kinetic energy spectra, and compensated density-weighted kinetic energy spectra plots. For comparative analysis, we obtain the high-resolution ILES and CS+RF solutions by using a parallel computing approach using the Open Message Passing Interface (MPI) framework [91,92]. A detailed discussion on the MPI methodology implemented in our study can be found in [49]. Using both high- and coarse-resolution simulation results, the scaling behaviors of the kinetic energy spectra plots are also investigated in this section.

4.1. Two-Dimensional RTI Test Problem: Case Setup

In our numerical experiments, we use a two-dimensional implementation of RTI using the aforementioned numerical schemes. In general, RTI arises at the interface of two fluids when a dense fluid is supported above a comparatively lower density fluid in a gravitational field or stay in the presence of relative acceleration. Since it is found in numerous literature that many properties related to the RTI-induced flows such as the overall growth rate of RTI mixing, dissipation scales, velocity field, and so on, more or less depend on the initial conditions of the flow domain [14,93–95], we consider RTI with multi-mode or randomized perturbation and RTI with single-mode perturbation in our study. We first focus on the case of randomized initial perturbation where our computational domain is set $(x, y) \in [0, 0.5] \times [-0.375, 0.375]$ with the following initial conditions:

$$\rho(x, y) = \begin{cases} 1.0, & \text{if } |y| \leq 0 \\ 2.0, & \text{if } |y| > 0 \end{cases} \tag{30}$$

$$u(x, y) = 0, \tag{31}$$

$$v(x, y) = \frac{\lambda \alpha}{2} \left[1 + \cos(2\pi y / L_y) \right], \tag{32}$$

$$p(x, y) = 2.5 - \rho y. \tag{33}$$

Here, L_y is set 0.75 and the amplitude of the perturbation is set at $\lambda = 0.01$ and α is a random number with a value in between 0 and 1. Since λ is updating itself at each grid point, we note that it is an implicit function of both x and y. On the other hand, for the single-mode RTI, the computational domain is set $(x, y) \in [0, 0.5] \times [-0.75, 0.75]$ with the following initial conditions:

$$\rho(x, y) = \begin{cases} 1.0, & \text{if } |y| \leq 0 \\ 2.0, & \text{if } |y| > 0 \end{cases} \tag{34}$$

$$u(x, y) = 0, \tag{35}$$

$$v(x, y) = \frac{\lambda}{4} \left[1 + \cos(2\pi x / L_x) \right] \left[1 + \cos(2\pi y / L_y) \right], \tag{36}$$

$$p(x, y) = 2.5 - \rho y, \tag{37}$$

where L_x and L_y is set 0.5 and 1.5 respectively with the similar amplitude of the perturbation as the multi-mode case, $\lambda = 0.01$. The similar two-dimensional RTI test problem set up has been used in various studies related to RTI [96,97]. Figure 1 shows the schematic of the computational domain for both cases of the RTI test problem where it can be seen that we consider the normalized gravity is acting in vertically downward direction in our problem definitions.

We must note here that we apply the periodic boundary condition on the left and right boundaries, and the reflective boundary condition on the top and bottom boundaries of our computational domain for both test setup. To get a better understanding on the boundary conditions used in our test setup domain, we can consider an arbitrary two-dimensional domain illustrated in Figure 2. To apply the periodic and reflective boundary condition for our 7-point stencil scheme, we take three ghost points in each direction of the four boundaries of our computational domain. For periodic boundary condition on the left and right boundaries, the ghost point values of the time-dependent variables in vector q (from Equation (2)) can be computed by:

$$
\begin{aligned}
(\rho)_{0,j} &= (\rho)_{N_x,j}, & (\rho u)_{0,j} &= (\rho u)_{N_x,j}, & (\rho v)_{0,j} &= (\rho v)_{N_x,j}, & (\rho e)_{0,j} &= (\rho e)_{N_x,j}, \\
(\rho)_{-1,j} &= (\rho)_{N_x-1,j}, & (\rho u)_{-1,j} &= (\rho u)_{N_x-1,j}, & (\rho v)_{-1,j} &= (\rho v)_{N_x-1,j}, & (\rho e)_{-1,j} &= (\rho e)_{N_x-1,j}, \\
(\rho)_{-2,j} &= (\rho)_{N_x-2,j}, & (\rho u)_{-2,j} &= (\rho u)_{N_x-2,j}, & (\rho v)_{-2,j} &= (\rho v)_{N_x-2,j}, & (\rho e)_{-2,j} &= (\rho e)_{N_x-2,j}, \\
(\rho)_{N_x+1,j} &= (\rho)_{1,j}, & (\rho u)_{N_x+1,j} &= (\rho u)_{1,j}, & (\rho v)_{N_x+1,j} &= (\rho v)_{1,j}, & (\rho e)_{N_x+1,j} &= (\rho e)_{1,j}, \\
(\rho)_{N_x+2,j} &= (\rho)_{2,j}, & (\rho u)_{N_x+2,j} &= (\rho u)_{2,j}, & (\rho v)_{N_x+2,j} &= (\rho v)_{2,j}, & (\rho e)_{N_x+2,j} &= (\rho e)_{2,j}, \\
(\rho)_{N_x+3,j} &= (\rho)_{3,j}, & (\rho u)_{N_x+3,j} &= (\rho u)_{3,j}, & (\rho v)_{N_x+3,j} &= (\rho v)_{3,j}, & (\rho e)_{N_x+3,j} &= (\rho e)_{3,j},
\end{aligned}
\tag{38}
$$

where $j = -2, -1, ..., N_y + 3$. On the other hand, our approximation for the reflective boundary condition is as follows:

$$
\begin{aligned}
(\rho)_{i,0} &= (\rho)_{i,1}, & (\rho u)_{i,0} &= (\rho u)_{i,1}, & (\rho v)_{i,0} &= -(\rho v)_{i,1}, & (\rho e)_{i,0} &= (\rho e)_{i,1}, \\
(\rho)_{i,-1} &= (\rho)_{i,2}, & (\rho u)_{i,-1} &= (\rho u)_{i,2}, & (\rho v)_{i,-1} &= -(\rho v)_{i,2}, & (\rho e)_{i,-1} &= (\rho e)_{i,2}, \\
(\rho)_{i,-2} &= (\rho)_{i,3}, & (\rho u)_{i,-2} &= (\rho u)_{i,3}, & (\rho v)_{i,-2} &= -(\rho v)_{i,3}, & (\rho e)_{i,-2} &= (\rho e)_{i,3}, \\
(\rho)_{i,N_y+1} &= (\rho)_{i,N_y}, & (\rho u)_{i,N_y+1} &= (\rho u)_{i,N_y}, & (\rho v)_{i,N_y+1} &= -(\rho v)_{i,N_y}, & (\rho e)_{i,N_y+1} &= (\rho e)_{i,N_y}, \\
(\rho)_{i,N_y+2} &= (\rho)_{i,N_y-1}, & (\rho u)_{i,N_y+2} &= (\rho u)_{i,N_y-1}, & (\rho v)_{i,N_y+2} &= -(\rho v)_{i,N_y-1}, & (\rho e)_{i,N_y+2} &= (\rho e)_{i,N_y-1}, \\
(\rho)_{i,N_y+3} &= (\rho)_{i,N_y-2}, & (\rho u)_{i,N_y+3} &= (\rho u)_{i,N_y-2}, & (\rho v)_{i,N_y+3} &= -(\rho v)_{i,N_y-2}, & (\rho e)_{i,N_y+3} &= (\rho e)_{i,N_y-2},
\end{aligned}
\tag{39}
$$

where $i = 1, 2, ..., N_x$. For the parallelization, we do the domain decomposition in the y-direction and update the ghost points of the local domain by transferring the information from the adjacent domain. Although we implement our reflective boundary conditions as defined by Equation (39), we note that the boundary conditions are often applied on the velocity rather than momentum. We stress that simulations of unsteady compressible flows require an accurate control of wave reflections from the boundaries of the computational domain since such waves may propagate from the boundary and interact with the flow [98]. We plan to implement more accurate characteristics-based boundary conditions in our future studies.

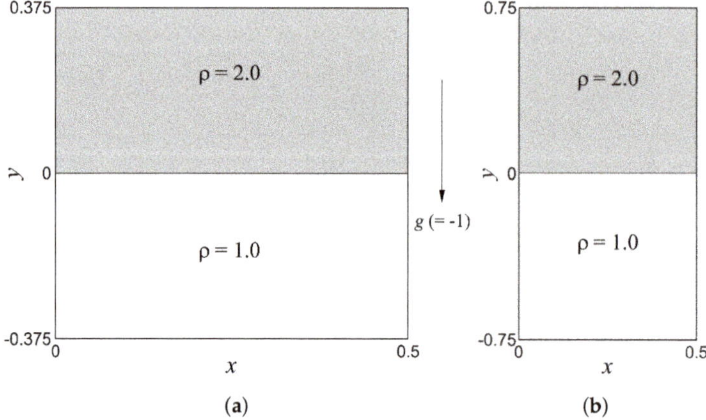

Figure 1. Computational domain and initial conditions for the (**a**) multi-mode and (**b**) single-mode Rayleigh–Taylor instability (RTI) test case. Please note that we apply periodic boundary condition on left and right boundaries, and reflective boundary condition on top and bottom boundaries.

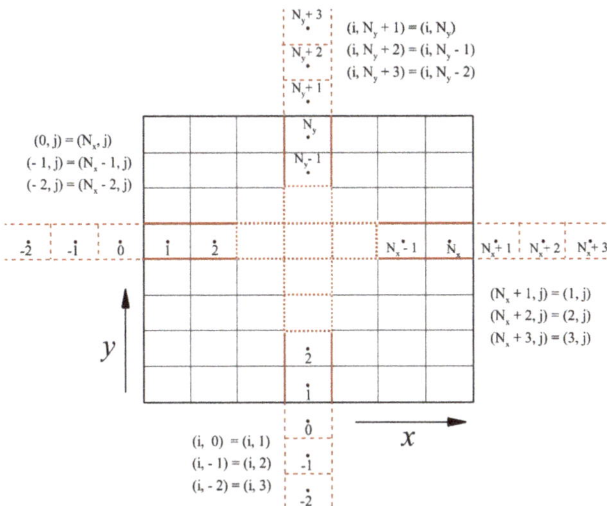

Figure 2. Illustration of the periodic boundary condition (left and right boundaries) and the reflective boundary condition (top and bottom boundaries) on an arbitrary two-dimensional domain.

4.2. RTI with Random (Multi-Mode) Perturbation

The nonlinear evolution of the Rayleigh–Taylor instability from multi-mode initial perturbations is studied based on the density field contour and density-weighted kinetic energy spectra to assess the performance of the underlying modeling schemes. Figure 3 shows the time evolution of the density field at high resolution using the ILES-Roe scheme. As it is shown in [99] that the DNS and ILES give similar results for global properties of RTI-induced mixing, we use the high-resolution results of ILES schemes to avoid the higher computational cost of DNS. Also, it is apparent in Figure 3 that several fine scale structures are captured using the ILES-Roe scheme because of its capability to resolve the smaller scales in high wavenumber region. It can be observed that the mixing growth rate is uniform along the interface at $t = 1.6$ with multiple modes. It has been seen before in [21] where authors found a uniform growth of the mixing region initially for an idealized initial condition whereas the experimental results of same test condition show the presence of dominant scale at the same time. In our study, even though there are some dominant scales or modes present at $t = 4.0$, there are a considerable amount of unmixed region can be seen at the same time which indicates the slow mixing rate for this initial condition. In Figure 4), we present the density contour plots for coarse-resolutions obtained by different ILES-Riemann solver combinations at final time of our simulation, i.e., at $t = 4.0$. As we can see, a clear difference in the growth of scales as well as mixing for both solvers. Since the Rusanov solver is more dissipative than the Roe solver [100], it is expected to have different evolution of the scales in the flow field. Similarly, we can observe different flow field evolution of CS+RF scheme for different filtering strength, σ in Figure 5. Since the higher value of σ adds more dissipation, this solver induces different amount of perturbation in the flow field with the evolution of time than the solver with lower value of σ. Hence, we plot the density-weighted kinetic energy spectra to get a better view in the performance of different solvers [27,101–104]. To include these density effects, we define the energy spectrum built on density-weighted velocity vector which can be expressed as:

$$\omega = (u_\rho, v_\rho) \doteq (\sqrt{\rho}u, \sqrt{\rho}v), \tag{40}$$

where the density-weighted velocity components are

$$u_\rho(x,y,t) = \sqrt{\rho(x,y,t)}u(x,y,t), \tag{41}$$

$$v_\rho(x,y,t) = \sqrt{\rho(x,y,t)}v(x,y,t). \tag{42}$$

We then can calculate the density-weighted kinetic spectra by following expressions:

$$E(k,t) = \frac{1}{N_y}\sum_{j=1}^{N_y}\frac{1}{2}\left(|\hat{u}_\rho(k,y_j,t)|^2 + |\hat{v}_\rho(k,y_j,t)|^2\right), \tag{43}$$

where k refers to the wavenumber along x-direction. We obtain the Fourier coefficients using a standard FFT algorithm [105]

$$\hat{u}_\rho(k,y_j,t) = \sum_{i=1}^{N_x} u_\rho(x_i,y_j,t)\exp^{ikx_i}, \tag{44}$$

$$\hat{v}_\rho(k,y_j,t) = \sum_{i=1}^{N_x} v_\rho(x_i,y_j,t)\exp^{ikx_i}, \tag{45}$$

where i refers to unit imaginary number, and (x_i, y_j) determines the Cartesian grid. Since our domain is periodic only in x-direction, we note that our spectra calculations are averaged in y-direction as illustrated in Equation (43). The other statistical measures investigated in our study are the classical kinetic energy spectra and power density spectra. The energy spectra can be calculated using the following definition in wavenumber space [106]:

$$E(k,t) = \frac{1}{N_y}\sum_{j=1}^{N_y}\frac{1}{2}\left(|\hat{u}(k,y_j,t)|^2 + |\hat{v}(k,y_j,t)|^2\right), \tag{46}$$

where the velocity components $\hat{u}$ and $\hat{v}$ can be computed using a similar fast Fourier transform algorithm presented in Equation (44). To quantify the effect of the scale content of density field, we use the power spectrum that reflects the average packaging of density over different scales at any given time in the simulation. This may be given by the following expression:

$$P(k,t) = \frac{1}{N_y}\sum_{j=1}^{N_y}\frac{1}{2}\left(|\hat{\rho}(k,y_j,t)|^2\right), \tag{47}$$

where $\hat{\rho}$ is the Fourier coefficients of the density field.

For the validation of our spectra plots, we follow the well-established theory for two-dimensional RTI systems [14,43,44]. In his seminal paper, Chertkov [43] proposed a phenomenological theory corresponding to the Bolgiano scaling [107] which can be abstracted to $k^{-7/5}$ scaling law for density or temperature and $k^{-11/5}$ scaling law for velocity. In Figure 6, we can see that the spectra plots (on the left) obtained by the ILES and CS+RF schemes are showing a clear inertial subrange with the $k^{-11/5}$ scaling. However, it is apparent that the CS+RF scheme results are the most aligned with the $k^{-11/5}$ reference line. Moreover, we present the kinetic energy spectra plots without density weighting in Figure 7 which supports the conclusions of the density-weighted energy spectra plots. To validate further, we plot the regular and compensated power density spectra in Figure 8 where it can be seen that the density spectra for CS+RF scheme are following the $k^{-7/5}$ scaling law.

The time evolution of the spectra shows similar statistical trends for all schemes. Therefore, we will only focus on the results at final time in our subsequent analyses. To compare the dissipation characteristics of the schemes, we place the density-weighted spectra of both ILES schemes in a single plot as well as for both CS+RF scheme with different filtering strength σ in Figure 9. It can be observed that the Rusanov solver is more dissipative than the Roe solver and the higher σ value adds more dissipation. These findings are consistent with the previously found results in the literature as well. Also, the spectra are following the reference $k^{-11/5}$ scaling. The density-weighted spectra plots compensated by $k^{11/5}$ for the CS+RF scheme with different filtering strength show that all the lines are flat above the axis line. However, the kinetic energy spectra plots without the density weighting in Figure 10 exhibit a similar trend as the density-weighted ones. On the other hand, the power density spectra plots in Figure 11 show that the $k^{-7/5}$ scaling law is maintained for both set of schemes. However, the compensated spectra plots indicate that the CS+RF scheme is more consistent with the scaling law than the ILES schemes. We present another set of density-weighted spectra plot varying grid resolution to show a comparison between the ILES and CS+RF schemes in Figure 12. It is apparent that the CS+RF captures more scales in the inertial subrange than the ILES schemes. However, the CS+RF scheme reaches the effective grid cut-off scales earlier than the ILES schemes. It is because the CS+RF solvers do the filtering once at the end of the simulation whereas the ILES solvers implicitly adds dissipation throughout the simulation. As a result, ILES schemes capture wide range of scale at high wavenumber even though they resolve comparatively less scales in the inertial subrange. Some key points can be seen from Figure 12 that the $\sigma = 1.0$ solver is the most dissipative among all solvers considered in this study, and $\sigma = 0.4$ solver captures more scales in the inertial subrange than the other solvers. Also, ILES-Roe solver resolves the highest range of scales in high wavenumber for both coarse and high resolution which explains the appearance of very fine small-scale structures in the density field contour plot obtained by ILES-Roe solver.

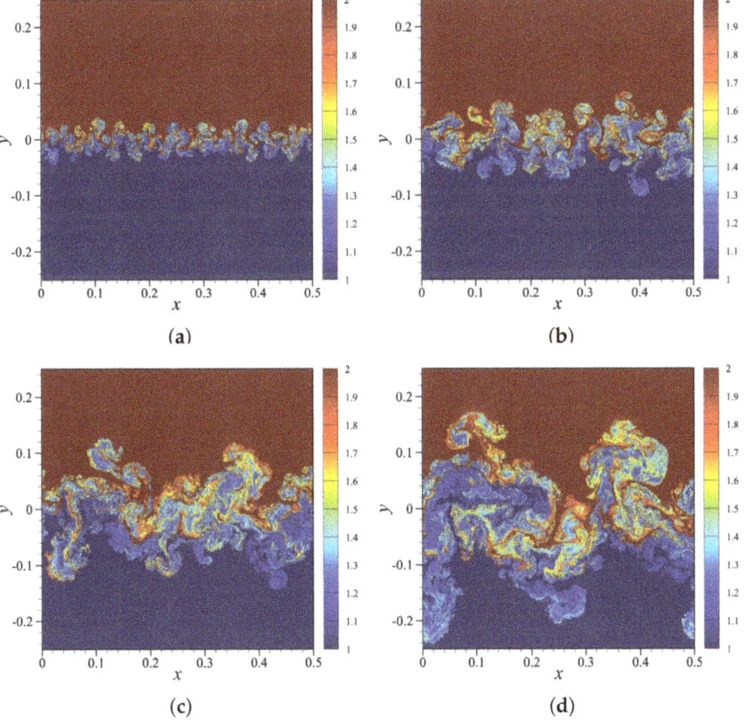

Figure 3. Time evolution of density field for the RTI problem with multi-mode perturbation: (**a**) $t = 1.6$; (**b**) $t = 2.4$; (**c**) $t = 3.2$; (**d**) $t = 4.0$. Results are obtained by the ILES-Roe scheme at a resolution of 16384×24576.

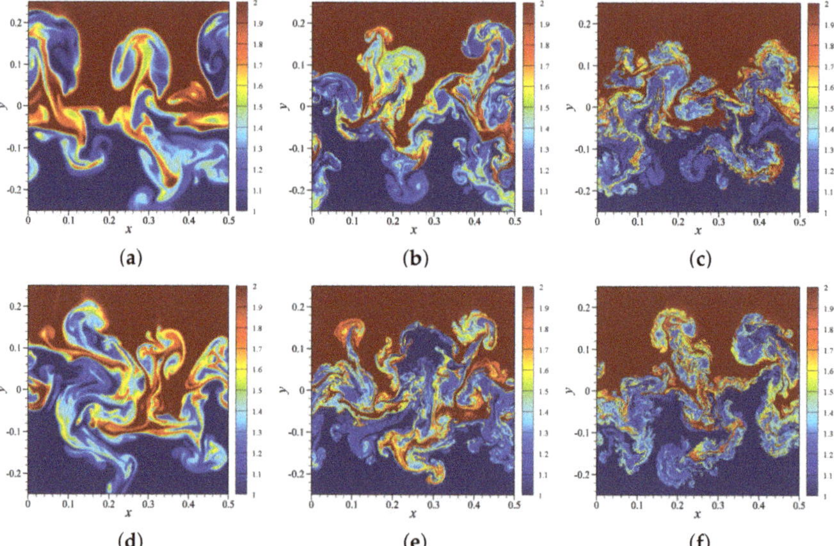

Figure 4. Density contours for the RTI problem with multi-mode perturbation at $t = 4.0$ for different coarse grid resolutions and ILES schemes: (**a**) ILES-Rusanov with 256×384 resolution; (**b**) ILES-Rusanov with 1024×1536 resolution; (**c**) ILES-Rusanov with 4096×6144 resolution; (**d**) ILES-Roe with 256×384 resolution; (**e**) ILES-Roe with 1024×1536 resolution; (**f**) ILES-Roe with 4096×6144 resolution. The gravity is directed in a vertically downward direction.

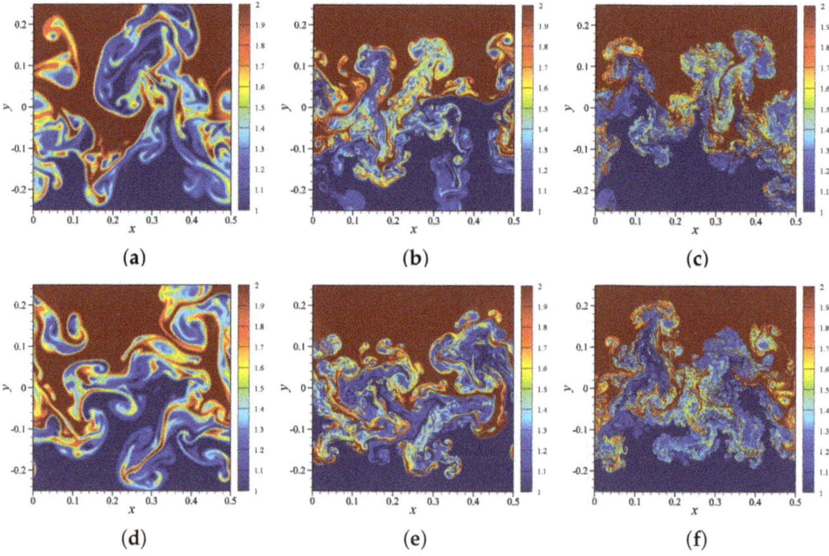

Figure 5. Density contours for the RTI problem with multi-mode perturbation at $t = 4.0$ for different coarse grid resolutions and filtering strength, σ of CS+RF schemes: (**a**) $\sigma = 1.0$ with 256×384 resolution; (**b**) $\sigma = 1.0$ with 1024×1536 resolution; (**c**) $\sigma = 1.0$ with 4096×6144 resolution; (**d**) $\sigma = 0.4$ with 256×384 resolution; (**e**) $\sigma = 0.4$ with 1024×1536 resolution; (**f**) $\sigma = 0.4$ with 4096×6144 resolution. The gravity is directed in a vertically downward direction.

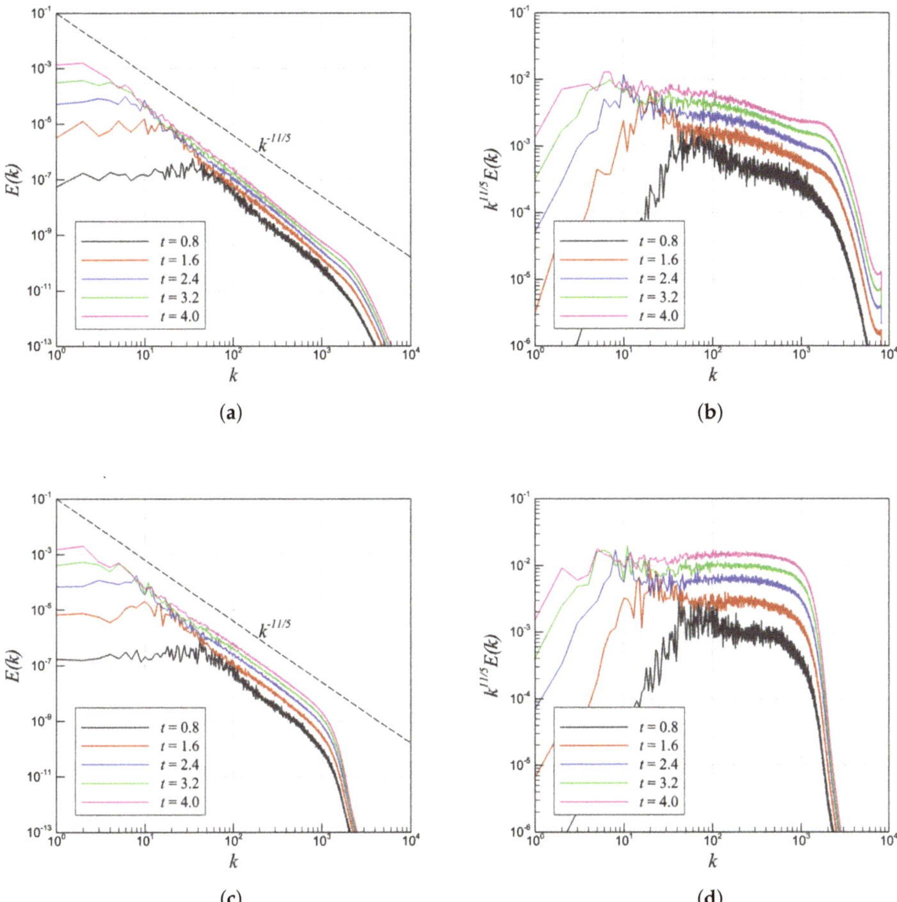

Figure 6. Time evolution of density-weighted kinetic energy spectra and compensated density-weighted kinetic energy spectra for the RTI problem with multi-mode perturbation obtained using different modeling approaches at a resolution of 16384×24576; (**a**) density-weighted spectra using ILES-Roe solver; (**b**) compensated density-weighted spectra using ILES-Roe solver; (**c**) density-weighted spectra using CS+RF ($\sigma = 1.0$) solver; (**d**) compensated density-weighted spectra using CS+RF ($\sigma = 1.0$) solver.

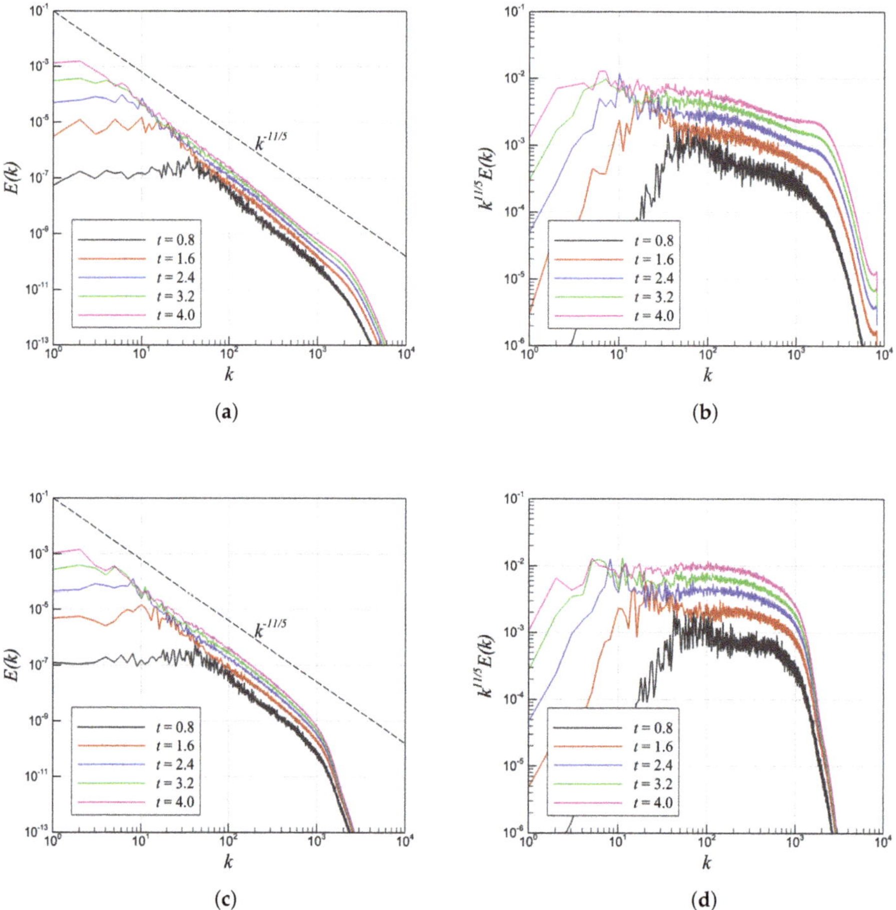

Figure 7. Time evolution of kinetic energy spectra and compensated kinetic energy spectra for the RTI problem with multi-mode perturbation obtained using different modeling approaches at a resolution of 16384×24576; (**a**) kinetic energy spectra using ILES-Roe solver; (**b**) compensated kinetic energy spectra using ILES-Roe solver; (**c**) kinetic energy spectra using CS+RF ($\sigma = 1.0$) solver; (**d**) compensated kinetic energy spectra using CS+RF ($\sigma = 1.0$) solver.

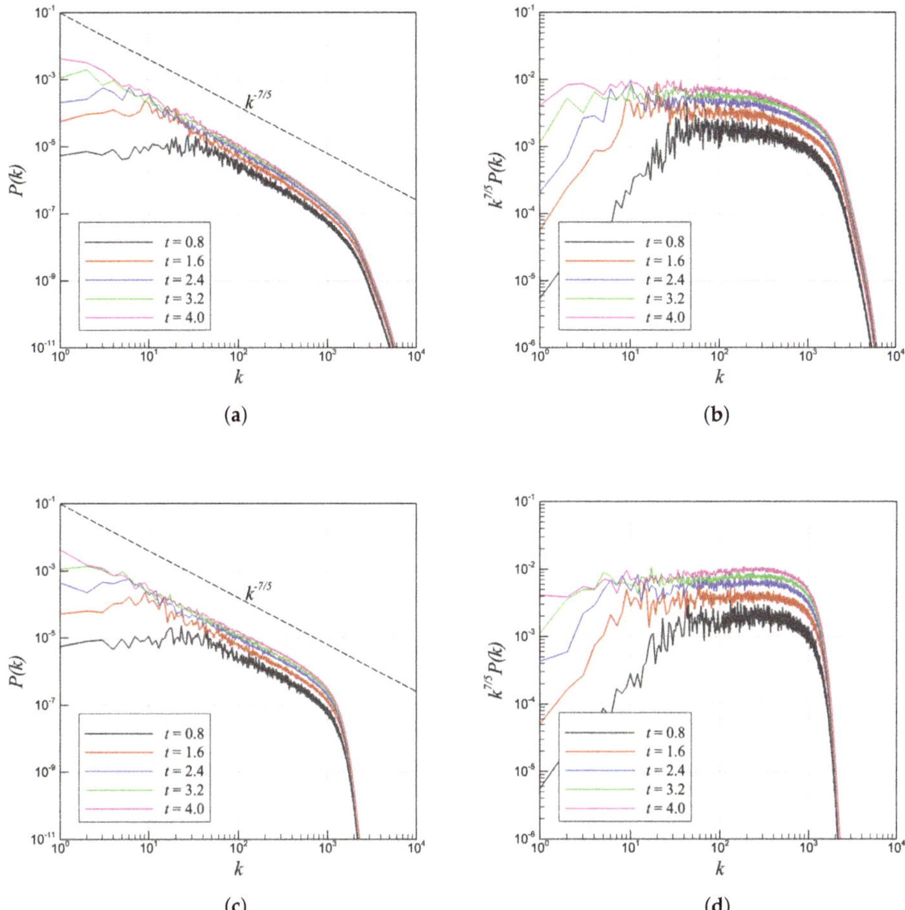

Figure 8. Time evolution of power density spectra and compensated power density spectra for the RTI problem with multi-mode perturbation obtained using different modeling approaches at a resolution of 16384×24576; (**a**) power density spectra using ILES-Roe solver; (**b**) compensated power density spectra using ILES-Roe solver; (**c**) power density spectra using CS+RF ($\sigma = 1.0$) solver; (**d**) compensated power density spectra using CS+RF ($\sigma = 1.0$) solver.

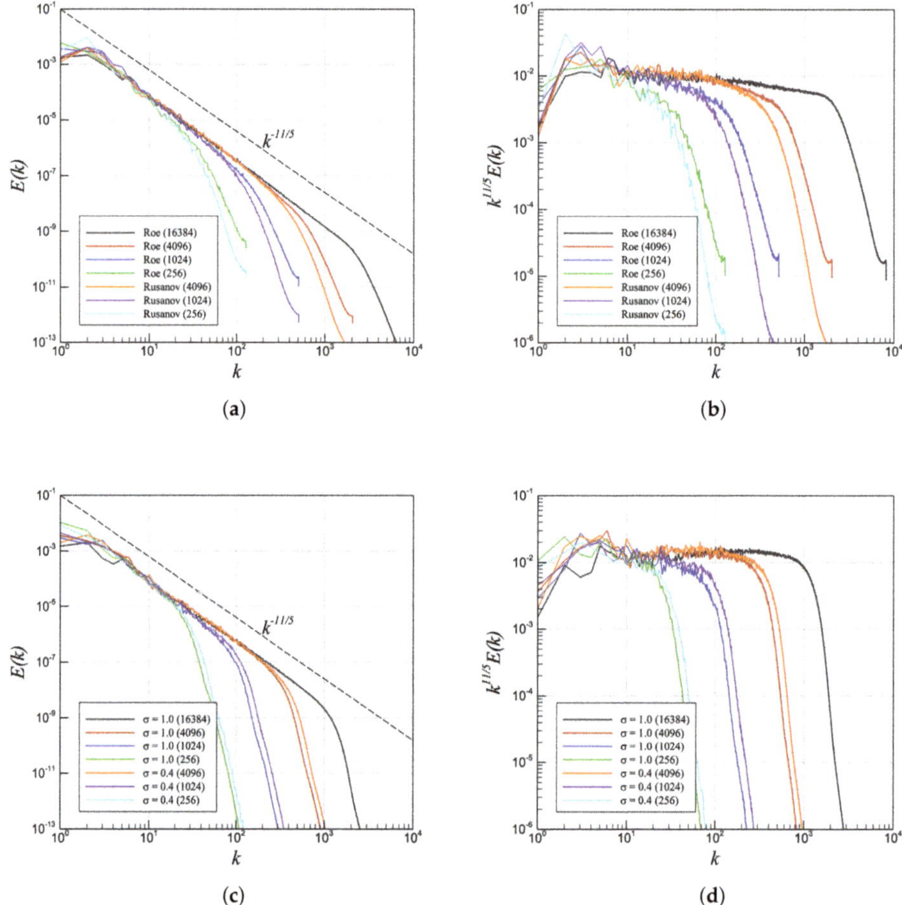

Figure 9. Comparison of ILES (ILES-Roe and ILES-Rusanov) models and CS+RF ($\sigma = 1.0$ and $\sigma = 0.4$) models for the RTI problem with multi-mode perturbation showing the density-weighted kinetic energy spectra and compensated density-weighted kinetic energy spectra at different resolutions; (**a**) density-weighted spectra using ILES solvers; (**b**) compensated density-weighted spectra using ILES solvers; (**c**) density-weighted spectra using CS+RF solvers; (**d**) compensated density-weighted spectra using CS+RF solvers.

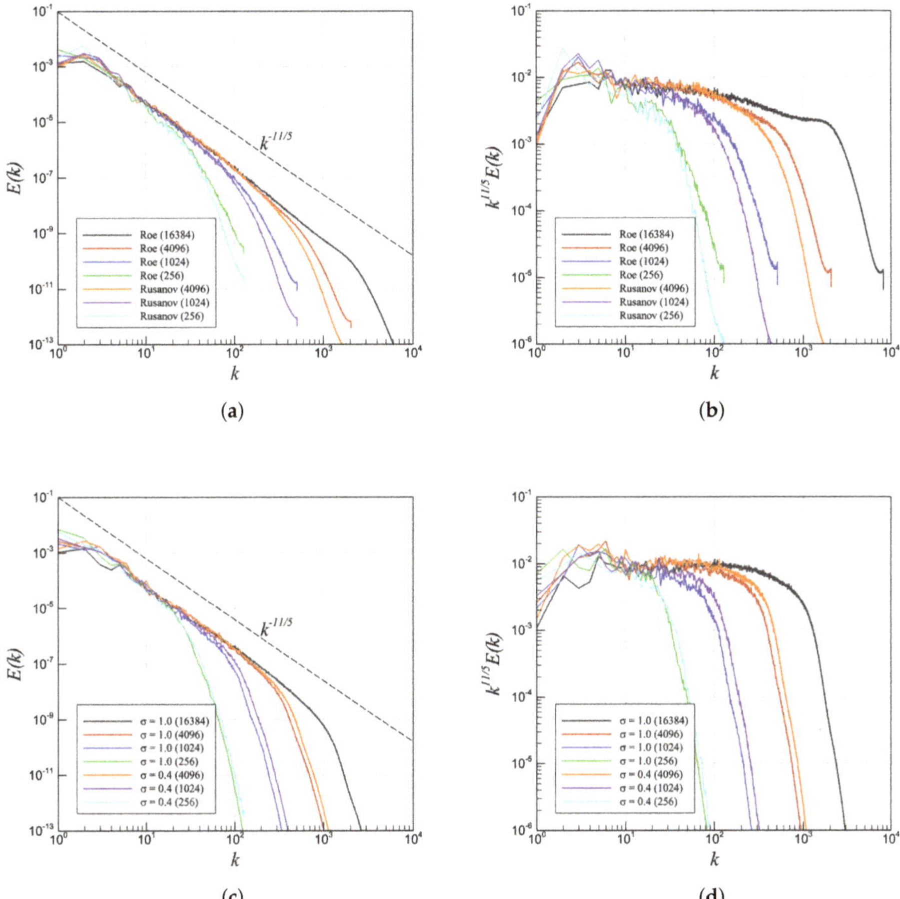

Figure 10. Comparison of ILES (ILES-Roe and ILES-Rusanov) models and CS+RF ($\sigma = 1.0$ and $\sigma = 0.4$) models for the RTI problem with multi-mode perturbation showing the kinetic energy spectra and compensated kinetic energy spectra at different resolutions; (**a**) kinetic energy spectra using ILES solvers; (**b**) compensated kinetic energy spectra using ILES solvers; (**c**) kinetic energy spectra using CS+RF solvers; (**d**) compensated kinetic energy spectra using CS+RF solvers.

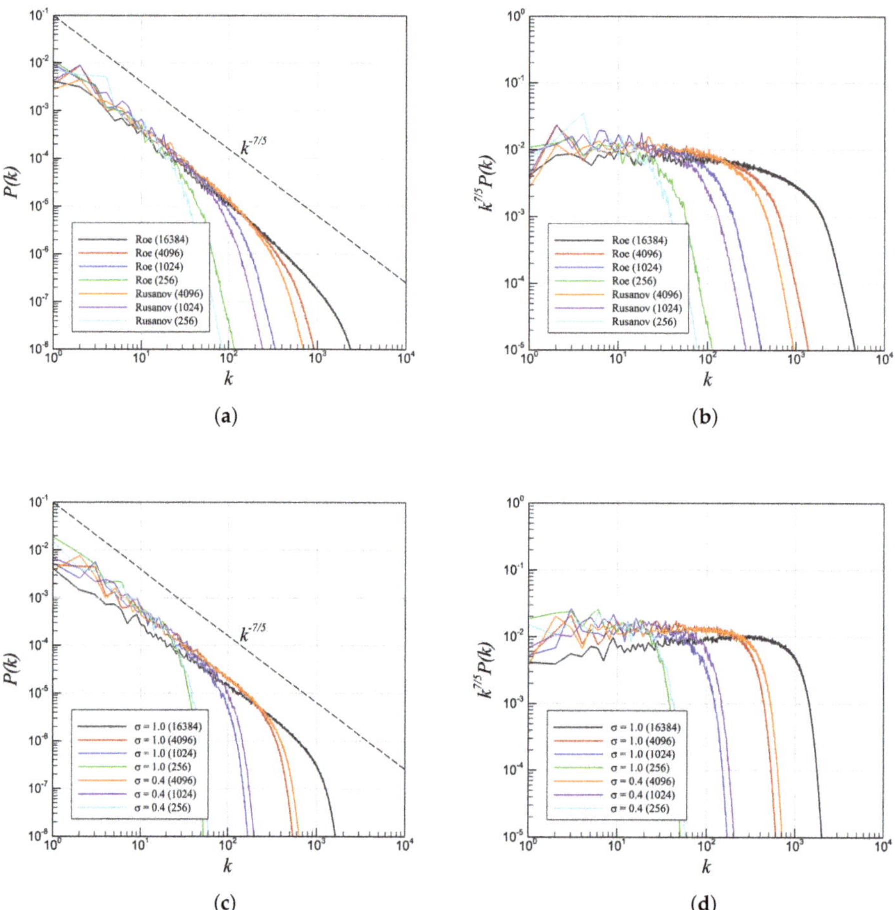

Figure 11. Comparison of ILES (ILES-Roe and ILES-Rusanov) models and CS+RF ($\sigma = 1.0$ and $\sigma = 0.4$) models for the RTI problem with multi-mode perturbation showing the power density spectra and compensated power density spectra at different resolutions; (**a**) power density spectra using ILES solvers; (**b**) compensated power density spectra using ILES solvers; (**c**) power density spectra using CS+RF solvers; (**d**) compensated power density spectra using CS+RF solvers.

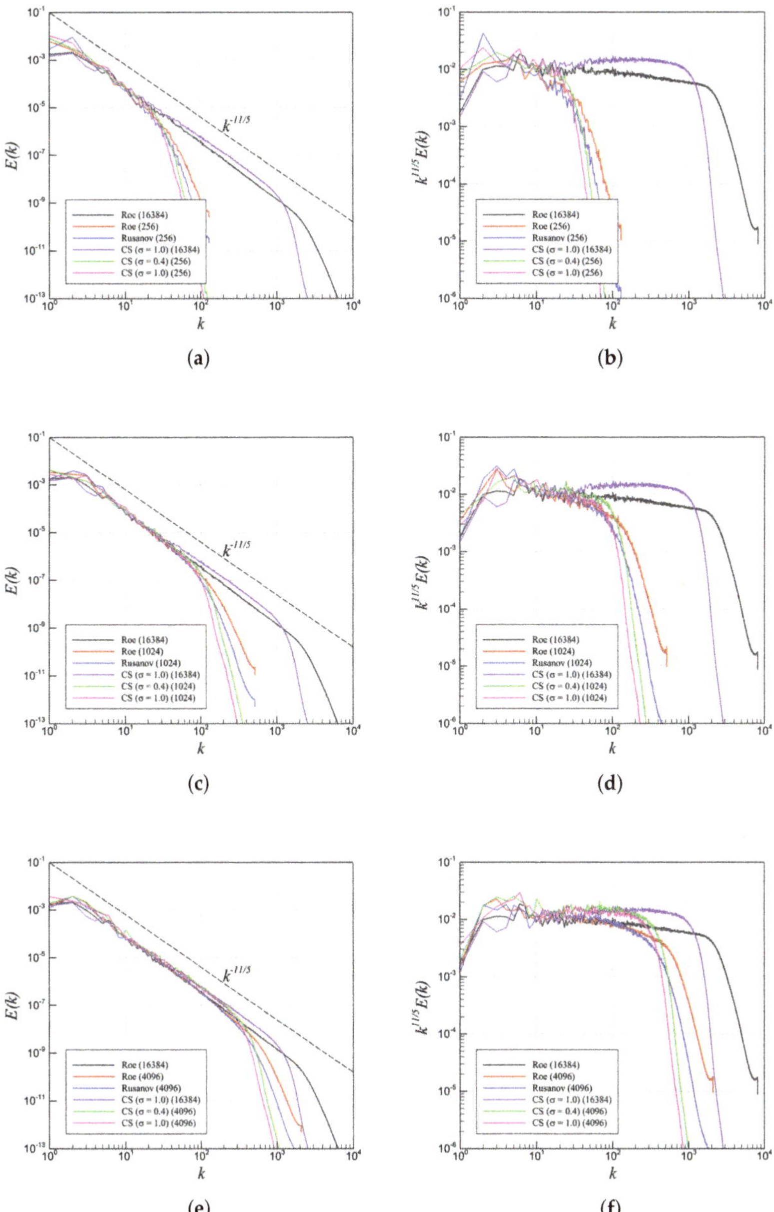

Figure 12. Comparison between ILES and CS+RF models for the RTI problem with multi-mode perturbation at different resolutions; (**a**) density-weighted spectra at 256×384 resolution; (**b**) compensated density-weighted spectra at 256×384 resolution; (**c**) density-weighted spectra at 1024×1536 resolution; (**d**) compensated density-weighted spectra at 1024×1536 resolution; (**e**) density-weighted spectra at 4096×6144 resolution; (**f**) compensated density-weighted spectra at 4096×6144 resolution.

4.3. RTI with Single-Mode Perturbation

Numerical simulation of flows with RTI is comparatively challenging because the instability grows from the small scales of the flow field [19,108]. Since the analytical modeling can be done for single-mode RTI, the numerical study of the RTI with the single-mode perturbation setup has been started very early [33,109] and still being studied extensively to understand and explain the nature of RTI-induced flows [20,31,32,110–113]. For our analyses of the single-mode perturbation case, we first present the time evolution of the density field results obtained by ILES-Roe solver at a high resolution of 8192×24576 in Figure 13. It is observed in many studies that the tips of the spikes of a single-mode RTI-induced flow always maintains the symmetry [14]. Yet in our simulation, the line of symmetry within the spike of the single-mode RTI in Figure 13 is seen broken at time $t = 4.5$. This phenomenon was observed and well explained by Ramaprabhu et al. [31] at late-time of the RTI flow simulation which the authors referred to as "chaotic mixing" at the late-time regime. With simulations of the Euler equations, it can be also seen in [96] that the less dissipative schemes show this interface breaking up, while the more dissipative schemes suppress the instability. In our high-resolution simulation, the presence of small-scale structures can be seen in very early stage which lead to secondary instability, i.e., the KH vortex formation as well as chaotic mixing. However, for small or modest grid resolutions, the numerical viscosity suppresses the small-scale structures and preserves the symmetry which can be seen in Figures 14–17. In Figure 14, we present the state of the density field at $t = 2.7$ (top row) and $t = 4.5$ (bottom row) for the simulations by using the ILES-Rusanov solver at different grid resolutions. It is apparent that the 256×768 and 1024×3072 resolution results hold the symmetry. But the 4096×12288 result shows the development of smaller scales at $t = 2.7$ which leads to the loss of symmetry at final time, $t = 4.5$. Similar conclusions can be made for the ILES-Roe solver results in Figure 15. However, the loss of symmetry can be observed even in 1024×3072 resolution simulation for the ILES-Roe scheme since the ILES-Roe solver is less dissipative compared to the ILES-Rusanov solver. For both CS+RF schemes in Figures 16 and 17, the symmetry holds for lower resolutions and breaks for higher resolution. Since there is no physical viscosity in Euler simulations, we note that the breakup of the interface and the loss of symmetry can be due to the numerics. When increasing the resolution, the loss of symmetry in RTI problems have been also demonstrated in the literature (e.g., see [55,114,115]). Similar observations can be seen when we use the higher-order numerical schemes. We also refer to [116,117] for an illustration of symmetry breaking and increasing mixing in Richtmyer–Meshkov instability problems for solving Euler equations.

Figure 18 presents the density field plots at the final time, $t = 4.5$, to get a comparative idea between the performance of the ILES schemes. We can observe in Figure 18 that the symmetry is maintained in lower resolutions, but starts to break with the increase of the resolution for both ILES solvers. If we look at the 4096×12288 resolution results for both ILES solvers, we can see that the ILES-Roe scheme result is more deviated from the symmetry than the ILES-Rusanov scheme because of the dissipative behavior of the ILES-Rusanov scheme. Based on these findings, we can say our two-dimensional simulation results are consistent with the findings in [31] for three-dimensional RTI case. Additionally, the dimensionless Atwood number defined as:

$$A = \frac{\rho_2 - \rho_1}{\rho_2 + \rho_1},$$

(48)

is set as $A \sim 0.33$ in our case, and it is lower than 0.6, which indicates the formation of reacceleration phase in the flow field due to the secondary KH instabilities. As suggested in the literature [31], these secondary instabilities can be responsible for the change in the usual behavior of the spikes in single-mode RTI flows. These findings are also supported by the works of Liska and Wendroff [96] where they showed that the less dissipative schemes result in an interface break up while the instability might be suppressed by high dissipative schemes. The same observations can be found at final time in Figure 19 that the higher resolution results start to break the symmetry of the spike for both CS+RF schemes. However, the density fields obtained by the CS+RF and ILES schemes seem different due to different amount of dissipation added to the system by different solvers which would eventually lead to different evolution of the flow fields. To get a more precise understanding on the simulation results, we next focus on the density-weighted kinetic energy spectra plots. We can say from the time evolution of the kinetic energy spectra plot in Figure 20 that the trends of the spectra are similar at late-stage of the simulation. We can observe that the density-weighted spectra analysis for ILES-Roe scheme shows an inertial subrange following $k^{-11/5}$ scaling law. On the other hand, the kinetic energy spectra for CS+RF scheme follow the $k^{-11/5}$ scaling in Figure 21. To validate our findings further, we present the power density spectra plots for both ILES-Roe and CS+RF ($\sigma = 1.0$) schemes in Figure 22. The power density spectra display a good alignment with the $k^{-7/5}$ reference line. Since the time evolution of the field for both schemes follow a similar trend, we can consider the solutions at final time for rest of our analysis.

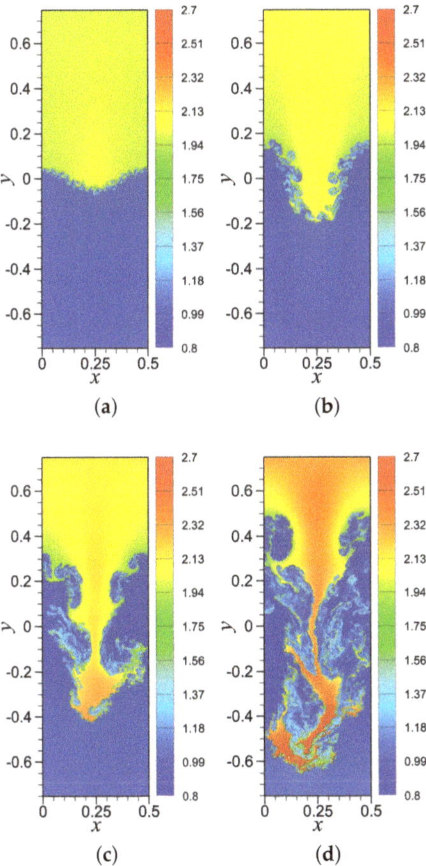

Figure 13. Time evolution of density field for the RTI problem with single-mode perturbation: (**a**) $t = 1.8$; (**b**) $t = 2.7$; (**c**) $t = 3.6$; (**d**) $t = 4.5$. Results are obtained by the ILES-Roe scheme at a resolution of 8192×24576.

Figure 23 shows that the ILES-Rusanov solver is more dissipative than the ILES-Roe solver as expected and the CS+RF scheme with $\sigma = 1.0$ is more dissipative than the $\sigma = 0.4$ solver. One interesting point can be noticed that the density-weighted spectra for the CS+RF scheme tend to deviate from the reference $k^{-11/5}$ line at high wavenumber; however, the kinetic energy spectra in

Figure 24 and the density-weighted spectra in Figure 25 clearly show that the spectra for the CS+RF scheme follow the reference scaling laws. On the other hand, the ILES spectra also maintain the inertial subrange following the $k^{-11/5}$ and $k^{-7/5}$ laws. Finally, similar to the previous section of multi-mode RTI case, we find the CS+RF solver captures more scales in the inertial range than the ILES solvers as shown in Figure 26. However, the ILES solvers resolve more scales in the high wavenumber region. This explains the reason we have seen different density field evolution for different solvers and appearance of smaller scales in ILES solvers than the CS+RF solvers. Since the ILES-Roe solver is least dissipative among the other solvers, we can see in the density contour plots that the ILES-Roe solution deviates most from the symmetry.

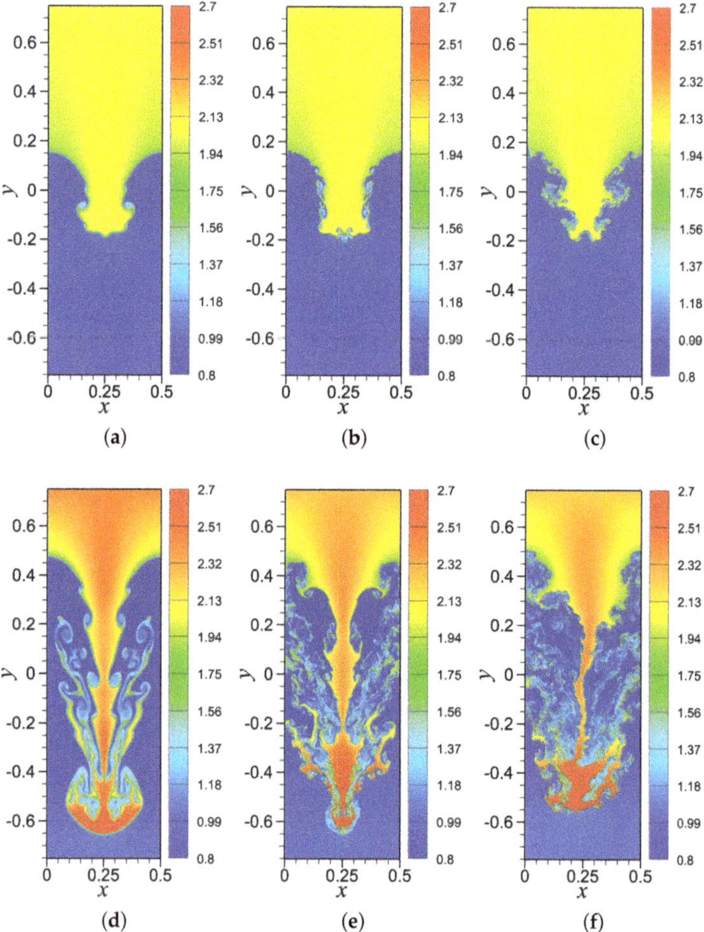

Figure 14. Time evolution of density field for the RTI problem with single-mode perturbation for different coarse grid resolutions using ILES-Rusanov scheme: (**a**) ILES-Rusanov with 256×768 resolution at $t = 2.7$; (**b**) ILES-Rusanov with 1024×3072 resolution at $t = 2.7$; and (**c**) ILES-Rusanov with 4096×12288 resolution at $t = 2.7$; (**d**) ILES-Rusanov with 256×768 resolution at $t = 4.5$; (**e**) ILES-Rusanov with 1024×3072 resolution at $t = 4.5$; and (**f**) ILES-Rusanov with 4096×12288 resolution at $t = 4.5$. The gravity is directed in a vertically downward direction.

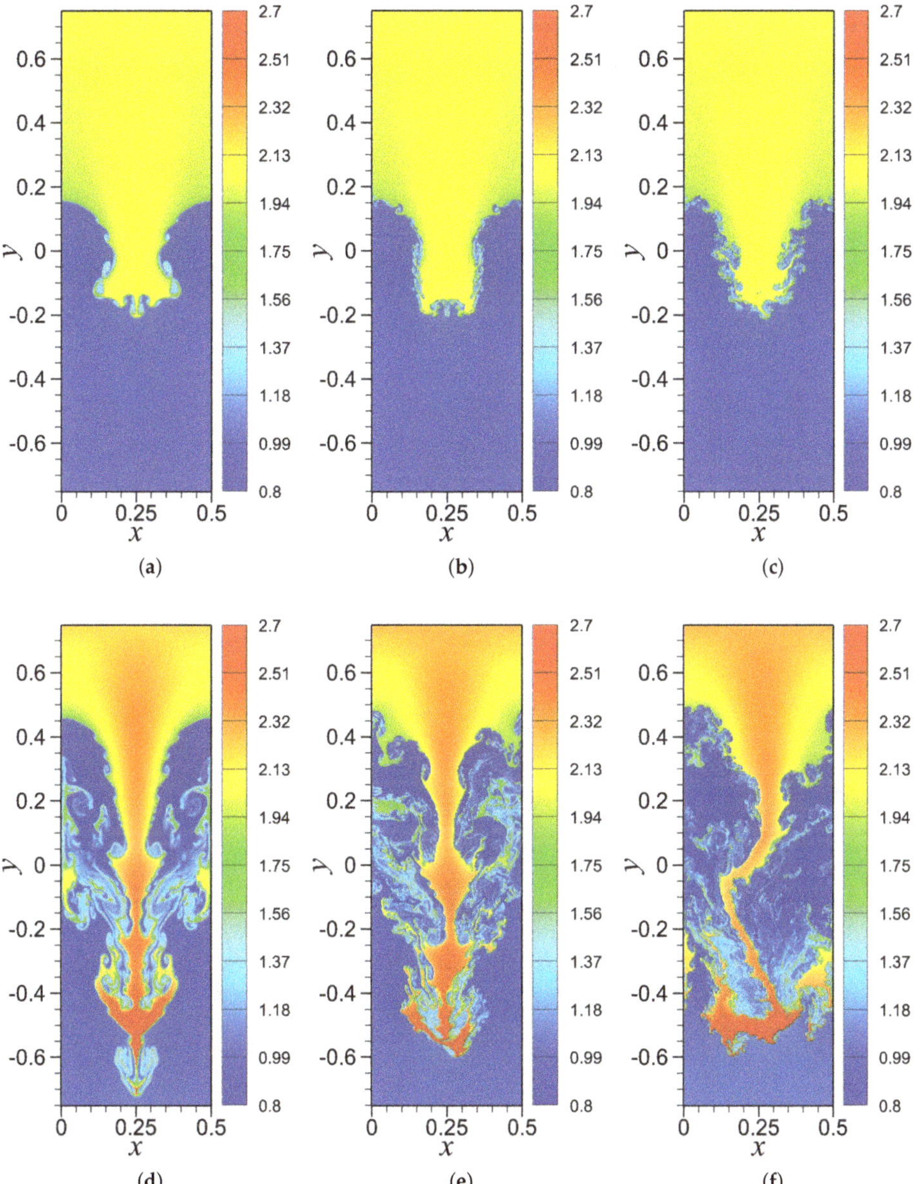

Figure 15. Time evolution of density field for the RTI problem with single-mode perturbation for different coarse grid resolutions using ILES-Roe scheme: (**a**) ILES-Roe with 256×768 resolution at $t = 2.7$; (**b**) ILES-Roe with 1024×3072 resolution at $t = 2.7$; and (**c**) ILES-Roe with 4096×12288 resolution at $t = 2.7$; (**d**) ILES-Roe with 256×768 resolution at $t = 4.5$; (**e**) ILES-Roe with 1024×3072 resolution at $t = 4.5$; and (**f**) ILES-Roe with 4096×12288 resolution at $t = 4.5$. The gravity is directed in a vertically downward direction.

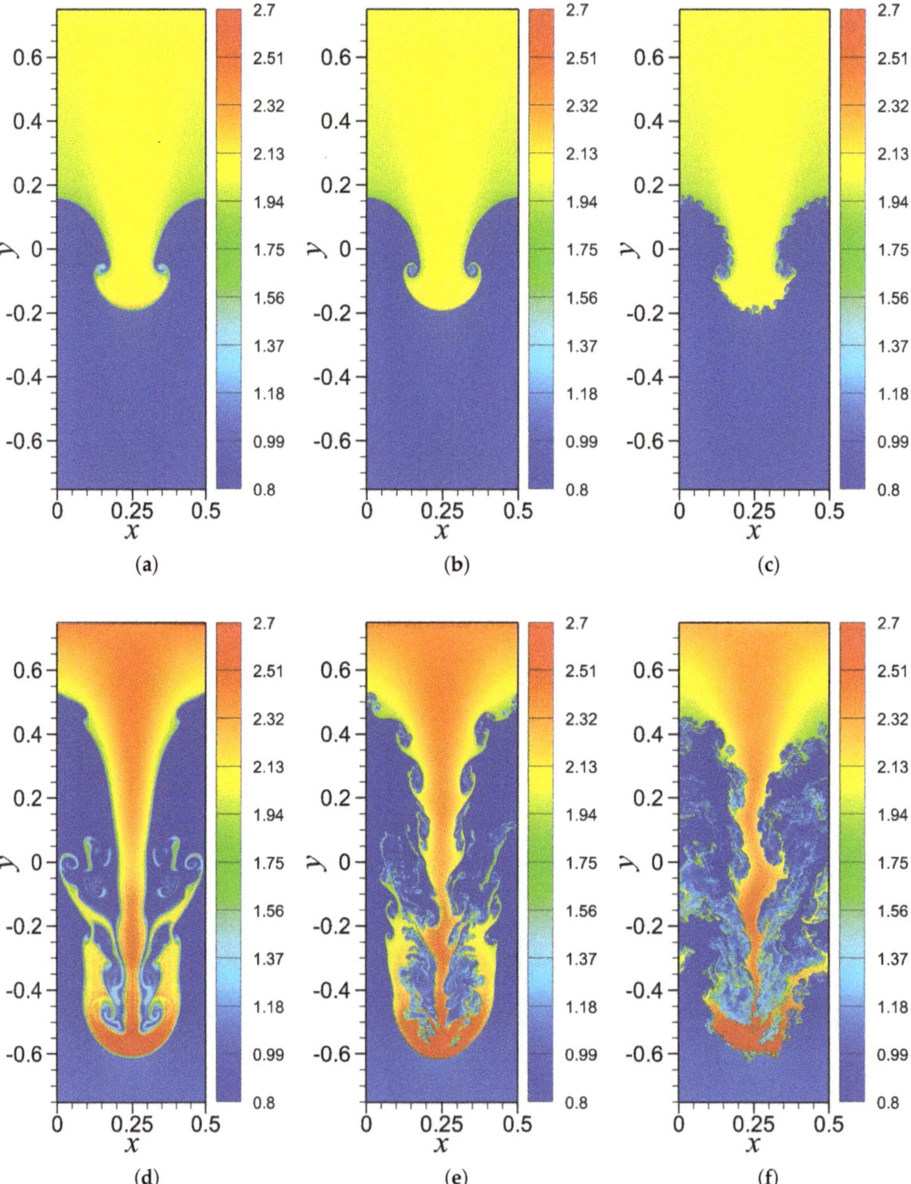

Figure 16. Time evolution of density field for the RTI problem with single-mode perturbation for different coarse grid resolutions using CS+RF scheme ($\sigma = 0.4$): (**a**) CS+RF scheme ($\sigma = 0.4$) with 256×768 resolution at $t = 2.7$; (**b**) CS+RF scheme ($\sigma = 0.4$) with 1024×3072 resolution at $t = 2.7$; and (**c**) CS+RF scheme ($\sigma = 0.4$) with 4096×12288 resolution at $t = 2.7$; (**d**) CS+RF scheme ($\sigma = 0.4$) with 256×768 resolution at $t = 4.5$; (**e**) CS+RF scheme ($\sigma = 0.4$) with 1024×3072 resolution at $t = 4.5$; and (**f**) CS+RF scheme ($\sigma = 0.4$) with 4096×12288 resolution at $t = 4.5$. The gravity is directed in a vertically downward direction.

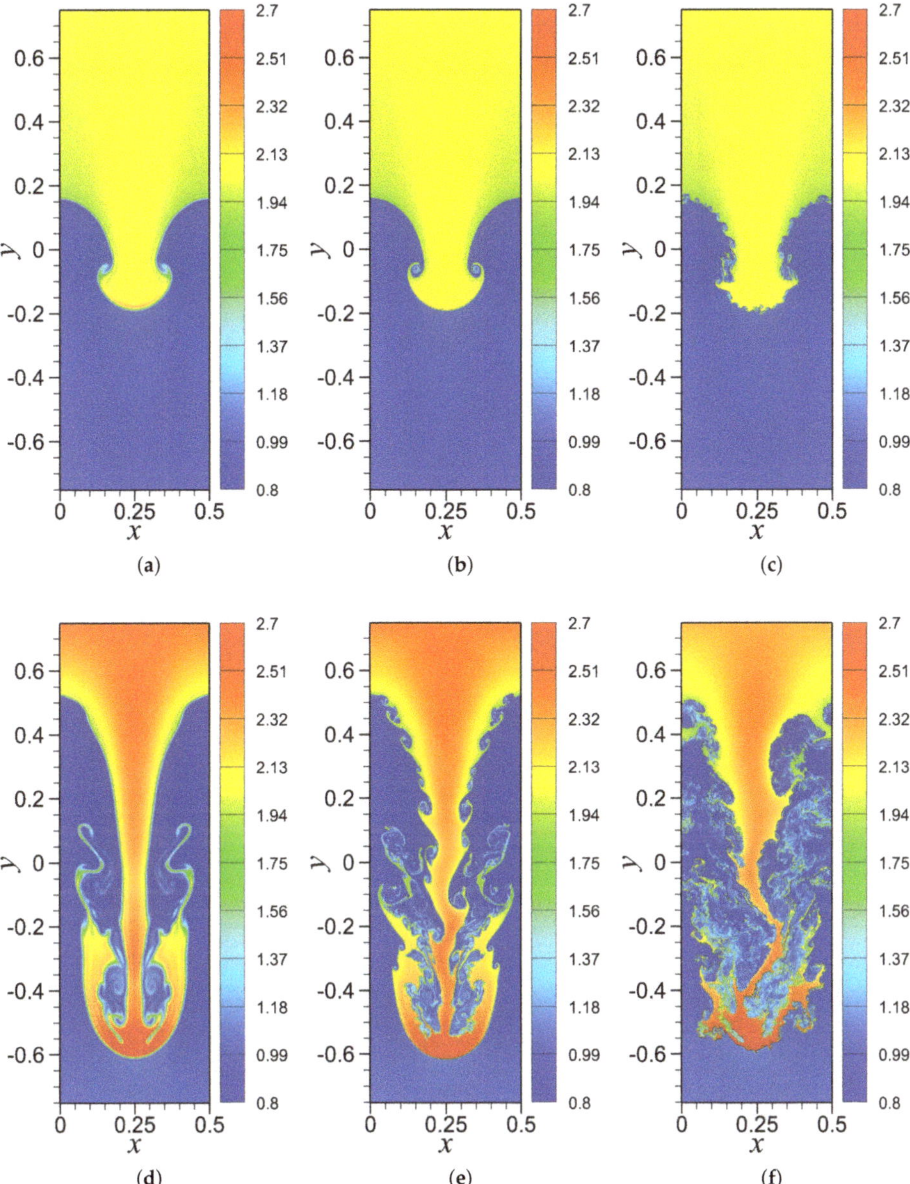

Figure 17. Time evolution of density field for the RTI problem with single-mode perturbation for different coarse grid resolutions using CS+RF scheme ($\sigma = 1.0$): (**a**) CS+RF scheme ($\sigma = 1.0$) with 256×768 resolution at $t = 2.7$; (**b**) CS+RF scheme ($\sigma = 1.0$) with 1024×3072 resolution at $t = 2.7$; and (**c**) CS+RF scheme ($\sigma = 1.0$) with 4096×12288 resolution at $t = 2.7$; (**d**) CS+RF scheme ($\sigma = 1.0$) with 256×768 resolution at $t = 4.5$; (**e**) CS+RF scheme ($\sigma = 1.0$) with 1024×3072 resolution at $t = 4.5$; and (**f**) CS+RF scheme ($\sigma = 1.0$) with 4096×12288 resolution at $t = 4.5$. The gravity is directed in a vertically downward direction.

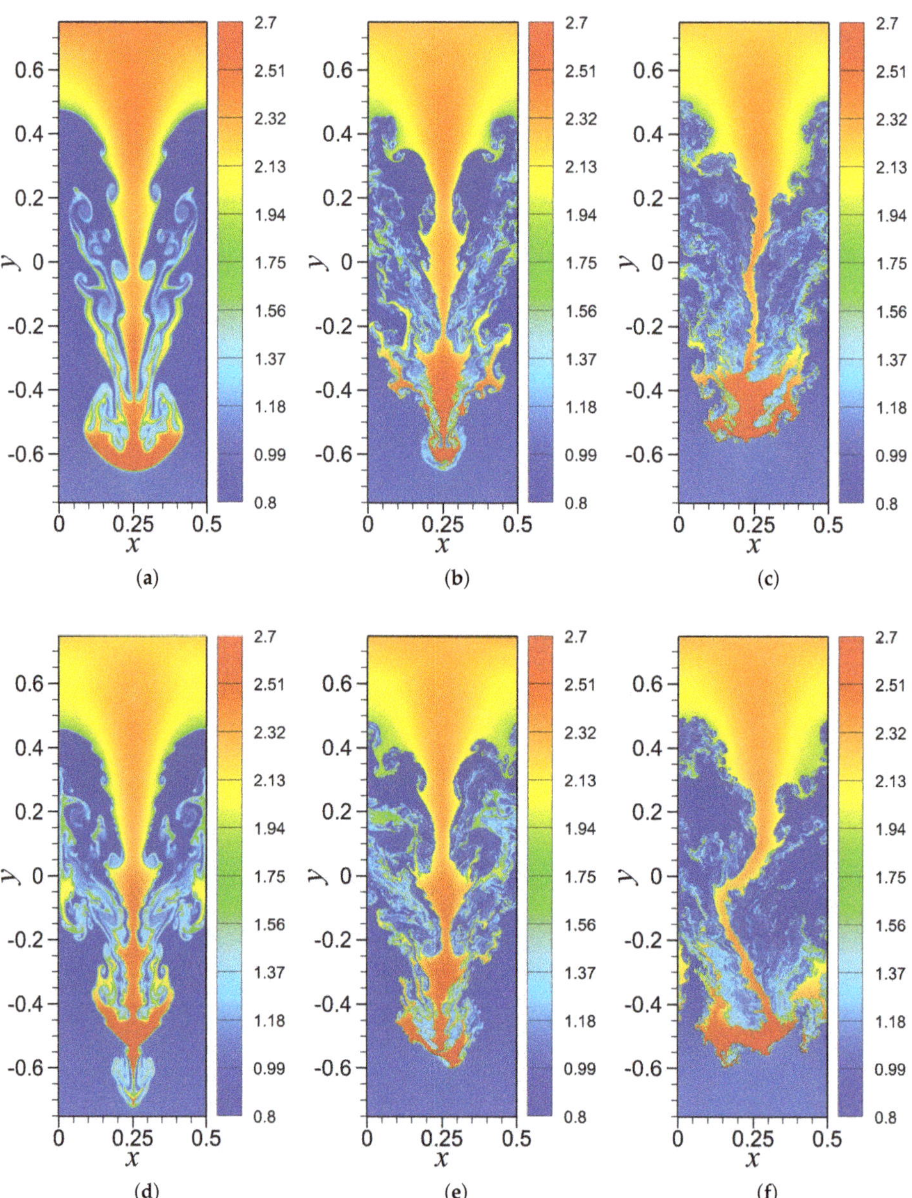

Figure 18. Density contours for the RTI problem with single-mode perturbation at $t = 4.5$ for different coarse grid resolutions and ILES schemes: (**a**) ILES-Rusanov with 256×768 resolution; (**b**) ILES-Rusanov with 1024×3072 resolution; (**c**) ILES-Rusanov with 4096×12288 resolution; (**d**) ILES-Roe with 256×768 resolution; (**e**) ILES-Roe with 1024×3072 resolution; (**f**) ILES-Roe with 4096×12288 resolution. The gravity is directed in a vertically downward direction.

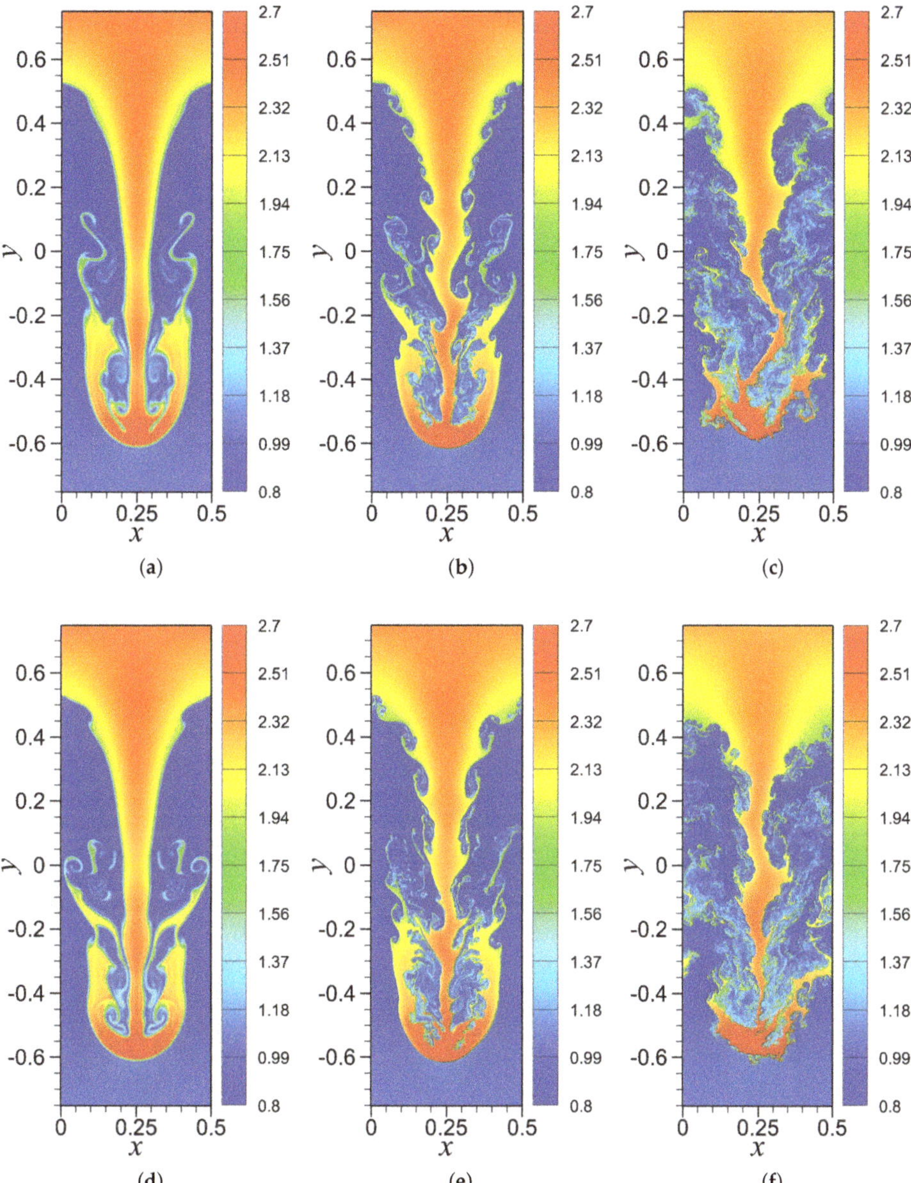

Figure 19. Density contours for the RTI problem with single-mode perturbation at $t = 4.5$ for different coarse grid resolutions and filtering strength, σ of CS+RF schemes: (**a**) $\sigma = 1.0$ with 256×768 resolution; (**b**) $\sigma = 1.0$ with 1024×3072 resolution; (**c**) $\sigma = 1.0$ with 4096×12288 resolution; (**d**) $\sigma = 0.4$ with 256×768 resolution; (**e**) $\sigma = 0.4$ with 1024×3072 resolution; (**f**) $\sigma = 0.4$ with 4096×12288 resolution. The gravity is directed in a vertically downward direction.

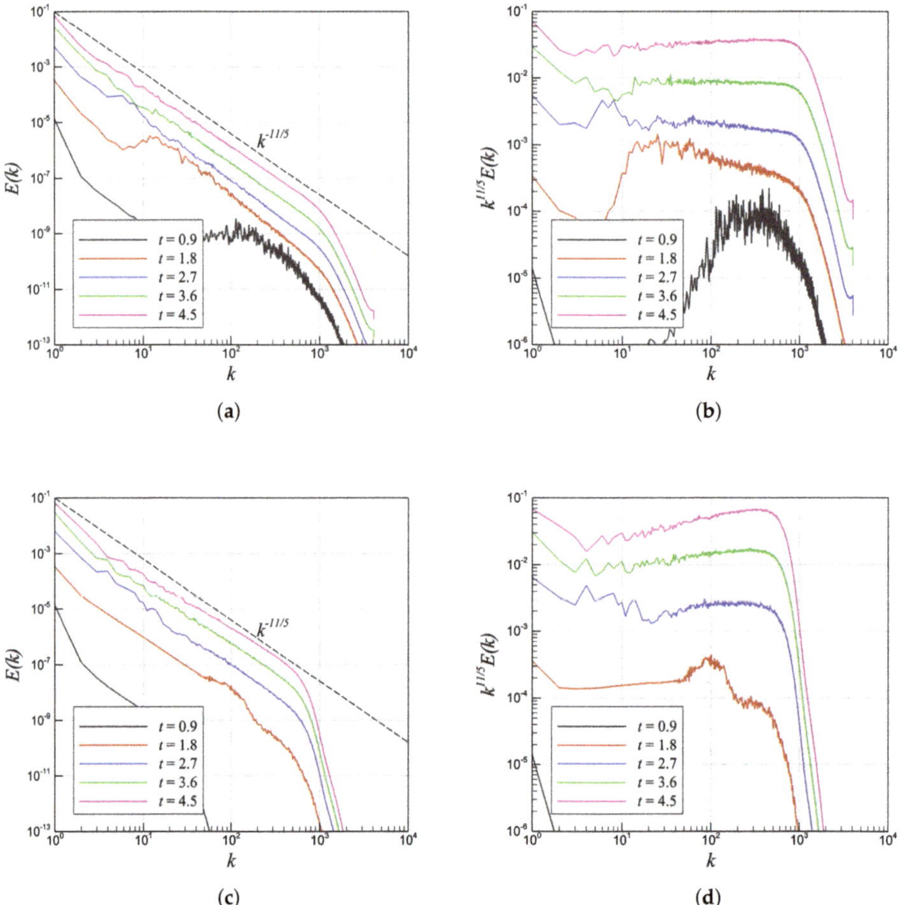

Figure 20. Time evolution of density-weighted kinetic energy spectra and compensated density-weighted kinetic energy spectra for the RTI problem with single-mode perturbation obtained using different modeling approaches at a resolution of 8192×24576; (**a**) density-weighted spectra using ILES-Roe solver; (**b**) compensated density-weighted spectra using ILES-Roe solver; (**c**) density-weighted spectra using CS+RF ($\sigma = 1.0$) solver; (**d**) compensated density-weighted spectra using CS+RF ($\sigma = 1.0$) solver.

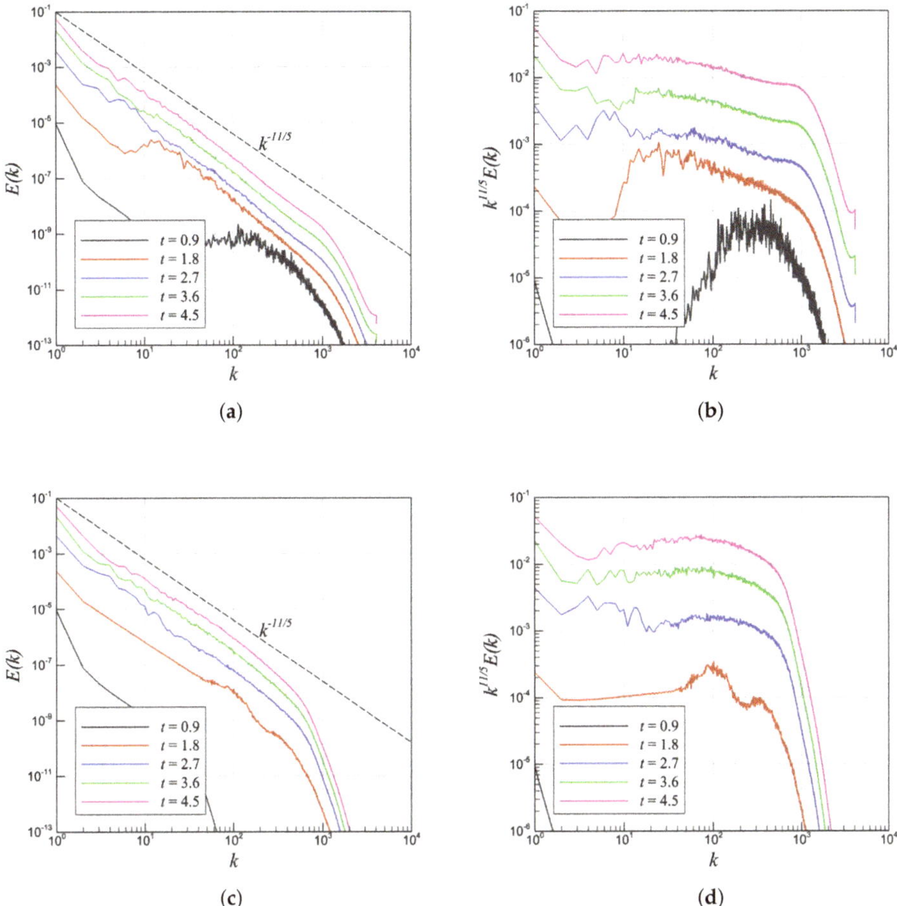

Figure 21. Time evolution of kinetic energy spectra and compensated kinetic energy spectra for the RTI problem with single-mode perturbation obtained using different modeling approaches at a resolution of 8192×24576; (**a**) kinetic energy spectra using ILES-Roe solver; (**b**) compensated kinetic energy spectra using ILES-Roe solver; (**c**) kinetic energy spectra using CS+RF ($\sigma = 1.0$) solver; (**d**) compensated kinetic energy spectra using CS+RF ($\sigma = 1.0$) solver.

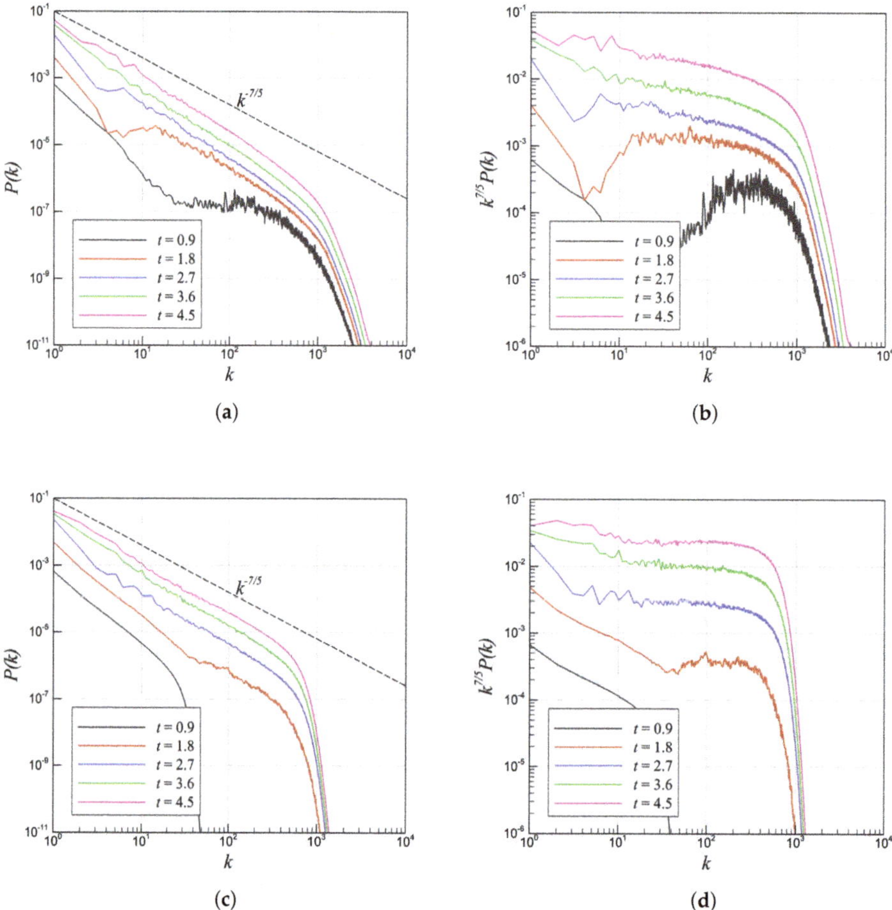

Figure 22. Time evolution of power density spectra and compensated power density spectra for the RTI problem with single-mode perturbation obtained using different modeling approaches at a resolution of 8192×24576; (**a**) power density spectra using ILES-Roe solver; (**b**) compensated power density spectra using ILES-Roe solver; (**c**) power density spectra using CS+RF ($\sigma = 1.0$) solver; (**d**) compensated power density spectra using CS+RF ($\sigma = 1.0$) solver.

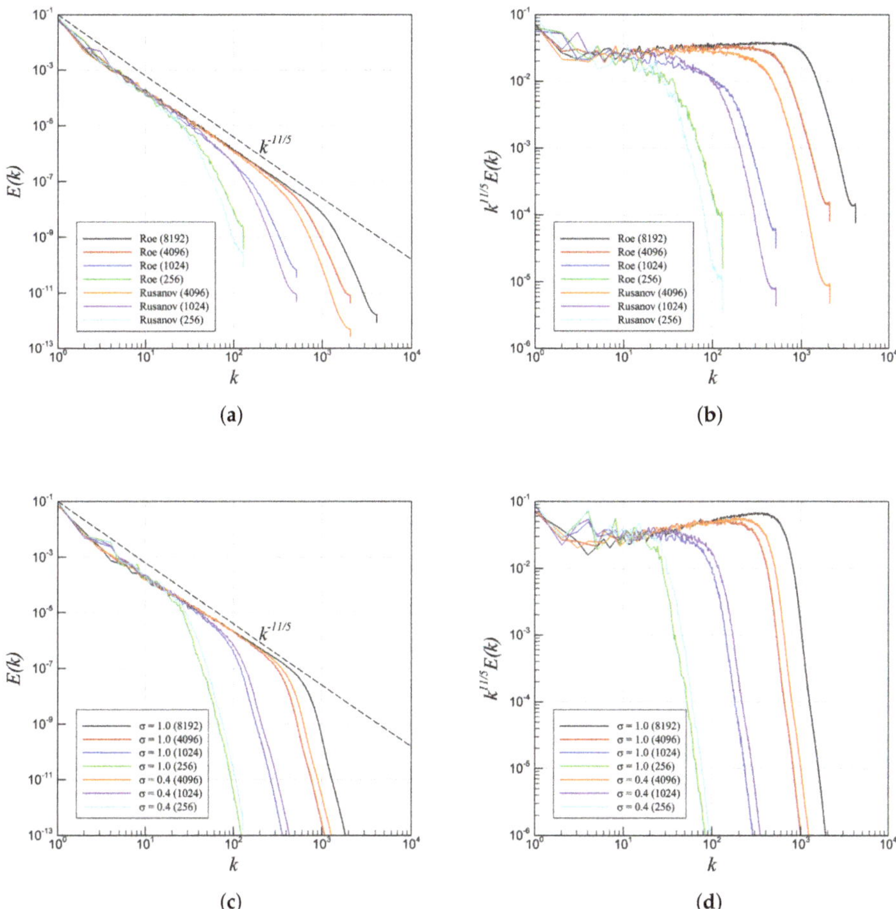

Figure 23. Comparison of ILES (ILES-Roe and ILES-Rusanov) models and CS+RF ($\sigma = 1.0$ and $\sigma = 0.4$) models at different resolutions for the RTI problem with single-mode perturbation showing the density-weighted kinetic energy spectra and compensated density-weighted kinetic energy spectra; (**a**) density-weighted spectra using ILES solvers; (**b**) compensated density-weighted spectra using ILES solvers; (**c**) density-weighted spectra using CS+RF solvers; (**d**) compensated density-weighted spectra using CS+RF solvers.

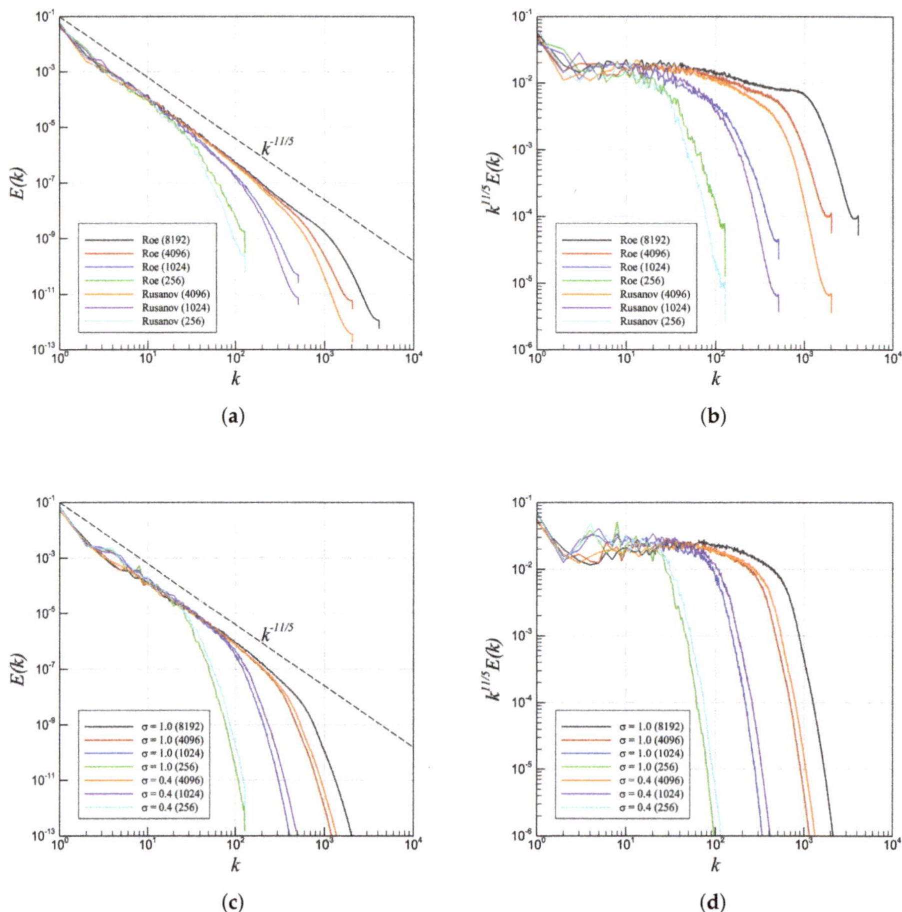

Figure 24. Comparison of ILES (ILES-Roe and ILES-Rusanov) models and CS+RF ($\sigma = 1.0$ and $\sigma = 0.4$) models at different resolutions for the RTI problem with single-mode perturbation showing the kinetic energy spectra and compensated kinetic energy spectra; (**a**) kinetic energy spectra using ILES solvers; (**b**) compensated kinetic energy spectra using ILES solvers; (**c**) kinetic energy spectra using CS+RF solvers; (**d**) compensated kinetic energy spectra using CS+RF solvers.

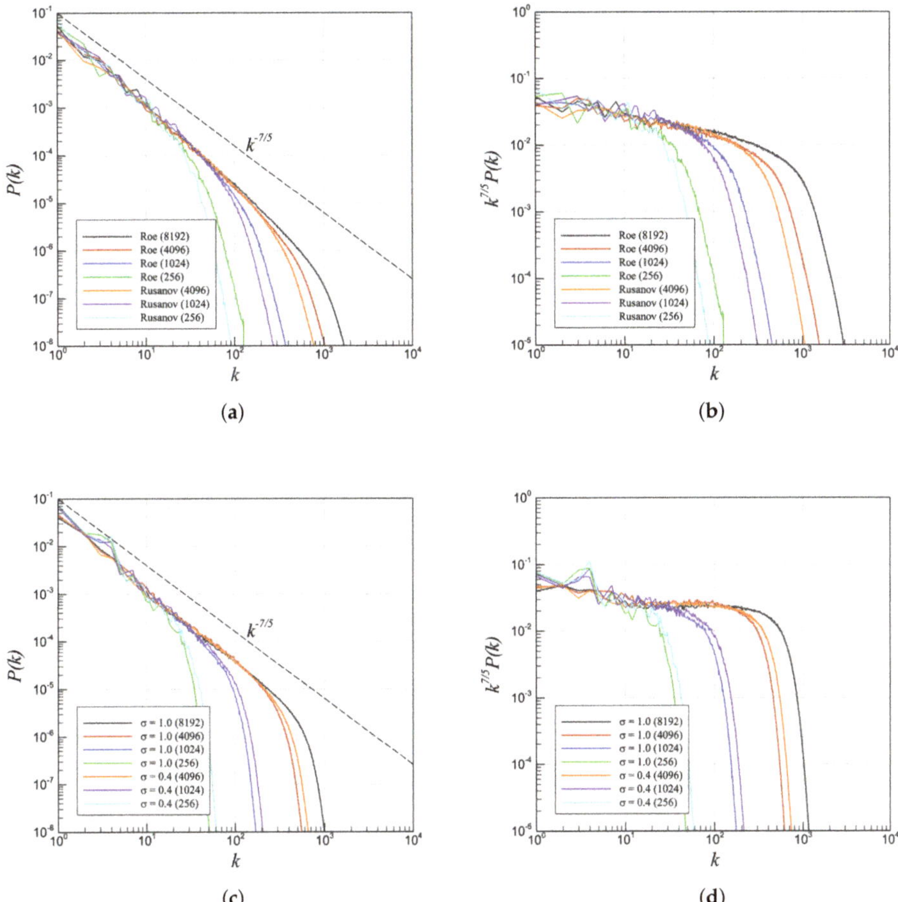

Figure 25. Comparison of ILES (ILES-Roe and ILES-Rusanov) models and CS+RF ($\sigma = 1.0$ and $\sigma = 0.4$) models at different resolutions for the RTI problem with single-mode perturbation showing the power density spectra and compensated power density spectra; (**a**) power density spectra using ILES solvers; (**b**) compensated power density spectra using ILES solvers; (**c**) power density spectra using CS+RF solvers; (**d**) compensated power density spectra using CS+RF solvers.

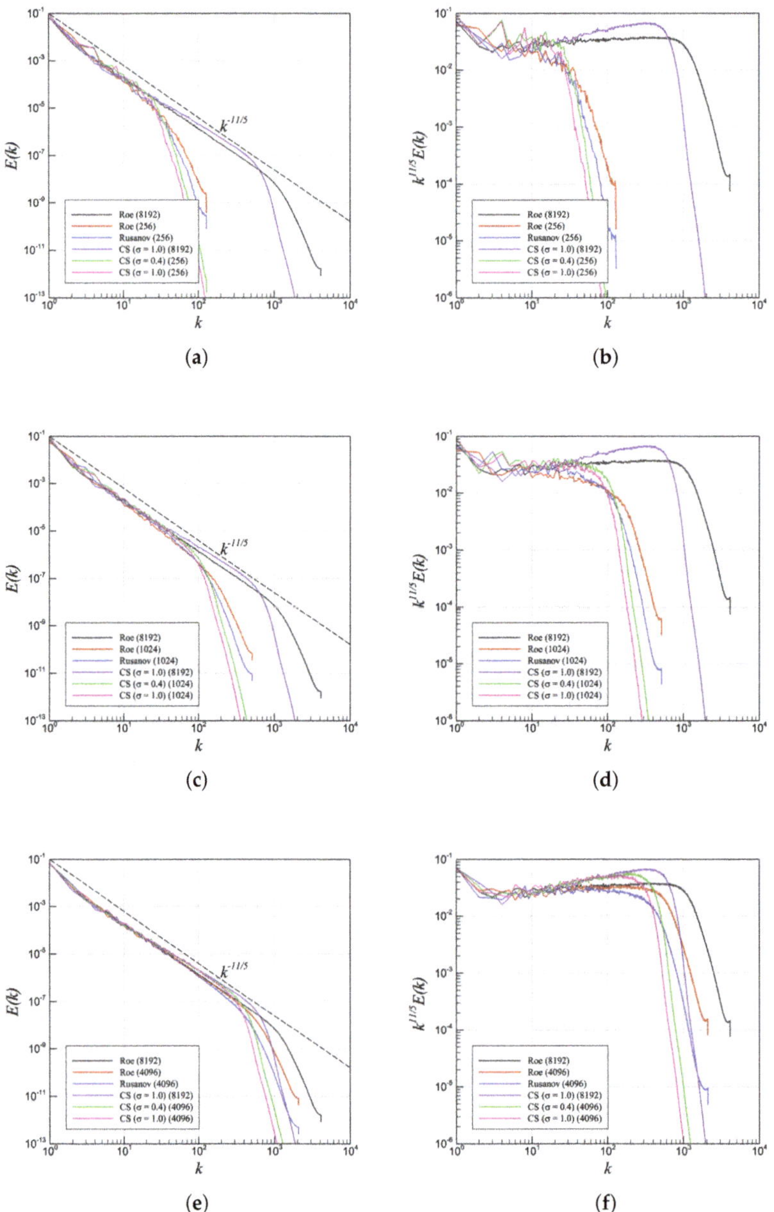

Figure 26. Comparison between ILES and CS+RF models for the RTI problem with single-mode perturbation at different resolutions; (**a**) density-weighted spectra at 256×768 resolution; (**b**) compensated density-weighted spectra at 256×768 resolution; (**c**) density-weighted spectra at 1024×3072 resolution; (**d**) compensated density-weighted spectra at 1024×3072 resolution; (**e**) density-weighted spectra at 4096×12288 resolution; (**f**) compensated density-weighted spectra at 4096×12288 resolution.

5. Summary and Conclusions

In this paper, we put an effort to show the performance of a relaxation filtering approach using central scheme (CS+RF) on resolving the flows resulting from Rayleigh–Taylor hydrodynamic instability, and compare the simulation results with the results obtained by two common ILES-Riemann solver schemes. To assess the performance of the solvers, we use the density field contours and different spectra plots. We further analyze the resolution capacity of both CS+RF and ILES schemes as well as the flow nature at high-resolution simulation using CS+RF and ILES schemes. To validate the observations from the field plots, we use the statistical tools, i.e., kinetic energy and power density spectra plots for both high and coarse resolutions which show consistency with the existing results in the literature. In our investigation, we consider the two-dimensional RTI test problem with two different initial conditions. From the simulation results of both cases, we come to this conclusion that the CS+RF schemes capture more scales in the inertial subregion whereas the ILES schemes resolve a wide range of scales in high wavenumber region. The ILES-Rusanov scheme is more dissipative than the ILES-Roe scheme, and hence, the ILES-Roe scheme tends to deviate more from symmetry in the spike of single-mode RTI case. On the other hand, it is also observed that the dissipation can be controlled by σ parameter for CS+RF scheme which also affect the perturbation as well as the evolution of the flow field. Furthermore, we observe that the kinetic energy spectra follow $k^{-11/5}$ scaling law for both multi-mode and single-mode RTI case whereas the power density spectra plots are seen to be more align to $k^{-7/5}$ line at different resolutions. Also, we observe a chaotic mixing at late-time stage for single-mode RTI case at high resolution. It is because of the formation of secondary KH instabilities at high-resolution simulation of single-mode RTI case. The higher numerical dissipation due to the coarser resolution suppresses the formation of secondary instabilities which is the reason behind the preservation of symmetry for our coarse-resolution simulation results. Overall, we believe the study of relaxation filtering approach using CS will be a good contribution to the numerical study of RTI-induced flows as well as for understanding the nature of the flow field with the evolution of the instability.

Author Contributions: Conceptualization, O.S.; Data curation, S.M.R. and O.S.; Formal analysis, S.M.R. and O.S.; Visualization, S.M.R. and O.S.; Methodology, O.S.; Supervision, O.S.; Writing—original draft, S.M.R.; and Writing—review and editing, S.M.R. and O.S.

Funding: This research received no external funding.

Acknowledgments: The computing for this project was performed by using resources from the High-Performance Computing Center (HPCC) at Oklahoma State University.

References

1. Taylor, G.I. The instability of liquid surfaces when accelerated in a direction perpendicular to their planes. I. *Proc. R. Soc. Lond. Ser. Math. Phys. Sci.* **1950**, *201*, 192–196.
2. Cui, A.; Street, R.L. Large-eddy simulation of coastal upwelling flow. *Environ. Fluid Mech.* **2004**, *4*, 197–223. [CrossRef]
3. Jevons, W.S. II. On the cirrous form of cloud. *Lond. Edinburgh Dublin Philos. Mag. J. Sci.* **1857**, *14*, 22–35. [CrossRef]
4. Bradley, P. The effect of mix on capsule yields as a function of shell thickness and gas fill. *Phys. Plasmas* **2014**, *21*, 062703. [CrossRef]
5. Bell, J.; Day, M.; Rendleman, C.; Woosley, S.; Zingale, M. Direct numerical simulations of type Ia supernovae flames. II. The Rayleigh-Taylor instability. *Astrophys. J.* **2004**, *608*, 883. [CrossRef]
6. Hillebrandt, W.; Niemeyer, J.C. Type Ia supernova explosion models. *Annu. Rev. Astron. Astrophys.* **2000**, *38*, 191–230. [CrossRef]
7. Racca, R.A.; Annett, C.H. Simple demonstration of Rayleigh-Taylor instability. *Am. J. Phys.* **1985**, *53*, 484–486. [CrossRef]

8. Chertkov, M.; Lebedev, V.; Vladimirova, N. Reactive Rayleigh–Taylor turbulence. *J. Fluid Mech.* **2009**, *633*, 1–16. [CrossRef]
9. Veynante, D.; Trouvé, A.; Bray, K.; Mantel, T. Gradient and counter-gradient scalar transport in turbulent premixed flames. *J. Fluid Mech.* **1997**, *332*, 263–293. [CrossRef]
10. Petrasso, R.D. Rayleigh's challenge endures. *Nature* **1994**, *367*, 217. [CrossRef]
11. Lewis, D.J. The instability of liquid surfaces when accelerated in a direction perpendicular to their planes. II. *Proc. R. Soc. Lond. Ser. Math. Phys. Sci.* **1950**, *202*, 81–96.
12. Lord, R. Investigation of the character of the equilibrium of an incompressible heavy fluid of variable density. *Sci. Pap.* **1900**; pp. 200–207.
13. Zhou, Y. Rayleigh–Taylor and Richtmyer–Meshkov instability induced flow, turbulence, and mixing. I. *Phys. Rep.* **2017**, *720–722*, 1–136.
14. Zhou, Y. Rayleigh–Taylor and Richtmyer–Meshkov instability induced flow, turbulence, and mixing. II. *Phys. Rep.* **2017**, *723–725*, 1–160. [CrossRef]
15. Piriz, A.; Cortazar, O.; Lopez Cela, J.; Tahir, N. The Rayleigh-Taylor instability. *Am. J. Phys.* **2006**, *74*, 1095–1098. [CrossRef]
16. Boffetta, G.; Mazzino, A. Incompressible Rayleigh–Taylor turbulence. *Annu. Rev. Fluid Mech.* **2017**, *49*, 119–143. [CrossRef]
17. Sharp, D.H. *Overview of Rayleigh-Taylor Instability*; Technical Report; Los Alamos National Lab.: Los Alamos, NM, USA, 1983.
18. Dimonte, G.; Youngs, D.; Dimits, A.; Weber, S.; Marinak, M.; Wunsch, S.; Garasi, C.; Robinson, A.; Andrews, M.; Ramaprabhu, P.; et al. A comparative study of the turbulent Rayleigh–Taylor instability using high-resolution three-dimensional numerical simulations: The Alpha-Group collaboration. *Phys. Fluids* **2004**, *16*, 1668–1693. [CrossRef]
19. Mueschke, N.J.; Schilling, O. Investigation of Rayleigh–Taylor turbulence and mixing using direct numerical simulation with experimentally measured initial conditions. I. Comparison to experimental data. *Phys. Fluids* **2009**, *21*, 014106. [CrossRef]
20. Wei, T.; Livescu, D. Late-time quadratic growth in single-mode Rayleigh-Taylor instability. *Phys. Rev. E* **2012**, *86*, 046405. [CrossRef]
21. Dalziel, S.; Linden, P.; Youngs, D. Self-similarity and internal structure of turbulence induced by Rayleigh–Taylor instability. *J. Fluid Mech.* **1999**, *399*, 1–48. [CrossRef]
22. Ristorcelli, J.; Clark, T. Rayleigh–Taylor turbulence: Self-similar analysis and direct numerical simulations. *J. Fluid Mech.* **2004**, *507*, 213–253. [CrossRef]
23. Xin, J.; Yan, R.; Wan, Z.H.; Sun, D.J.; Zheng, J.; Zhang, H.; Aluie, H.; Betti, R. Two mode coupling of the ablative Rayleigh-Taylor instabilities. *Phys. Plasmas* **2019**, *26*, 032703. [CrossRef]
24. Jun, B.I.; Norman, M.L.; Stone, J.M. A numerical study of Rayleigh-Taylor instability in magnetic fluids. *Astrophys. J.* **1995**, *453*, 332. [CrossRef]
25. Cabot, W. Comparison of two-and three-dimensional simulations of miscible Rayleigh-Taylor instability. *Phys. Fluids* **2006**, *18*, 045101. [CrossRef]
26. Anuchina, N.; Volkov, V.; Gordeychuk, V.; Es' kov, N.; Ilyutina, O.; Kozyrev, O. Numerical simulations of Rayleigh–Taylor and Richtmyer–Meshkov instability using MAH-3 code. *J. Comput. Appl. Math.* **2004**, *168*, 11–20. [CrossRef]
27. Zhao, D.; Aluie, H. Inviscid criterion for decomposing scales. *Phys. Rev. Fluids* **2018**, *3*, 054603. [CrossRef]
28. Livescu, D.; Ristorcelli, J. Buoyancy-driven variable-density turbulence. *J. Fluid Mech.* **2007**, *591*, 43–71. [CrossRef]
29. Kokkinakis, I.; Drikakis, D.; Youngs, D.; Williams, R. Two-equation and multi-fluid turbulence models for Rayleigh–Taylor mixing. *Int. J. Heat Fluid Flow* **2015**, *56*, 233–250. [CrossRef]
30. Youngs, D. Large eddy simulation and 1D/2D engineering models for Rayleigh-Taylor mixing. In Proceedings of the IWPCTM: 12th International Workshop on the Physics of Compressible Turbulent Mixing, Moscow, Russia, 12–17 July 2010; pp. 12–17.
31. Ramaprabhu, P.; Dimonte, G.; Woodward, P.; Fryer, C.; Rockefeller, G.; Muthuraman, K.; Lin, P.H.; Jayaraj, J. The late-time dynamics of the single-mode Rayleigh-Taylor instability. *Phys. Fluids* **2012**, *24*, 074107. [CrossRef]

32. Glimm, J.; Li, X.l.; Lin, A.D. Nonuniform approach to terminal velocity for single mode Rayleigh-Taylor instability. *Acta Math. Appl. Sin.* **2002**, *18*, 1–8. [CrossRef]
33. Daly, B.J. Numerical study of two fluid Rayleigh-Taylor instability. *Phys. Fluids* **1967**, *10*, 297–307. [CrossRef]
34. Kane, J.; Arnett, D.; Remington, B.; Glendinning, S.; Bazán, G.; Müller, E.; Fryxell, B.; Teyssier, R. Two-dimensional versus three-dimensional supernova hydrodynamic instability growth. *Astrophys. J.* **2000**, *528*, 989. [CrossRef]
35. Goncharov, V. Analytical model of nonlinear, single-mode, classical Rayleigh-Taylor instability at arbitrary Atwood numbers. *Phys. Rev. Lett.* **2002**, *88*, 134502. [CrossRef]
36. Young, Y.N.; Tufo, H.; Dubey, A.; Rosner, R. On the miscible Rayleigh–Taylor instability: Two and three dimensions. *J. Fluid Mech.* **2001**, *447*, 377–408. [CrossRef]
37. Youngs, D.L. Three-dimensional numerical simulation of turbulent mixing by Rayleigh–Taylor instability. *Phys. Fluids A Fluid Dyn.* **1991**, *3*, 1312–1320. [CrossRef]
38. Youngs, D.L. Numerical simulation of mixing by Rayleigh–Taylor and Richtmyer–Meshkov instabilities. *Laser Part. Beams* **1994**, *12*, 725–750. [CrossRef]
39. Shvarts, D.; Alon, U.; Ofer, D.; McCrory, R.; Verdon, C. Nonlinear evolution of multimode Rayleigh–Taylor instability in two and three dimensions. *Phys. Plasmas* **1995**, *2*, 2465–2472. [CrossRef]
40. Oron, D.; Arazi, L.; Kartoon, D.; Rikanati, A.; Alon, U.; Shvarts, D. Dimensionality dependence of the Rayleigh–Taylor and Richtmyer–Meshkov instability late-time scaling laws. *Phys. Plasmas* **2001**, *8*, 2883–2889. [CrossRef]
41. Zhou, Y. A scaling analysis of turbulent flows driven by Rayleigh–Taylor and Richtmyer–Meshkov instabilities. *Phys. Fluids* **2001**, *13*, 538–543. [CrossRef]
42. Shvarts, D.; Oron, D.; Kartoon, D.; Rikanati, A.; Sadot, O.; Srebro, Y.; Yedvab, Y.; Ofer, D.; Levin, A.; Sarid, E.; et al. Scaling laws of nonlinear Rayleigh-Taylor and Richtmyer-Meshkov instabilities in two and three dimensions (IFSA 1999). In *Edward Teller Lectures: Lasers and Inertial Fusion Energy*; World Scientific: Singapore, 2016; pp. 253–260.
43. Chertkov, M. Phenomenology of Rayleigh-Taylor turbulence. *Phys. Rev. Lett.* **2003**, *91*, 115001. [CrossRef]
44. Celani, A.; Mazzino, A.; Vozella, L. Rayleigh-Taylor turbulence in two dimensions. *Phys. Rev. Lett.* **2006**, *96*, 134504. [CrossRef]
45. Margolin, L. The reality of artificial viscosity. *Shock Waves* **2019**, *29*, 27–35. [CrossRef]
46. Brehm, C.; Barad, M.F.; Housman, J.A.; Kiris, C.C. A comparison of higher-order finite-difference shock capturing schemes. *Comput. Fluids* **2015**, *122*, 184–208. [CrossRef]
47. Margolin, L.G.; Rider, W.J.; Grinstein, F.F. Modeling turbulent flow with implicit LES. *J. Turbul.* **2006**, *7*, N15. [CrossRef]
48. Zhou, Y.; Thornber, B. A comparison of three approaches to compute the effective Reynolds number of the implicit large-eddy simulations. *J. Fluids Eng.* **2016**, *138*, 070905. [CrossRef]
49. Maulik, R.; San, O. Resolution and energy dissipation characteristics of implicit LES and explicit filtering models for compressible turbulence. *Fluids* **2017**, *2*, 14. [CrossRef]
50. San, O.; Kara, K. Numerical assessments of high-order accurate shock capturing schemes: Kelvin–Helmholtz type vortical structures in high-resolutions. *Comput. Fluids* **2014**, *89*, 254–276. [CrossRef]
51. Titarev, V.A.; Toro, E.F. Finite-volume WENO schemes for three-dimensional conservation laws. *J. Comput. Phys.* **2004**, *201*, 238–260. [CrossRef]
52. Pirozzoli, S. Numerical methods for high-speed flows. *Annu. Rev. Fluid Mech.* **2011**, *43*, 163–194. [CrossRef]
53. Brehm, C. On consistent boundary closures for compact finite-difference WENO schemes. *J. Comput. Phys.* **2017**, *334*, 573–581. [CrossRef]
54. Wong, M.L.; Lele, S.K. High-order localized dissipation weighted compact nonlinear scheme for shock-and interface-capturing in compressible flows. *J. Comput. Phys.* **2017**, *339*, 179–209. [CrossRef]
55. Zhao, S.; Lardjane, N.; Fedioun, I. Comparison of improved finite-difference WENO schemes for the implicit large eddy simulation of turbulent non-reacting and reacting high-speed shear flows. *Comput. Fluids* **2014**, *95*, 74–87. [CrossRef]
56. Hu, X.; Wang, Q.; Adams, N.A. An adaptive central-upwind weighted essentially non-oscillatory scheme. *J. Comput. Phys.* **2010**, *229*, 8952–8965. [CrossRef]
57. Mathew, J.; Foysi, H.; Friedrich, R. A new approach to LES based on explicit filtering. *Int. J. Heat Fluid Flow* **2006**, *27*, 594–602. [CrossRef]

58. Lund, T. The use of explicit filters in large eddy simulation. *Comput. Math. Appl.* **2003**, *46*, 603–616. [CrossRef]

59. Berland, J.; Lafon, P.; Daude, F.; Crouzet, F.; Bogey, C.; Bailly, C. Filter shape dependence and effective scale separation in large-eddy simulations based on relaxation filtering. *Comput. Fluids* **2011**, *47*, 65–74. [CrossRef]

60. Bose, S.T.; Moin, P.; You, D. Grid-independent large-eddy simulation using explicit filtering. *Phys. Fluids* **2010**, *22*, 105103. [CrossRef]

61. Vasilyev, O.V.; Lund, T.S.; Moin, P. A general class of commutative filters for LES in complex geometries. *J. Comput. Phys.* **1998**, *146*, 82–104. [CrossRef]

62. San, O. Analysis of low-pass filters for approximate deconvolution closure modelling in one-dimensional decaying Burgers turbulence. *Int. J. Comput. Fluid Dyn.* **2016**, *30*, 20–37. [CrossRef]

63. Najafi-Yazdi, A.; Najafi-Yazdi, M.; Mongeau, L. A high resolution differential filter for large eddy simulation: Toward explicit filtering on unstructured grids. *J. Comput. Phys.* **2015**, *292*, 272–286. [CrossRef]

64. Shivamoggi, B.K. Spectral laws for the compressible isotropic turbulence. In *Instability, Transition, and Turbulence*; Springer: New York, NY, USA, 1992; pp. 524–534.

65. Shivamoggi, B.K. Multi-fractal aspects of the fine-scale structure of temperature fluctuations in isotropic turbulence. *Phys. A Stat. Mech. Its Appl.* **1995**, *221*, 460–477. [CrossRef]

66. Shivamoggi, B.K. Intermittency in the enstrophy cascade of two-dimensional fully developed turbulence: Universal features. *Ann. Phys.* **2008**, *323*, 444–460. [CrossRef]

67. Shivamoggi, B. Compressible turbulence: Multi-fractal scaling in the transition to the dissipative regime. *Phys. A Stat. Mech. Its Appl.* **2011**, *390*, 1534–1538. [CrossRef]

68. Sun, B. The temporal scaling laws of compressible turbulence. *Mod. Phys. Lett. B* **2016**, *30*, 1650297. [CrossRef]

69. Sun, B. Scaling laws of compressible turbulence. *Appl. Math. Mech.* **2017**, *38*, 765–778. [CrossRef]

70. Parent, B. Positivity-preserving high-resolution schemes for systems of conservation laws. *J. Comput. Phys.* **2012**, *231*, 173–189. [CrossRef]

71. Gottlieb, S.; Shu, C.W. Total variation diminishing Runge-Kutta schemes. *Math. Comput. Am. Math. Soc.* **1998**, *67*, 73–85. [CrossRef]

72. Godunov, S.K. A difference method for numerical calculation of discontinuous solutions of the equations of hydrodynamics. *Mat. Sb.* **1959**, *89*, 271–306.

73. Grinstein, F.F.; Margolin, L.G.; Rider, W.J. *Implicit Large Eddy Simulation: Computing Turbulent Fluid Dynamics*; Cambridge University Press: New York, NY, USA, 2007.

74. Denaro, F.M. What does finite volume-based implicit filtering really resolve in large-eddy simulations? *J. Comput. Phys.* **2011**, *230*, 3849–3883. [CrossRef]

75. Liu, X.D.; Osher, S.; Chan, T. Weighted essentially non-oscillatory schemes. *J. Comput. Phys.* **1994**, *115*, 200–212. [CrossRef]

76. Harten, A.; Engquist, B.; Osher, S.; Chakravarthy, S.R. Uniformly high order accurate essentially non-oscillatory schemes, III. In *Upwind and High-Resolution Schemes*; Springer: New York, NY, USA, 1987; pp. 218–290.

77. Shu, C.W.; Osher, S. Efficient implementation of essentially non-oscillatory shock-capturing schemes. *J. Comput. Phys.* **1988**, *77*, 439–471. [CrossRef]

78. Jiang, G.S.; Shu, C.W. Efficient implementation of weighted ENO schemes. *J. Comput. Phys.* **1996**, *126*, 202–228. [CrossRef]

79. Henrick, A.K.; Aslam, T.D.; Powers, J.M. Mapped weighted essentially non-oscillatory schemes: Achieving optimal order near critical points. *J. Comput. Phys.* **2005**, *207*, 542–567. [CrossRef]

80. Borges, R.; Carmona, M.; Costa, B.; Don, W.S. An improved weighted essentially non-oscillatory scheme for hyperbolic conservation laws. *J. Comput. Phys.* **2008**, *227*, 3191–3211. [CrossRef]

81. Roe, P.L. Approximate Riemann solvers, parameter vectors, and difference schemes. *J. Comput. Phys.* **1981**, *43*, 357–372. [CrossRef]

82. Harten, A. High resolution schemes for hyperbolic conservation laws. *J. Comput. Phys.* **1983**, *49*, 357–393. [CrossRef]

83. Rusanov, V.V. The calculation of the interaction of non-stationary shock waves with barriers. *Zhurnal Vychislitel'Noi Mat. Mat. Fiz.* **1961**, *1*, 267–279.

84. Lax, P.D. Weak solutions of nonlinear hyperbolic equations and their numerical computation. *Commun. Pure Appl. Math.* **1954**, *7*, 159–193. [CrossRef]

85. Toro, E.F. *Riemann Solvers and Numerical Methods for Fluid Dynamics: A Practical Introduction*; Springer Science & Business Media: Berlin, Germany, 2013.

86. Hyman, J.M.; Knapp, R.J.; Scovel, J.C. High order finite volume approximations of differential operators on nonuniform grids. *Phys. D Nonlinear Phenom.* **1992**, *60*, 112–138. [CrossRef]

87. Fauconnier, D.; Bogey, C.; Dick, E. On the performance of relaxation filtering for large-eddy simulation. *J. Turbul.* **2013**, *14*, 22–49. [CrossRef]

88. Bogey, C.; Bailly, C. Computation of a high Reynolds number jet and its radiated noise using large eddy simulation based on explicit filtering. *Comput. Fluids* **2006**, *35*, 1344–1358. [CrossRef]

89. Bull, J.R.; Jameson, A. Explicit filtering and exact reconstruction of the sub-filter stresses in large eddy simulation. *J. Comput. Phys.* **2016**, *306*, 117–136. [CrossRef]

90. San, O. A dynamic eddy-viscosity closure model for large eddy simulations of two-dimensional decaying turbulence. *Int. J. Comput. Fluid Dyn.* **2014**, *28*, 363–382. [CrossRef]

91. Gropp, W.; Lusk, E.; Doss, N.; Skjellum, A. A high-performance, portable implementation of the MPI message passing interface standard. *Parallel Comput.* **1996**, *22*, 789–828. [CrossRef]

92. Gropp, W.D.; Gropp, W.; Lusk, E.; Skjellum, A.; Lusk, E. *Using MPI: Portable Parallel Programming with the Message-Passing Interface*; MIT Press: Cambridge, MA, USA, 1999; Volume 1.

93. Banerjee, A.; Andrews, M.J. 3D simulations to investigate initial condition effects on the growth of Rayleigh–Taylor mixing. *Int. J. Heat Mass Transf.* **2009**, *52*, 3906–3917. [CrossRef]

94. Vladimirova, N.; Chertkov, M. Self-similarity and universality in Rayleigh–Taylor, Boussinesq turbulence. *Phys. Fluids* **2009**, *21*, 015102. [CrossRef]

95. Xu, Z.; Duo-Wang, T. Direct Numerical Simulation of the Rayleigh- Taylor Instability with the Spectral Element Method. *Chin. Phys. Lett.* **2009**, *26*, 084703. [CrossRef]

96. Liska, R.; Wendroff, B. Comparison of several difference schemes on 1D and 2D test problems for the Euler equations. *Siam J. Sci. Comput.* **2003**, *25*, 995–1017. [CrossRef]

97. Stone, J.M.; Gardiner, T.A.; Teuben, P.; Hawley, J.F.; Simon, J.B. Athena: A new code for astrophysical MHD. *Astrophys. Suppl. Ser.* **2008**, *178*, 137–177. [CrossRef]

98. Poinsot, T.J.; Lele, S.K. Boundary conditions for direct simulations of compressible viscous flows. *J. Comput. Phys.* **1992**, *101*, 104–129. [CrossRef]

99. Youngs, D.L. Rayleigh–Taylor mixing: Direct numerical simulation and implicit large eddy simulation. *Phys. Scr.* **2017**, *92*, 074006. [CrossRef]

100. San, O.; Kara, K. Evaluation of Riemann flux solvers for WENO reconstruction schemes: Kelvin–Helmholtz instability. *Comput. Fluids* **2015**, *117*, 24–41. [CrossRef]

101. Lele, S.K. Compressibility effects on turbulence. *Annu. Rev. Fluid Mech.* **1994**, *26*, 211–254. [CrossRef]

102. Kritsuk, A.G.; Norman, M.L.; Padoan, P.; Wagner, R. The statistics of supersonic isothermal turbulence. *Astrophys. J.* **2007**, *665*, 416. [CrossRef]

103. Aluie, H. Scale decomposition in compressible turbulence. *Phys. D Nonlinear Phenom.* **2013**, *247*, 54–65. [CrossRef]

104. San, O.; Maulik, R. Stratified Kelvin–Helmholtz turbulence of compressible shear flows. *Nonlinear Process. Geophys.* **2018**, *25*. [CrossRef]

105. Press, W.H.; Teukolsky, S.A.; Vetterling, W.T.; Flannery, B.P. *Numerical Recipes in FORTRAN: The Art of Scientific Computing*, 2nd ed.; Cambridge University Press: New York, NY, USA, 1992.

106. Kida, S.; Orszag, S.A. Energy and spectral dynamics in forced compressible turbulence. *J. Sci. Comput.* **1990**, *5*, 85–125. [CrossRef]

107. Bolgiano, R., Jr. Turbulent spectra in a stably stratified atmosphere. *J. Geophys. Res.* **1959**, *64*, 2226–2229. [CrossRef]

108. Livescu, D.; Wei, T.; Petersen, M. *Direct Numerical Simulations of Rayleigh-Taylor Instability*; Conference Series; IOP Publishing: Bristol, UK, 2011; Volume 318, p. 082007.

109. Youngs, D.L. Numerical simulation of turbulent mixing by Rayleigh-Taylor instability. *Phys. D Nonlinear Phenom.* **1984**, *12*, 32–44. [CrossRef]

110. Ramaprabhu, P.; Dimonte, G.; Andrews, M. A numerical study of the influence of initial perturbations on the turbulent Rayleigh–Taylor instability. *J. Fluid Mech.* **2005**, *536*, 285–319. [CrossRef]

111. Shadloo, M.; Zainali, A.; Yildiz, M. Simulation of single mode Rayleigh–Taylor instability by SPH method. *Comput. Mech.* **2013**, *51*, 699–715. [CrossRef]

112. Lee, H.G.; Kim, K.; Kim, J. On the long time simulation of the Rayleigh–Taylor instability. *Int. J. Numer. Methods Eng.* **2011**, *85*, 1633–1647. [CrossRef]

113. Livescu, D.; Ristorcelli, J.; Petersen, M.; Gore, R. New phenomena in variable-density Rayleigh–Taylor turbulence. *Phys. Scr.* **2010**, *2010*, 014015. [CrossRef]

114. Shi, J.; Zhang, Y.T.; Shu, C.W. Resolution of high order WENO schemes for complicated flow structures. *J. Comput. Phys.* **2003**, *186*, 690–696. [CrossRef]

115. Fu, L.; Hu, X.Y.; Adams, N.A. A family of high-order targeted ENO schemes for compressible-fluid simulations. *J. Comput. Phys.* **2016**, *305*, 333–359. [CrossRef]

116. Latini, M.; Schilling, O.; Don, W.S. Effects of WENO flux reconstruction order and spatial resolution on reshocked two-dimensional Richtmyer–Meshkov instability. *J. Comput. Phys.* **2007**, *221*, 805–836. [CrossRef]

117. Tritschler, V.; Hu, X.; Hickel, S.; Adams, N. Numerical simulation of a Richtmyer–Meshkov instability with an adaptive central-upwind sixth-order WENO scheme. *Phys. Scr.* **2013**, *2013*, 014016. [CrossRef]

Article

Turbulence Model Assessment in Compressible Flows around Complex Geometries with Unstructured Grids

Guillermo Araya

High Performance Computing and Visualization Laboratory, Department of Mechanical Engineering, University of Puerto Rico, Mayaguez, PR 00681, USA; araya@mailaps.org

Received: 23 February 2019; Accepted: 15 April 2019; Published: 28 April 2019

Abstract: One of the key factors in simulating realistic wall-bounded flows at high Reynolds numbers is the selection of an appropriate turbulence model for the steady Reynolds Averaged Navier–Stokes equations (RANS) equations. In this investigation, the performance of several turbulence models was explored for the simulation of steady, compressible, turbulent flow on complex geometries (concave and convex surface curvatures) and unstructured grids. The turbulence models considered were the Spalart–Allmaras model, the Wilcox k-ω model and the Menter shear stress transport (SST) model. The FLITE3D flow solver was employed, which utilizes a stabilized finite volume method with discontinuity capturing. A numerical benchmarking of the different models was performed for classical Computational Fluid Dynamic (CFD) cases, such as supersonic flow over an isothermal flat plate, transonic flow over the RAE2822 airfoil, the ONERA M6 wing and a generic F15 aircraft configuration. Validation was performed by means of available experimental data from the literature as well as high spatial/temporal resolution Direct Numerical Simulation (DNS). For attached or mildly separated flows, the performance of all turbulence models was consistent. However, the contrary was observed in separated flows with recirculation zones. Particularly, the Menter SST model showed the best compromise between accurately describing the physics of the flow and numerical stability.

Keywords: Navier–Stokes equations (RANS); Direct Numerical Simulation (DNS); turbulence models; compressible flow

---

1. Introduction

Wall-bounded turbulent flows at high Reynolds numbers are mainly characterized by a wide range of time and length scales (Pope [1]). In Direct Numerical Simulation (DNS), the full Navier–Stokes (NS) equations are generally solved to capture all time scales and eddies in the flow. This is an accurate approach that supplies extensive information, but it demands the use of a very fine mesh, and significant computational resources, to properly capture features in the order of the Kolmogorov scales [2]. Filtered Navier–Stokes equations, with an additional sub-grid scale stress term, are employed in Large Eddy Simulation (LES), in which large scales or eddies are directly resolved and the effects of the small scales are modeled [3]. The LES approach still requires high mesh resolution in boundary layer flows, particularly, in the near-wall region. Furthermore, the use of hybrid approaches (RANS-LES) to overcome the fine resolution needed in wall-bounded flows is an emerging and promising area [4]; nevertheless, there is still much ground to cover for industrial applications. From the perspective of performing turbulent industrial flow simulations, the workhorse is still the use of the Reynolds Averaged Navier–Stokes equations (RANS), which are obtained by time averaging the full NS equations. In this approach, almost the full power spectra of velocity fluctuations and turbulent kinetic energy are modeled, while capturing the very large turbulent scales of low frequencies. These equations require a model or closure to compute the Reynolds stresses, which arise from the convective terms of the NS equations after applying the time averaging process. The selection of an appropriate turbulence model is crucial to accurately represent important physical aspects of the flow, such as boundary layer separation and shock–boundary layer interaction [5]. Catalano and Amato [5] tested five different turbulence models in 2D and 3D classical aerodynamic applications: Spalart–Allmaras, Myong and Kasagi k-ε, Wilcox k-ω, Kok turbulent/non-turbulent (TNT), and Menter SST, where k is the turbulent kinetic energy, ε is the turbulent dissipation and ω is the specific rate of dissipation. They concluded that the Menter shear stress transport (SST) model exhibited the best balance between the physical capabilities and the numerical robustness for the transonic and high-lift flows explored. Knight et al. [6] performed a review of recent CFD simulations of shockwave turbulent boundary layer interactions. In addition, Knight et al. [6] tested five different configurations: 2D compression corner, 2D shock impingement, 2D expansion-compression corner, 3D single fin and 3D double fin in structured and unstructured grids by considering several two-equation turbulence models. In general, they reported a fair agreement between RANS results to experiments, except for heat transfer predictions of strong separated shock wave turbulent boundary layer interactions where turbulence models still need to make some improvements. In addition, Manzari et al. [7] successfully tested the k-ω turbulence model in 3D compressible flows on unstructured tetrahedral grids. Furthermore, the use of multi-grid techniques for solution acceleration in 3D turbulent flows on unstructured meshes constitutes a challenging problem [8].

In the present study, the one-equation Spalart–Allmaras model and the two-equation Wilcox k-ω and Menter SST turbulence models were implemented within an innovative unstructured mesh finite volume flow solver. We computed and validated not only global quantities (e.g., lift, drag and moment coefficient) but also local and boundary layer parameters such as displacement/momentum thickness, pressure coefficient, skin friction coefficient and streamwise velocity profiles. Consequently, this study represents an extensive and thorough analysis of popular turbulence models in large scale systems at experimental Reynolds numbers in compressible flows.

2. Governing Equations

The unsteady compressible Navier–Stokes equations are expressed, over a 3D domain $\Omega \subset \Re^3$ with closed surface Γ, in the integral form

$$\int_\Omega \frac{\partial U_i}{\partial t} dx + \int_{\partial \Omega} F_{ij} n_j dx = \int_{\partial \Omega} G_{ij} n_j dx, \tag{1}$$

where $\vec{n} = (n_1, n_2, n_3)$ is the unit normal vector to Γ. In addition, the unknown vector of conservative variables is expressed as

$$U_i = \begin{bmatrix} \rho \\ \rho u_i \\ \rho \epsilon \end{bmatrix} \tag{2}$$

where ρ is the fluid density, u_i denotes the ith component of the velocity vector and ϵ is the specific total energy and the inviscid and viscous flux vectors are expressed as

$$F_{ij} = \begin{bmatrix} \rho u_j \\ \rho u_i u_j + p \delta_{ij} \\ u_i(\rho \epsilon + p) \end{bmatrix} \qquad G_{ij} = \begin{bmatrix} 0 \\ \tau_{ij} \\ u_k \tau_{kj} - q_j \end{bmatrix} \tag{3}$$

respectively. The quantity

$$\tau_{ij} = -\frac{2}{3} \mu \frac{\partial u_k}{\partial x_k} \delta_{ij} + \mu \left(\frac{\partial u_i}{\partial x_j} + \frac{\partial u_j}{\partial x_i} \right), \tag{4}$$

is the deviatoric stress tensor, where μ is the dynamic viscosity and δ_{ij} is the Kronecker delta. The quantity $q_j = -k \partial T / \partial x_j$ is the heat flux, where k is the thermal conductivity and T is the absolute temperature. The viscosity varies with temperature according to Sutherlands's law and the

Prandtl number is assumed to be constant and equal to 0.72. In addition, the medium is assumed to be calorically perfect.

3. Solution Procedure

3.1. Hybrid Unstructured Mesh Generation

The computational domain was represented by an unstructured hybrid mesh for viscous problems. The process of mesh generation began with the discretization of the domain boundary into a set of triangular or quadrilateral meshes that satisfy a mesh control function specified by the user [9]. Additionally, the distribution of the mesh parameters [10], such as spacing, stretching and direction of stretching, were described by a background mesh as well as point, line and planar sources. For viscous flows, the boundary layers were generated using the advancing layer method [11] and the rest of the computational domain was filled with tetrahedral elements using a Delaunay incremental Bowyer Watson point insertion [12]. Hybrid unstructured meshes were constructed by merging certain elements of this tetrahedral mesh in the boundary layers.

3.2. Spatial Discretization

The discretization of the governing equations can be performed in a number of different ways. A computationally efficient approach was implemented, consisting on a cell vertex finite volume method. This involved the identification of a dual mesh, with the medial dual being constructed as an assembly of triangular facet, Γ_I^K, where each facet was formed by connecting edge midpoints, element and face centroids in the basic mesh, in such a way that only one node was contained within each dual mesh cell. Hence, the dual mesh cells formed the control volumes for the finite volume process. When hybrid meshes are employed, the method for constructing the median dual has to be modified to ensure that no node lies outside its corresponding control volume. To perform the numerical integration of the fluxes, a set of coefficients was calculated for each edge using the dual mesh segment associated with the edge. The values of the internal edge coefficients, C_j^{IJ}, and the boundary edge coefficients, D_j^{IJ}, are defined as follows,

$$C_j^{IJ} = \sum_{K \in \Gamma_{IJ}} A_{\Gamma_I^K} n_j^{\Gamma_I^K}, \qquad D_j^{IJ} = \sum_{K \in \Gamma_{IJ}^B} A_{\Gamma_I^K} n_j^{\Gamma_I^K}, \tag{5}$$

where $A_{\Gamma_I^K}$ is the area of facet Γ_I^K, $n_j^{\Gamma_I^K}$ is the outward unit normal vector of the facet, Γ_{IJ}^B is the set of dual mesh faces on the computational boundary touching the edge between nodes I and J and $n_j^{\Gamma_I^K}$ denotes the normal of the facet in the outward direction of the computational domain. The numerical integration of the fluxes over the dual mesh segment associated with an edge was performed by assuming the flux to be constant, and equal to its approximated value at the midpoint of the edge, i.e., a form of mid point quadrature. The calculation of a surface integral for the inviscid flux over the control volume surface for node I is defined as follows,

$$\int_{\partial \Omega_I} \mathbf{F}_j n_j dx \approx \sum_{J \in \Lambda_I} \frac{C_j^{IJ}}{2} \left(\mathbf{F}_j^I + \mathbf{F}_j^J \right) + \sum_{J \in \Lambda_I^B} D_j^{IJ} \mathbf{F}_j^I, \tag{6}$$

where Λ_I denotes the set of nodes connected to node I by an edge and Λ_I^B denotes the set of nodes connected to node I by an edge on the computational boundary. Thus, the last term is non-zero only in a boundary node. Similar formula can be implemented for the viscous fluxes. The use of edge based data structure has become widely used due to its efficiency in terms of memory and CPU requirements, compared to the traditional element based data structure, especially in three dimensions.

The resulting discretization is basically central difference in character. Therefore, the addition of a stabilizing dissipation is required for practical flow simulations. This was achieved by replacing the physical flux function by a consistent numerical flux function, such as the Jameson–Schmidt–Turkel (JST) flux function [13] or the Harten–Lax–van Leer-Contact (HLLC) solver [14]. Discontinuity capturing may be accomplished by the use of an additional harmonic term in regions of high pressure gradients, identified by using a pressure switch.

3.3. Relaxation Scheme

The FLITE3D flow solver [15] can simulate both unsteady and steady problems. However, in this investigation, a steady flow analysis was performed. Therefore, the relaxation scheme employed consisted of a three-stage Runge–Kutta approach with local time stepping. The viscous and artificial dissipation terms were only computed one time at every time step to reduce the computational requirements of the scheme. The stability limit of this scheme or maximum CFL is 1.8, where CFL is the Courant–Friedrich–Levy number. The local time step values for the momentum and energy equations were computed based on inviscid and viscous term contributions [16]. Similarly, local time steps were implemented for the solution of the corresponding transport equations of the turbulence model approaches (either one- or two-equation models in the present study) with contributions of the inviscid and viscous parts as well as a source term correction. More details can be found in [16].

4. Solver Parallelization

A parallel implementation of the flow solver was achieved by using the single program multiple data concept in conjunction with standard Message Passing Interface (MPI) routines for message passing. In the parallel implementation, at the start of each time step, the interface nodes of the sub-domain with the lower number obtained contributions from the interface edges. These partially updated nodal contributions were then broadcasted to the corresponding interface nodes in the neighboring higher-number domains. A loop over the interior edges was followed by the receipt of the interface nodal contributions and the subsequent updating of all nodal values. The procedure was completed by sending the updated interface nodal values from the higher domain back to the corresponding interface nodes of the lower domain. In addition, the computation and communication processes took place concurrently. The employed approach was a global procedure in which the agglomeration was performed across the parallel domain boundaries. Global agglomeration ensures equivalence with a sequential approach for any number of domain, but increases the communication load in the inner–grid mapping [8,16].

5. Turbulence Modeling in RANS

To obtain the compressible RANS equations, unsteady equations (Equation (1)) were averaged in time to smooth the instantaneous turbulent fluctuations in the flow field, while still allowing the capture of the time-dependency in the large time scales of interest. In many engineering problems, this assumption is valid, but this averaging procedure breaks down if the time scale of the physical phenomena of relevance is similar to that of the turbulence itself. For compressible flows, the density weighted Favre averaging procedure is mostly employed [16]. The Favre averaging procedure applied to Equations (1) generates the extra convective term:

$$\tau_{ij}^{R} = -\overline{\rho u_i'' u_j''},\tag{7}$$

which is the Favre averaged Reynolds stress tensor. The most straightforward approach is to associate the unknown Reynolds stresses with the computed mean flow quantities by means of a turbulence model or closure. If the Boussinesq hypothesis is applied, this results in a linear relationship to the mean flow strain tensor through the eddy viscosity μ_t [17],

$$\tau_{ij}^R = \mu_t \left(\frac{\partial U_i}{\partial x_j} + \frac{\partial U_j}{\partial x_i} - \frac{2}{3} \frac{\partial U_k}{\partial x_k} \delta_{ij} \right) - \frac{2}{3} \rho k \delta_{ij}, \tag{8}$$

where k is the turbulent kinetic energy. The eddy viscosity depends on the velocity and the length scales of the turbulent eddies, i.e., $\mu_t \sim k^{1/2} \ell$, where ℓ is the turbulence length scale. In this study, one- and two-transport-equation models were considered, in which additional partial differential equations were solved to describe the transport of the eddy viscosity. Consequently, nonlocal and history effects on μ_t were taken into account. In the one-equation turbulence model of Spalart–Allmaras [18], a combination of the turbulent scales, through the viscosity-like variable, $\tilde{\nu}$, is obtained by solving an empirical transport equation. Two-equation turbulence models are complete, because transport equations are solved for both turbulent scales, i.e., the velocity and the length scale. The original k-ω model [17] exhibits a freestream dependency of ω, which is generally not present in the k-ϵ model. Menter [19] combined the advantages of both models by means of blending functions, which permits the switching from k-ω, close to a wall, to k-ϵ, when approaching the edge of a boundary layer. A further improvement by Menter [19] is a modification to the eddy viscosity, based on the idea of the Johnson–King model, which establishes that the transport of the main turbulent shear stresses is crucial in the simulations of strong Adverse Pressure Gradient (APG) flows. This new approach is called the Menter shear–stress transport model (SST). In particular, the Menter SST turbulence model is well-known for its good performance in boundary layer flows subjected to APG, with eventual separation.

5.1. Turbulent Transport Equations

In this section, a brief description of the governing equations for the three turbulence models is presented. The differential equations are presented in the normalized form. A convenient scaling guarantees a unit order of all variables, which decreases round-off errors of calculations. The reader is referred to Appendix B of Sørensen's thesis [16] for the scaling and the normalized version of momentum and energy equations employed in the present study; however, the normalization symbol * is dropped here for simplicity. Furthermore, the dimensionless Spalart–Allmaras transport equation [18] is expressed as;

$$\underbrace{\frac{1}{St_\infty} \frac{\partial \tilde{\nu}}{\partial t}}_{\text{transient term}} + \underbrace{U_j \frac{\partial \tilde{\nu}}{\partial x_j}}_{\text{conv. term}} = \underbrace{c_{b1}[1 - f_{v1}]S\tilde{\nu}}_{\text{production term}} + \underbrace{\frac{1}{Re_\infty \sigma} \left\{ \frac{\partial}{\partial x_j} \left[(\nu + \tilde{\nu}) \frac{\partial \tilde{\nu}}{\partial x_j} \right] + c_{b2} \frac{\partial \tilde{\nu}}{\partial x_j} \frac{\partial \tilde{\nu}}{\partial x_j} \right\}}_{\text{diffusion term}}$$

$$\underbrace{- \frac{1}{Re_\infty} \left(c_{w1} f_w - \frac{c_{b1}}{\kappa^2} f_{t2} \right) \left(\frac{\tilde{\nu}}{d} \right)^2}_{\text{dissipation term}} + \underbrace{Re_\infty f_{t1} (\Delta U)^2,}_{\text{transition term}} \tag{9}$$

In Equation (9), the normalized quantity $\tilde{\nu}$ is related to the eddy viscosity by

$$\mu_t = \rho \tilde{\nu} \frac{\chi^3}{\chi^3 + c_{v1}^3}, \tag{10}$$

where ν is the molecular kinematic viscosity of the fluid, $\chi = \tilde{\nu}/\nu$, and c_{v1} is equal to 7.1.

Additionally, $St_\infty = U_\infty t/L$ and $Re_\infty = \rho U_\infty L/\mu_\infty$ are the Strouhal and Reynolds numbers, respectively; U_j represents the time Favre-averaged velocity; and μ_∞ is the freestream dynamic viscosity. The rest of the parameters and constants are defined as follows:

$$S = |\omega| + \frac{\tilde{v}}{Re_\infty \kappa^2 d^2} f_{v2}, \tag{11}$$

$$f_{v2} = 1 - \frac{\chi}{1 + \chi f_{v1}}, \tag{12}$$

$$f_\omega = g \left(\frac{1 + c_{w3}^6}{g^6 + c_{w3}^6} \right)^{\frac{1}{6}}, \tag{13}$$

$$g = r + c_{w2} r \left(r^5 - 1 \right), \tag{14}$$

$$r = \min \left(\frac{\tilde{v}}{Re_\infty S \kappa^2 d^2}, 10 \right), \tag{15}$$

$$f_{t2} = c_{t3} exp(-c_{t4}\chi^2), \tag{16}$$

$$f_{t1} = c_{t1} g_t exp \left[-c_{t2} \frac{\omega_t^2}{(\Delta U)^2} \left(d^2 + g_t^2 d_t^2 \right) \right], \tag{17}$$

$$g_t = \min \left(0.1, \frac{\Delta U}{\omega_t \Delta x} \right), \tag{18}$$

where $|\omega|$ is the vorticity magnitude, d is the distance from a given point to the nearest wall, $c_{b1} = 0.1355$, $\sigma = 2/3$, $c_{b2} = 0.622$, $\kappa = 0.41$, $c_{w1} = c_{b1}/\kappa^2 + (1 + c_{b2})/\sigma$, $c_{w2} = 0.3$, $c_{w3} = 2$, $c_{t1} = 1$, $c_{t2} = 2$, $c_{t3} = 1.1$, and $c_{t4} = 2$. Moreover, the variable ΔU is the difference in velocity between the point and the associated trigger point, d_t indicates the closest distance to such a trigger point or curve, ω_t is the vorticity magnitude at the associated trigger point, and Δx is the surface grid spacing at this point. The convective term is conveniently rewritten in conservative form,

$$U_j \frac{\partial \tilde{v}}{\partial x_j} = \frac{\partial (U_j \tilde{v})}{\partial x_j} - \tilde{v} \frac{\partial U_j}{\partial x_j}, \tag{19}$$

which introduces an extra source term. Nevertheless, the contribution of this source term is negligible [20]; thus, it is ignored in the present formulation.

Furthermore, the corresponding normalized transport equations in the Wilcox k-ω model [17] for the turbulent kinetic energy, k, and the specific dissipation rate, ω, in compressible flows read as follows:

$$\underbrace{\frac{1}{St_\infty} \frac{\partial(\rho k)}{\partial t}}_{\text{transient term}} + \underbrace{\frac{\partial(\rho U_j k)}{\partial x_j}}_{\text{conv. term}} = \underbrace{\tau_{ij} \frac{\partial U_j}{\partial x_j}}_{\text{production term}} - \underbrace{\beta_k \rho \omega k}_{\text{dissipation term}} + \underbrace{\frac{1}{Re_\infty} \frac{\partial}{\partial x_j} \left[\left(\mu + \frac{\mu_t}{\sigma_k} \right) \frac{\partial k}{\partial x_j} \right]}_{\text{diffusion term}} \tag{20}$$

$$\underbrace{\frac{1}{St_\infty} \frac{\partial(\rho \omega)}{\partial t}}_{\text{transient term}} + \underbrace{\frac{\partial(\rho U_j \omega)}{\partial x_j}}_{\text{conv. term}} = \underbrace{\alpha \frac{\omega}{k} \tau_{ij} \frac{\partial U_j}{\partial x_j}}_{\text{production term}} - \underbrace{\beta_\omega \rho \omega^2}_{\text{dissipation term}} + \underbrace{\frac{1}{Re_\infty} \frac{\partial}{\partial x_j} \left[\left(\mu + \frac{\mu_t}{\sigma_\omega} \right) \frac{\partial \omega}{\partial x_j} \right]}_{\text{diffusion term}}, \tag{21}$$

where $\beta_k = 0.09$, $\beta_\omega = 3/40$, $\sigma_k = 2$, $\sigma_\omega = 2$ and $\alpha = 5/9$. The normalized eddy viscosity is defined as:

$$\mu_t = Re_\infty \rho k/\omega. \tag{22}$$

In the Menter SST model for the ω equation, an extra term is considered in Equation (21), which is called the cross-diffusion term,

$$\underbrace{2(1-F_1)\frac{\rho\sigma_{\omega 2}}{\omega}\frac{\partial k}{\partial x_j}\frac{\partial \omega}{\partial x_j}}_{\text{cross-diffusion term}},\tag{23}$$

where $\sigma_{\omega 2} = 0.856$. F_1 is a blending function defined as,

$$F_1 = \tanh\left\{\left[\min\left(\max\left(\frac{\sqrt{k}}{\beta_k\omega y},\frac{500\nu}{y^2\omega Re_\infty}\right),\frac{4\rho\sigma_{\omega 2}k}{CD_{k\omega}y^2}\right)\right]\right\},\tag{24}$$

and

$$CD_{k\omega} = \max\left(\frac{2\rho\sigma_{\omega 2}}{\omega}\frac{\partial k}{\partial x_j}\frac{\partial \omega}{\partial x_j},10^{-10}\right).\tag{25}$$

Thus, the blending function F_1 generates values close to one far from the wall (k-ϵ model) and almost zero values near the edge of the boundary layer (k-ω model). More details can be found in [19]. The main differences of Menter SST equations with respect to the Wilcox k-ω equations [17] can be summarized as follows: (i) the consideration of a cross-diffusion term in the ω equation (k-ϵ model), which makes the model insensitive to the freestream boundary condition of ω; (ii) the implementation of a stress limiter for the maximum value of ω as well as a production limiter to impede the build-up of turbulence in stagnation zones; and (iii) the application of a blending function to compute the corresponding constants of the k-ϵ and k-ω models.

5.2. Discretization of the Turbulent Transport Equations

The turbulence model equations involve the solution of partial differential equations, which are discretized in a similar manner as the governing equations of the flow. Nevertheless, to avoid instabilities from the convective term, a first order upwind discretization was used plus the addition of a term to introduce adequate dissipation for stabilization purposes:

$$\int_{\partial\Omega_I} U_j\, tu\, n_j\, dx \approx \sum_{J\in\Lambda_I}\frac{C_j^{IJ}}{2}\left(U_j^I\, tu_I + U_j^J\, tu_J\right) - \left(C_j^{IJ}\frac{U_j^J\, tu^J - U_j^I\, tu^I}{tu_J - tu_I}\right)(tu_J - tu_I) +$$

$$\sum_{J\in\Lambda_I^B} D_j^{IJ}\mathbf{F}_j^I,\tag{26}$$

where tu stands for turbulence unknowns ($\tilde{\nu}$, k and ω). Furthermore, the volume integrals were calculated using the midpoint rule and the gradients appearing in the model were calculated as follows:

$$\int_{\Omega_I}\frac{\partial U_i}{\partial x_j}\,d\mathbf{x} = \int_{\partial\Omega_I} U_i\, n_j\, dx,\tag{27}$$

Equation (27) in discrete form reads,

$$\frac{\partial U_i}{\partial x_j}\Big|^I \approx \partial_j^h U_i^I \equiv \frac{1}{V_I}\left[\sum_{J\in\Lambda_I}\frac{C_j^{IJ}}{2}\left(U_i^I + U_i^J\right) + \sum_{J\in\Lambda_I^B} D_j^{IJ} U_i^I\right],\tag{28}$$

where V_I is the volume of the control volume. Equations (27) and (28) are expressed only for the velocity but can be applied to any flow parameter. The second-order diffusion term was calculated using the compact stencil form according to Equation (3.38) in [16], where the gradient along the edges are evaluated by means of the compact finite difference scheme.

6. Initial and Boundary Conditions

Generally speaking, all run tests were started from freestream conditions. Due to the similarity of the two-equation models, the steady solution by the Wilcox k-ω model was used as starting condition for the Menter SST model to save computational resources. The boundary conditions for the flow parameters (i.e., velocity, density and total energy) are discussed in [15]. In this section, focus is given to the corresponding boundary conditions of the turbulence variables. Furthermore, the boundary conditions employed in this investigation were classified as: solid wall, inflow, outflow, symmetry and engine boundaries.

6.1. Solid Wall Boundary

In the Spalart–Allmaras model, the eddy viscosity was set to zero since no velocity fluctuations were present at such locations. Similarly, the turbulent kinetic energy, k, was assigned a zero value at the wall. The specific dissipation rate, ω, did not possess a natural boundary condition at the wall. However, based on the asymptotic solution given by Wilcox [17], the following value for ω was prescribed according to the implemented normalization:

$$\omega_0 = \frac{6\nu_w}{\beta y_w^2 Re_L}, \tag{29}$$

where ν_w is the laminar kinematic viscosity at the wall, β is a constant ($= 3/40$), y_w is the local first off-wall point and Re_L is the Reynolds number.

6.2. Inflow and Outflow Boundaries

In the Spalart–Allmaras turbulence model, the independent variable, $\tilde{\nu}$, was set to ten percent of the laminar viscosity at the inflow and outflow. In a similar way and based on recommendations by Spalart and Rumsey [21], the following values for k_∞/U_∞^2 and $\omega_\infty/(U_\infty/L)$ were adopted in the two-equation turbulence models at these boundaries: 1×10^{-6} and 5, respectively.

6.3. Symmetry Boundary

A symmetry condition is defined as a boundary where the normal velocity component is zero, being the density and total energy normal derivative zero, as well. Furthermore, a symmetry condition can also be compared to an inviscid wall (streamline). Hence, the turbulent eddy viscosity was set to zero at this location in turbulence models. If the configuration of the problem possesses a symmetry plane, this represents a very convenient way to significantly reduce the computational resources required.

6.4. Engine Boundary

Some test cases in the present study considered jet engines, such as the F15 aircraft. The internal parts of the engines were assumed to be outside the computational domain. Therefore, it was necessary to prescribe the corresponding boundary conditions at the engine inlets and outlets. For engine inlets, the mass flow, m_e, is defined as follows:

$$m_e = -\int_{\Gamma_e} \rho u_j n_j^e dx, \tag{30}$$

where Γ_e is the inlet engine surface and n_j^e is the corresponding unit normal. Furthermore, the independent variables of the turbulence model equations were prescribed fixed values at engine inlets and outlets. In general, freestream values for the turbulence variables were set at the engine inflow and outflow.

7. Results and Discussion

The Spalart–Allmaras, Wilcox k-ω and Menter SST turbulence models were implemented in the FLITE3D flow solver [15]. The results of classical aerodynamic test cases are presented in this section.

7.1. Supersonic Flat Plate

Although the flat plate shows a very simple geometry without a streamwise pressure gradient, it is very appropriate to evaluate the performance of any turbulence model due to the extensive experimental data, theoretical/empirical correlations and high-resolution numerical data available from the literature. In this study, a supersonic flat plate at a freestream Mach number, M_∞, equal to 2 was computed. The Reynolds number, based on the streamwise x-coordinate was 1×10^7 per unit length. The computational domain dimensions shown in Figure 1a were 10.4 m $\times$ 0.78 m $\times$ 0.78 m along the x-streamwise, y-vertical and z-spanwise directions, respectively. The boundary layer thickness by the end of the flat plate was $\delta_{ref} \approx 0.06$ m, therefore, the computational box in terms of the reference boundary layer thickness was $173\delta_{ref} \times 13\delta_{ref} \times 13\delta_{ref}$. Hence, the computational box was wide enough to eliminate any influence from the lateral faces on the flow statistics. Furthermore, numerical results shown here were taken from the vicinity of the central longitudinal plane, far from lateral surfaces. Additionally, the height of the computational domain was sufficient ($\sim 13\delta_{ref}$) to allow a natural streamwise developing of the turbulent boundary layer without interference of the top boundary condition. The hybrid mesh consisted of 1.1 million tetrahedral and 0.5 million prismatic elements. The boundary layer consisted of 24 layers with the closest point to the wall located at a distance of approximately $y^+ \approx 0.7$ in wall units. A close-up of the boundary layer elements can be observed in Figure 1b. A region with a bottom symmetry condition was imposed upstream of the flat plate leading edge. Downstream of the flat plate (where the no-slip condition was assumed for the velocity field), a freestream condition was prescribed. Similarly, at inflow, outflow and top surfaces, freestream values were imposed. An isothermal wall condition Was assumed for the temperature field with a wall to freestream temperature ratio (T_w/T_∞) equal to 1.75. This was a 3D case with a symmetry condition assumed in lateral faces. Iso-contours of the turbulent kinetic energy, $k^* = k/U_\infty^2$, are shown in Figure 1c) from the Menter SST model. The starting and ending points of the flat plate were represented by two cross-sectional YZ cutting planes. The full $X-$length of the flat plate was about $87\delta_{ref}$. Figure 1c presents the natural evolution of k^* in the direction of the flow, which was mainly responsible for triggering turbulence.

Figure 2 shows the streamwise variation of the skin friction coefficient, C_f, as a function of Re_x. Generally speaking, the corresponding values obtained by Spalart–Allmaras, Wilcox k-ω and Menter SST turbulence models depicted an excellent agreement with theoretical correlations from White [22] and Schlichting [23], particularly by the end of the flat plate. However, the Menter SST model exhibited a shorter transition and the C_f profile quickly tended to realistic fully-turbulent values downstream from the leading edge. Additionally, Direct Numerical Simulation (DNS) data from Araya and Jansen [24] are also included for a supersonic isothermal flat plate ($T_w/T_\infty = 2.25$) at a freestream Mach number, M_∞, of 2.5 and low Reynolds numbers. Again, the Menter SST turbulence model exhibited a good agreement of C_f with high spatial/temporal numerical results. Generally speaking, the mean streamwise velocity along the boundary layer followed the DNS profile of Araya and Jansen [24] and the 1/7 power-law distribution in the outer region (i.e., $y/\delta > 0.1$), as shown in Figure 3a for $Re_x = 3.4 \times 10^7$. Nevertheless, all velocity profiles computed from turbulence models depicted some deviations in the inner region (where the 1/7 power law did not work properly) from the DNS profile. The thermal boundary layer plays a crucial role in the transport phenomena of compressible wall-bounded flows; consequently, its accurate understanding and modeling are very important. In other words, the coupled behavior of the velocity and thermal field should be properly represented. In Figure 3b, the computed temperature distributions, T/T_∞, as a function of the mean streamwise velocity, U/U_∞, are plotted at a streamwise station where $Re_x = 3.4 \times 10^7$. For comparison, the theoretical Crocco–Busemann relation (see page 502 in White [22]) is also included.

The Crocco–Busemann relation assumes a linear variation of the total enthalpy across the boundary layer in zero pressure gradient with a unitary turbulent Prandtl number. Furthermore, present RANS thermal profiles showed good agreement with the quadratic Crocco–Busemann relationship. It is worth mentioning that the Crocco–Busemann relation is a function of the local wall-temperature; therefore, a slightly different theoretical profile was obtained for each turbulence model.

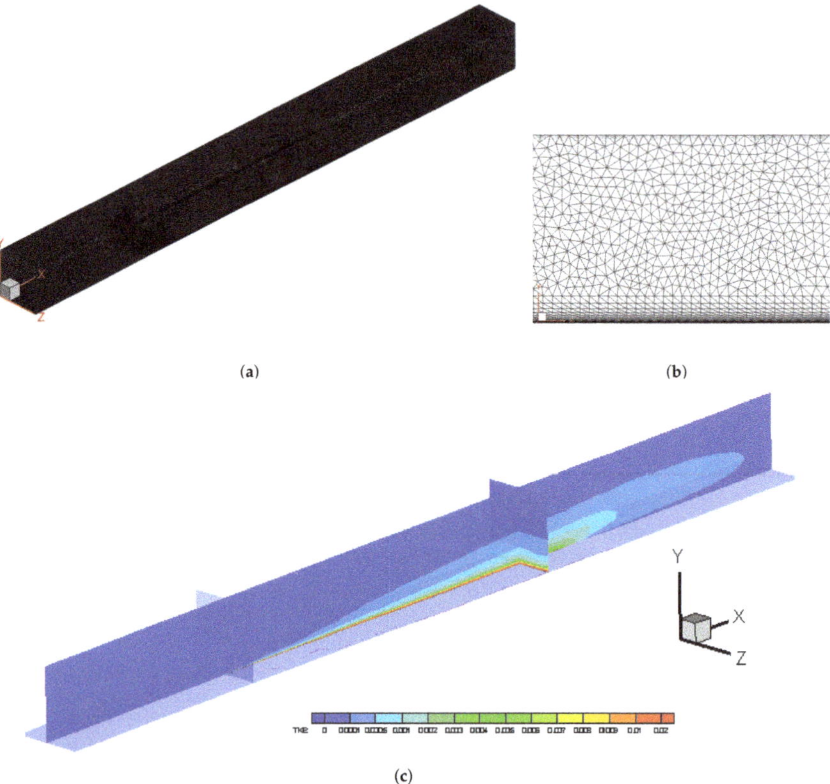

(a)

(b)

(c)

Figure 1. Mesh schematic (**a**); close-up of the near wall region (**b**); and iso-contours of turbulent kinetic energy from Menter SST (**c**) in the supersonic flat plate.

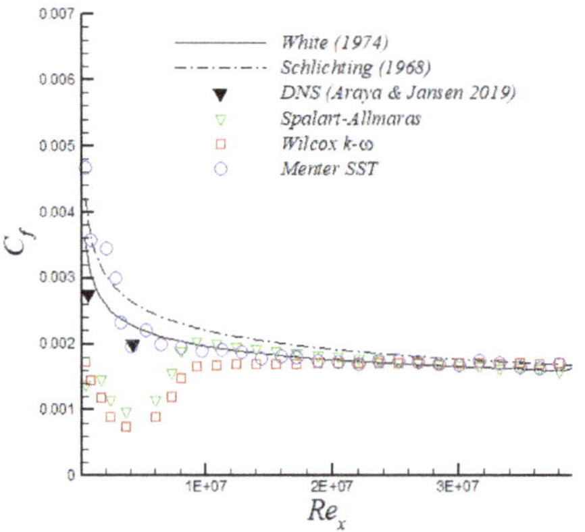

Figure 2. Skin friction coefficient as a function of Re_x.

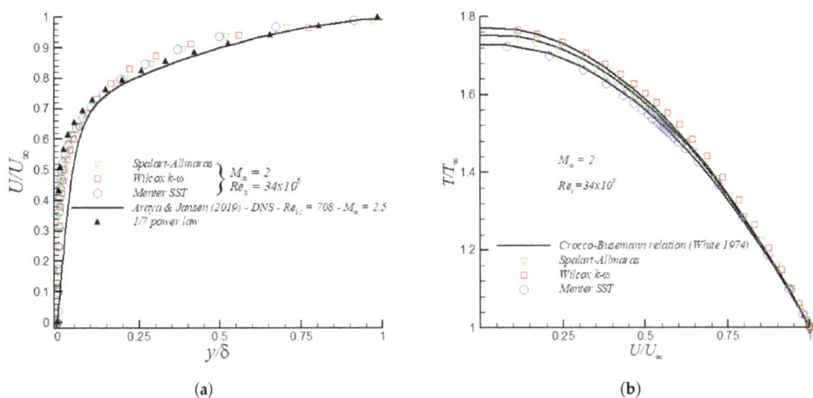

Figure 3. Mean streamwise velocity (**a**); and temperature distribution (**b**) in the supersonic flat plate.

7.2. Transonic RAE 2822 Airfoil

This evaluation test consisted on the simulation of steady turbulent transonic flow around the RAE2822 airfoil, which is the standard test Case 9 [25] for turbulence modeling. The freestream Mach number was $M_\infty = 0.73$, the Reynolds number was 6.5×10^6 and the angle of attack was $\alpha = 2.79°$. This problem was solved using a 3D hybrid and unstructured mesh. The mesh consisted of 4.2 million tetrahedral elements, 1.9 million prismatic elements and 2 thousand pyramid elements. The total number of elements was approximately 6.1 million. The boundary layer consisted of 40 layers with the closest point to the wall located at a distance of $y/c = 0.5 \times 10^{-6}$ or approximately $y^+ \approx 0.1–0.4$ in wall units.

Table 1 shows the computed values of lift, drag and moment coefficients for the Spalart–Allmaras, Wilcox k-ω and Menter SST turbulence models, together with experimental data from the AGARD AR–138 report [25] and other numerical simulations in 2D and structured meshes [26,27]. In Table 1, N/A stands for Not Available. For this simulation case, all turbulence models produced values of C_L within 11 lift counts (1 lift count = 0.001) with respect to experiments by Cook et al. [25]. Similarly, the predicted total drag coefficients, $C_{D\ total}$, showed good agreement with experiments with a maximum discrepancy of 1.5 drag counts (1 drag count = 0.0001), at most, for the Menter SST model; nevertheless, it generated the most accurate moment coefficient at 25% of the chord, $C_{m\ 25\%}$.

Table 1. Lift, drag and moment coefficients in RAE2822.

	SA	**k-ω**	**SST**	**Exp. [25]**	**Num. [26]**	**Num. [27]**
C_L	0.801	0.811	0.792	0.803	$0.794 - 0.824$	0.824
$C_{D\ total}$	0.0175	0.0178	0.0182	0.0168	$0.0164 - 0.0165$	0.0165
$C_{D\ friction}$	0.0054	0.0054	0.0064	N/A	$0.0054 - 0.0055$	0.0055
$C_{m\ 25\%}$	-0.1121	-0.1144	-0.1099	-0.099	-0.0874	N/A

In general, the three models are in good agreement with the experimental results and other simulation values, as shown in Table 1. Furthermore, the pressure (C_p) and skin friction (C_f) coefficients are depicted in Figure 4, together with the corresponding experimental data from the AGARD AR–138 report [25]. The predicted values of C_p in the upper surface looked very similar in all turbulence models. The weak shock at $x/c \approx 0.05$ was not fully captured by turbulence models with discrepancies in the order of 9% in peaks of C_p. Downstream, a quasi-zero pressure gradient zone with almost constant values of C_p and about half-chord in length was observed. The second and stronger shock located at $x/c \approx 0.55$ was accurately captured by all models. Perhaps, the Menter SST slightly separated from the experimental values by the end of the shock (i.e., $x/c \approx 0.57$). The strong shock was characterized by a sharp increase of wall pressure. The presence of a very strong APG induced a small flow separation zone, which is discussed below. Beyond this zone of "pockets" with supersonic flow, the wall pressure kept increasing towards the trailing edge, but at a moderate rate. In addition, the numerical predictions of C_p on the lower surface were almost identical to the experimental values given by all turbulence models. The skin friction coefficient is defined as $C_f = \frac{\tau_w}{1/2\rho_e U_e^2}$, where τ_w is the wall shear stress, and ρ_e and U_e are the density and velocity at the edge of the boundary layer, respectively. From the results in Figure 4b, it was inferred that the closest values to experiments were produced by the Menter SST model in the upper side, particularly at the strong shock location (i.e., $x/c \approx 0.55$). Furthermore, the Spalart–Allmaras and Wilcox k-ω models were observed to perform similarly; however, the Wilcox k-ω was the only turbulence model that predicted a shock-induced separation due to the presence of a very strong APG, with a small recirculating zone around $0.55 < x/c < 0.60$, as shown in Figure 4b. All models slightly underpredicted the experimental value of the skin friction on the lower surface. Figure 5 shows iso-contours of the Mach number for the Wilcox k-ω model. This transonic regime was characterized by the formation of regions or "pockets" with supersonic flow. Since the flow must return to its freestream conditions by the time of reaching the trailing edge, this caused the formation of a strong shock by $x/c \approx 0.55$ where subsonic flow was observed beyond that point.

During the mesh generation, a good resolution was ensured by clustering the first-off wall point well inside the linear viscous layer (i.e., $y^+ < 4$) to accurately predict the skin friction coefficient. Accordingly, the distance distribution of the first off-wall point in wall units, Δz_w^+, along the RAE2822 airfoil and based on the Menter SST model is shown in Figure 6 together with the airfoil coordinates in the right vertical axis. It can be seen that Δz_w^+ was around 0.4 in the vicinity of the leading edge for the lower surface; however, Δz_w^+ was lower than 0.1 in the upper surface, where the flow faced more complex pressure changes. Consequently, the high spatial resolution of the employed mesh was demonstrated. In Figure 7, profiles of the mean streamwise velocity downstream of the strong shock in the upper surface (i.e., at $x/c = 0.65$ and 0.75) are depicted. Notice that the mean streamwise velocity was normalized by the local velocity at the edge of the boundary layer, U_e. In general, for velocity profiles shown in Figure 7a,b, the Menter SST model produced the most accurate velocity profiles of the three turbulence models tested in this study, when compared to experimental values from Cook et al. [25]. It is interesting to point out that both velocity profiles at $x/c = 0.65$ and 0.75 were located in a zone of very strong APG or increasing pressure, as shown in Figure 4a. Furthermore, the Shear Stress Transport (SST) formulation by Menter [19] focused on improving the performance of this turbulence model in adverse pressure gradient flows. Thus, the results depicted in Figure 7a,b support the previous statement. In addition, the corresponding displacement thickness, δ^*, momentum thickness, θ^*, and, shape factor $H\,(=\delta^*/\theta^*)$ were computed at these stations. These integral boundary layer parameters (i.e., δ^* and θ^*) are defined as follows for compressible flows,

$$\delta^* = \int_0^\delta \left(1 - \frac{\rho U}{\rho_e U_e}\right) d\mathbf{z}, \tag{31}$$

$$\theta^* = \int_0^\delta \frac{\rho U}{\rho_e U_e} \left(1 - \frac{U}{U_e}\right) d\mathbf{z}, \tag{32}$$

where subscripts e stands for values at the edge of the boundary layer. In Table 2, the computed values of δ^*, θ^* and H are exhibited for the three turbulence models together with experimental data from Cook et al. [25]. Furthermore, it was established that the three models analyzed in this investigation gave a similar level of accuracy on the calculation of the boundary layer parameters; nevertheless, the Menter SST was slightly superior to the other models with an average discrepancy of 5% with respect to experiments. In addition, the knowledge of the shape factor, H, in turbulent flows is important in assessing how strongly the APG is imposed. In fact, the higher is the value of H, the stronger is the APG. Similarly, it is well known that any flow subjected to strong deceleration is prone to separation, where most of the turbulence models find enormous difficulties. In this opportunity, we picked up two stations in the RAE2822 airfoil (i.e., $x/c = 0.65$ and 0.75) with strong APG or high shape factors. The idea was to evaluate the performance of the turbulence models in extreme conditions. Generally speaking, the Menter SST model showed the best performance in describing the physics of the flow subjected to strong APG.

Table 2. Boundary layer parameters at $x/c = 0.65$ and 0.75 in the RAE2822 airfoil.

Location	Parameter	SA	k-ω	SST	Exp. [25]
$x/c = 0.65$	δ^*	0.004782	0.005029	0.005304	0.004956
	θ^*	0.001816	0.001794	0.002010	0.002043
	H	2.633	2.803	2.638	2.426
$x/c = 0.75$	δ^*	0.006571	0.006630	0.007276	0.006884
	θ^*	0.003032	0.003013	0.003204	0.002999
	H	2.167	2.201	2.271	2.296

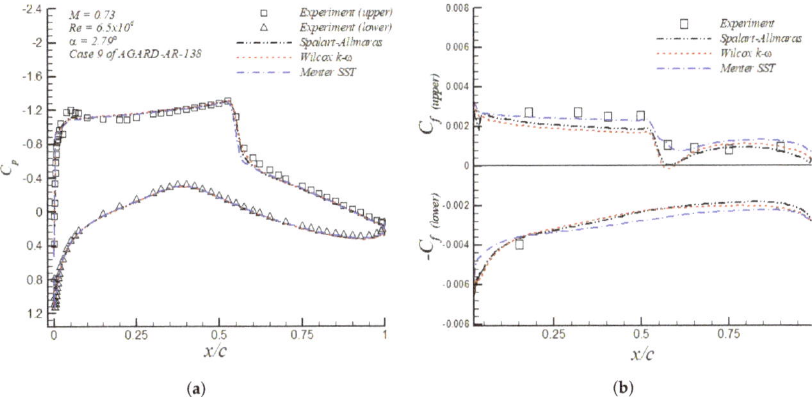

(a) (b)

Figure 4. Pressure (**a**); and skin friction (**b**) coefficient in the RAE2822 airfoil.

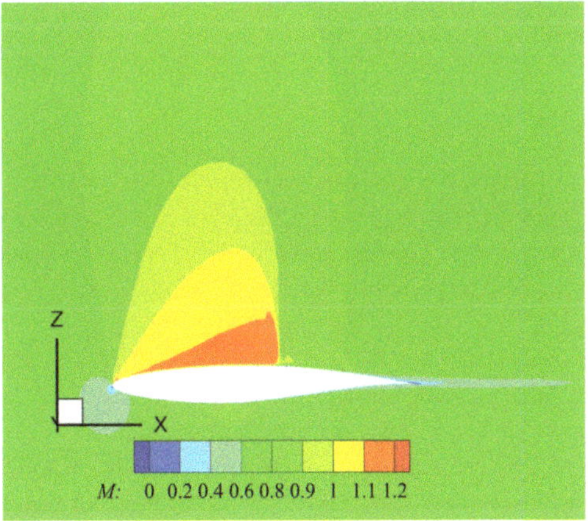

Figure 5. Iso-contours of Mach number (Wilcox k-ω model).

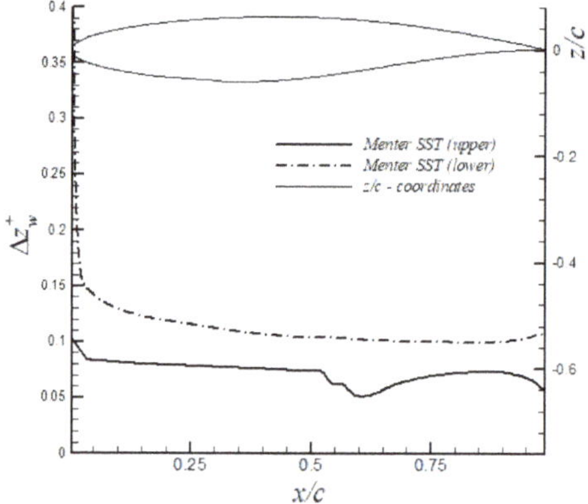

Figure 6. Distance distribution of the first off-wall point in wall units along the RAE airfoil for the Menter SST model.

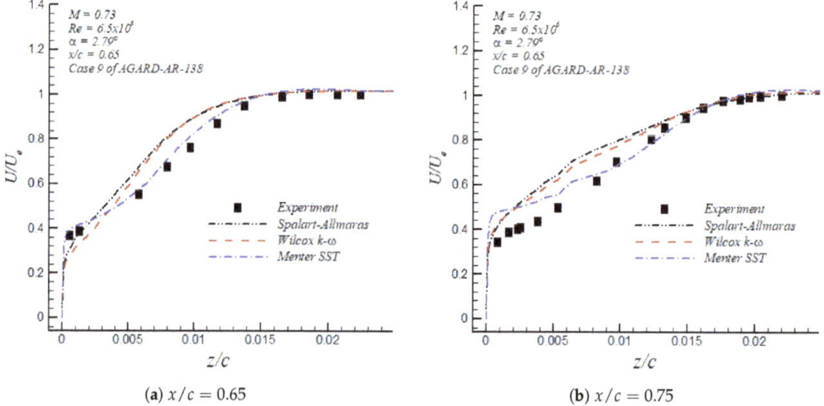

(a) $x/c = 0.65$

(b) $x/c = 0.75$

Figure 7. Mean streamwise velocity distribution in the RAE2822 airfoil.

7.3. ONERA M6 Wing

The third test case considered the simulation of the flow over the ONERA M6 wing [28], where the freestream Mach number was $M_\infty = 0.84$, the angle of attack was $\alpha = 3.06°$ and the Reynolds number based on the mean geometric chord was 1.2×10^7. The hybrid mesh employed consisted of 6.3 million tetrahedral elements, 2.5 million prisms and 11.6 thousand pyramids. The surface mesh consisted of 1.1 million triangular elements. The mesh had 35 viscous layers and the first off-wall point was located at $y^+ \approx 0.4$ in wall units.

The values of lift and drag coefficient for the Spalart–Allmaras, Wilcox k-ω and Menter SST turbulence models are shown in Table 3 together with numerical results from Le Moigne and Qin [29] and Nielsen and Anderson [30]. A maximum of nine lift counts deviation in the lift coefficient C_L between the three turbulence models was observed. In addition, a maximum of 14 drag counts difference in the drag coefficient C_D was computed, of which nine drag counts were due to the friction contribution to the drag coefficient. Furthermore, the present numerical results of lift and drag coefficients are in good agreement with numerical values by Le Moigne and Qin [29] and Nielsen and Anderson [30], who employed the Baldwin–Lomax and Spalart–Allmaras turbulence model, respectively. The pressure coefficient profiles C_p at different sections across the wing span are depicted in Figure 8. All three models captured reasonably well the corresponding location of both shocks at each wing span section, given by abrupt modifications on C_p. The most significant discrepancies were observed at $y/(b/2) = 0.8$, where none of the three models were able to appropriately represent the shock within $0.25 < x/c < 0.35$. However, a similar behavior of the Spalart–Allmaras and Wilcox k-ω turbulence models at $y/(b/2) = 0.8$ was reported by Huang et al. [31], Jakirlić et al. [32] and Nielsen and Anderson [30]. The skin friction coefficient C_f is plotted in Figure 9 at spanwise stations $y/(b/2) = 0.9$ and 0.95. These figures show the good agreement among all turbulence models about the location of the flow separation point $x/c \sim 0.25$ and 0.2 for $y/(b/2) = 0.9$ and 0.95, respectively. The downstream recovery of C_f was different in all turbulence models, with the Wilcox k-ω and Menter SST turbulence models predicting similar reattachment lengths and Spalart–Allmaras producing the largest reattachment length. The Menter SST model induced higher values for the skin friction downstream of the reattachment point, which was consistent with the largest values of $C_{D\,friction}$ shown in Table 3.

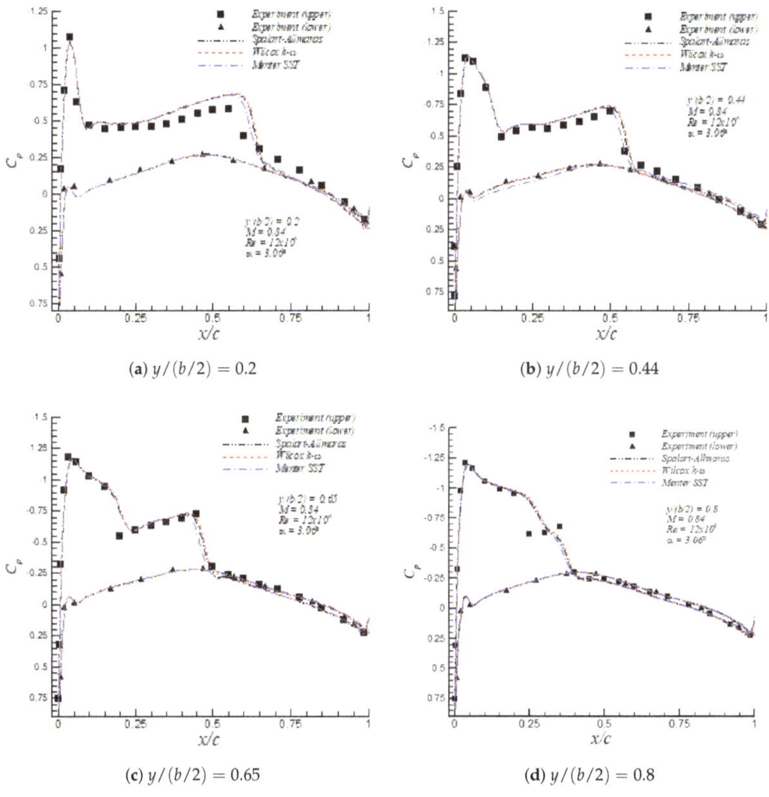

(**a**) $y/(b/2) = 0.2$

(**b**) $y/(b/2) = 0.44$

(**c**) $y/(b/2) = 0.65$

(**d**) $y/(b/2) = 0.8$

Figure 8. *Cont.*

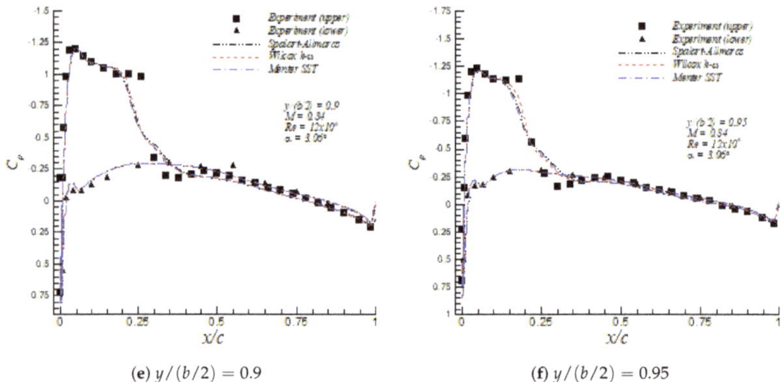

(**e**) $y/(b/2) = 0.9$ (**f**) $y/(b/2) = 0.95$

Figure 8. Pressure coefficient distributions in the ONERA M6 wing at $\alpha = 3.06°$.

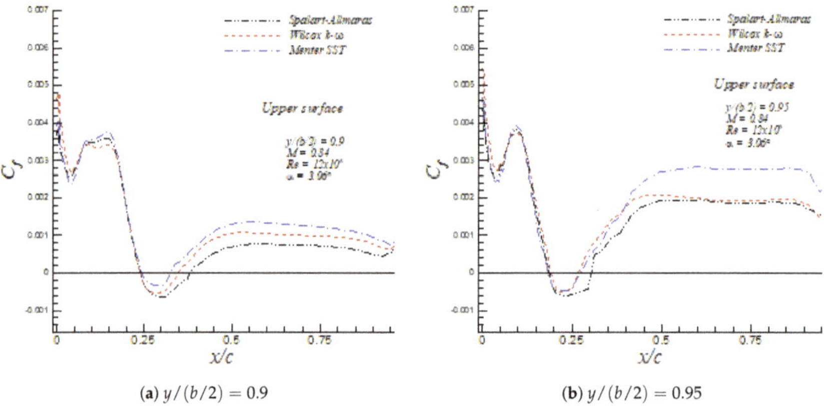

(**a**) $y/(b/2) = 0.9$ (**b**) $y/(b/2) = 0.95$

Figure 9. Skin friction distribution in the ONERA M6 wing at $\alpha = 3.06°$.

Table 3. Lift and drag coefficients in ONERA M6 wing at $\alpha = 3.06°$.

Aerodynamic Coeff.	SA	k-ω	SST	Num. [29]	Num. [30]
C_L	0.260	0.262	0.253	0.270	0.253
$C_{D\ total}$	0.0175	0.0179	0.0189	0.0174	0.0168
$C_{D\ friction}$	0.0048	0.0051	0.0057	0.0050	N/A

Figure 10 shows iso-surfaces of negative streamwise velocities normalized by the freestream velocity U_∞. The recirculating flow bubble in dark blue (identified by a white circle) predicted by the Spalart–Allmaras model was slightly larger than that of predicted by the Menter SST model. Iso-contours of the streamwise velocity, normalized by the freestream velocity U_∞, are depicted in Figure 11 in Menter SST at $y/(b/2) = 0.9$, where a strong shock wave can be clearly observed around $x/c \sim 0.25$ of the upper surface. Furthermore, the interaction of this shock with the turbulent boundary layer provoked a recirculating zone represented by the dark blue area.

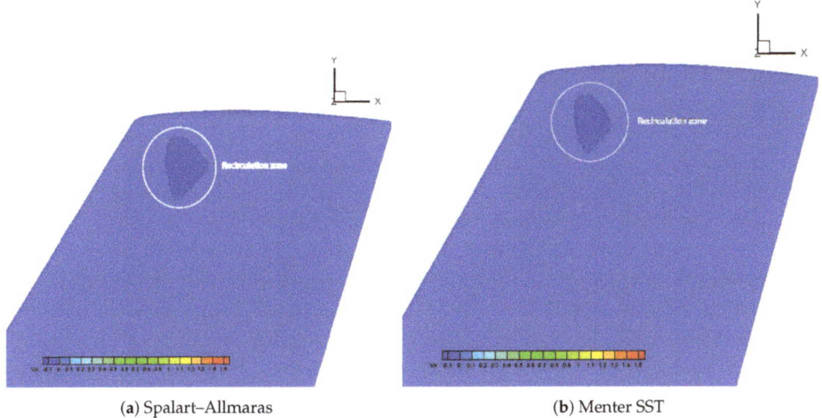

(**a**) Spalart–Allmaras (**b**) Menter SST

Figure 10. Iso-surfaces of streamwise velocity over the upper surface in the ONERA M6 wing at $\alpha = 3.06°$ (flow from left to right).

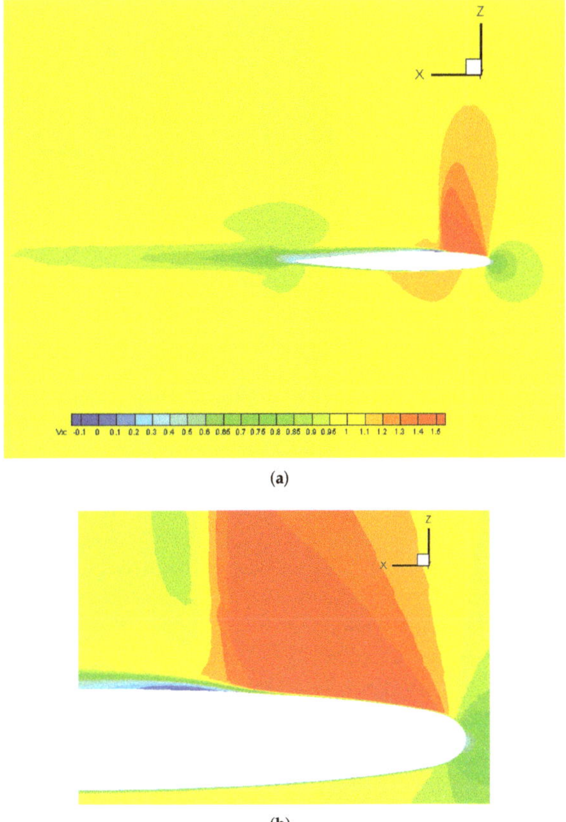

(**a**)

(**b**)

Figure 11. Iso-contours of streamwise velocity in Menter SST at $y/(b/2) = 0.9$ (**a**) and zoom over the leading edge (**b**) in the ONERA M6 wing at $\alpha = 3.06°$ (flow from right to left).

7.4. Generic F15 Aircraft Configuration

The last example corresponds to a generic F15 aircraft configuration, which represents a very complex and challenging geometry for CFD simulations with concave and convex surface curvatures. The freestream Mach number was $M_\infty = 0.9$, the Reynolds number, based on the airplane length ($L_x \sim 18$ m), was 3.58×10^8 and the angle of attack was five degrees. The hybrid mesh consisted of 12.5 million nodes and 49 million elements (36 million tetrahedra, 12.7 million prisms and 48.5 thousand pyramids). Due to the spanwise symmetry nature of the problem, only half of the model was used for the simulation. In Figure 12a, the employed surface mesh is plotted by mirroring the half-aircraft grid across the symmetry plane. In addition, details of the viscous layers can be seen by means of the cut plane at $x/L_x \approx 0.67$ in Figure 12b. Additionally, the boundary layer was made of 44 layers and the first off-wall point was prescribed at 1×10^{-5} m, which ensured that the location of this point was in the linear viscous layer for an accurate skin friction computation. A great deal of thought was given to the design of all computational meshes and dimensions in the present study to ensure grid-independent numerical results. The selected mesh spacing distribution, shown in this section, was the result of grid convergence study that was carried out to obtain grid independent results. The numerical simulations were run on 128 processors. Table 4 shows the lift coefficient, C_L, and the drag coefficient, C_D, for each turbulence model. The numerical results of C_L obtained by the three turbulence models were consistent with a maximum difference of seven lift counts. Furthermore, the largest discrepancy observed in the C_D coefficient was 24 drag counts between the Menter SST and Wilcox k-ω turbulence models.

Table 4. Lift and drag coefficients in F15 aircraft at $\alpha = 5.0°$.

Aerodynamic Coeff.	Spalart–Allmaras	Wilcox k-ω	Menter SST
C_L	0.181	0.174	0.175
$C_{D\ total}$	0.0397	0.0374	0.0398

In Figure 13a, iso-contours of the pressure coefficient C_p over the entire F15 aircraft configuration are depicted. Strong shock waves were observed on the canopy over the cock pit, leading edge of wings and between the vertical tails. The cut plane in Figure 13b at $x/L_x \approx 0.67$ depicted high values of C_p toward the wing tip. In addition, a top view of iso-contours of C_p are shown in Figure 14. The value of the pressure coefficient at four points (labeled as I–IV in Figure 14) were compared with flight and wind tunnel data from Webb et al. [33], as shown in Figure 15. In general, the present numerical results are in good agreement with the experimental wind tunnel data. Furthermore, it can be noticed that the pressure coefficients given by the Menter SST model showed a better match with experimental data than those calculated by Spalart–Allmaras and Wilcox k-ω models. In addition, the most significant discrepancies of the numerical results of C_p with experiments were found in the region $0.3 < x/L_x < 0.5$, just above the turbine intake. The discrepancy can be attributed to the difference between the estimated engine mass flow prescribed and the used engine mass flow during the flight test. This was emphasized by the significant scattering between the experimental data (in flight and wind tunnel), particularly, at Points I and II. In this zone, the different turbulence models predicted different separation regions, which are discussed below. Hence, this resulted in different local flow patterns that induced different pressure coefficients.

(**a**) Front view of the whole grid configuration.

(**b**) Cross-sectional cut at $x/L_x \approx 0.67$.

Figure 12. Surface mesh in the F15 aircraft configuration.

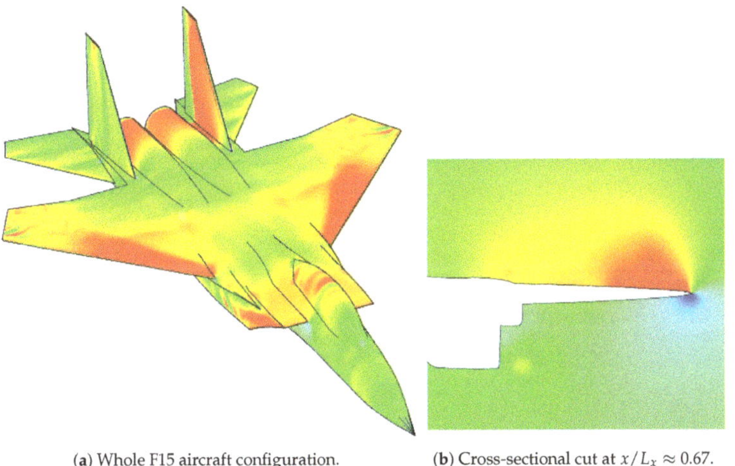

(**a**) Whole F15 aircraft configuration.

(**b**) Cross-sectional cut at $x/L_x \approx 0.67$.

Figure 13. Iso-contours of the pressure coefficient in the F15 aircraft configuration (C_p range of color contour as in Figure 14).

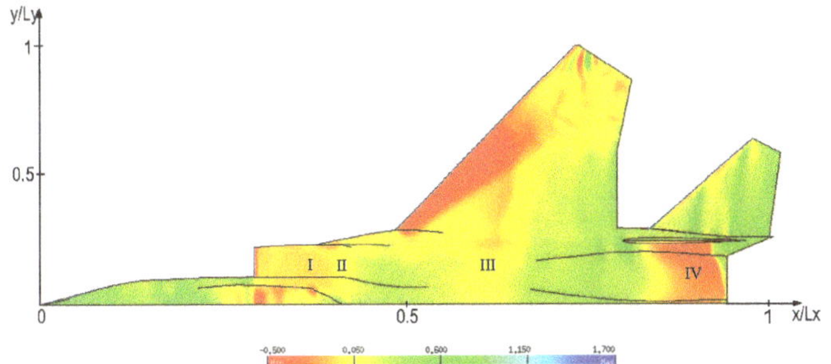

Figure 14. Iso-contours of pressure coefficient in the upper surface of the F15 aircraft.

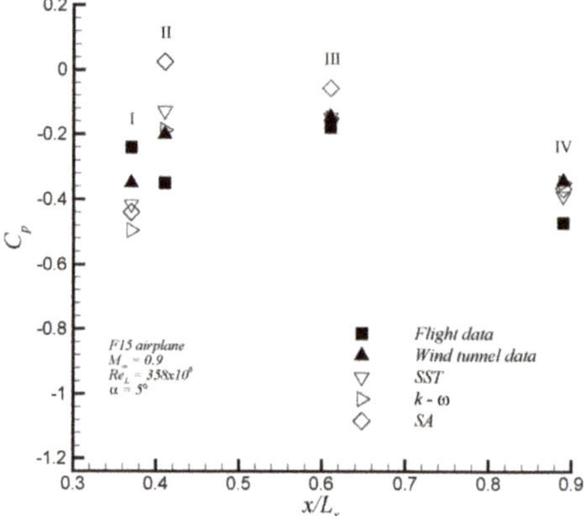

Figure 15. Variation of the pressure coefficient in the top surface along x/L_x of F15 aircraft for present numerical data (open symbols) experimental data (closed symbols).

The skin friction coefficient C_f at a spanwise location of $y = 1m$ or $y/L_y = 0.15$, where $L_y = 6.52$ m is the semi-span length of the aircraft, is plotted in Figure 16. The study of the drag coefficient due to friction showed that this section had the largest discrepancy between the three turbulence models, as observed in Figure 16a. Figure 16b indicates that the Wilcox k-ω model predicted two separation bubbles at $x/L_x \approx 0.325$ and 0.36, respectively, while the other two models predicted a single separated region (i.e., at $x/L_x \approx 0.35$) with the Menter SST predicting a much larger separation region than that computed by the Spalart–Allmaras model. Furthermore, the iso-contours of streamwise velocity (normalized by the freestream velocity U_∞) by the Menter SST model, as shown in Figure 17, indicated the presence of a lengthy recirculation zone with negative values of the velocity above the turbine intake, given by the shock wave-boundary layer interaction.

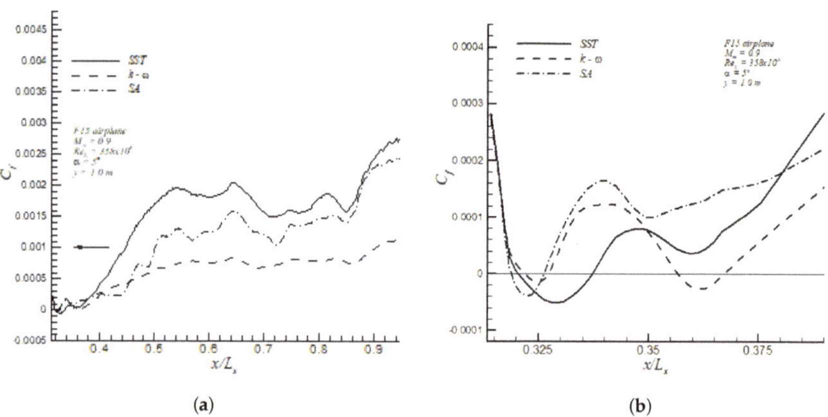

(a)

(b)

Figure 16. Skin friction coefficient in the upper surface of F15 at $y/L_y = 0.15$ (a); and zoom of the separated flow zone (b).

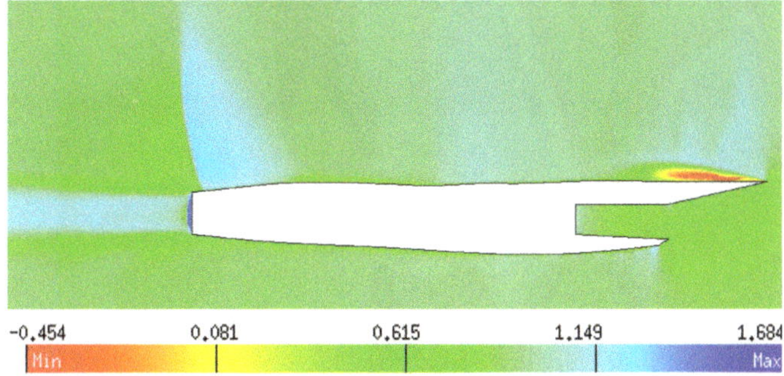

Figure 17. Iso-contours of the streamwise velocity at $y/L_y = 0.15$ (flow from right to left).

Figures 18 and 19 show the pressure coefficients on the upper and lower surfaces of the wing at two different spanwise locations, $y/L_y = 0.36$ and 0.59, respectively. In addition, the vertical coordinates, z, of the local profile are represented by the vertical axis on the right. It is shown in Figure 18a that, at $y/L_y = 0.36$, the C_p profiles predicted by the three turbulence models showed similar upper peak values in the vicinity of the leading edge, i.e., $C_p \approx -1$. Hence, it can be concluded that the pressure difference between the surface static pressure to the freestream static pressure was equal to the freestream dynamic pressure in absolute value at this location. On the other hand, the Wilcox k-ω model predicted a stronger second shock wave located at $x/L_x \approx 0.7$. In the lower side, a moderate APG zone was observed in $0.55 < x/L_x < 0.6$, as shown in Figure 18b, followed by a nearly ZPG region. Furthermore, Figure 19a shows similar peak values of C_p in the upper surface and close to the leading edge at station $y/L_y = 0.59$ by all three turbulence models. However, the Spalart–Allmaras model predicted a much stronger second shock located further downstream of the leading edge (at $x/L_x \approx 0.7$) than those predicted by the Menter SST and Wilcox k-ω model. All turbulence models exhibited similar C_p distributions in the lower surface, as shown in Figure 19b.

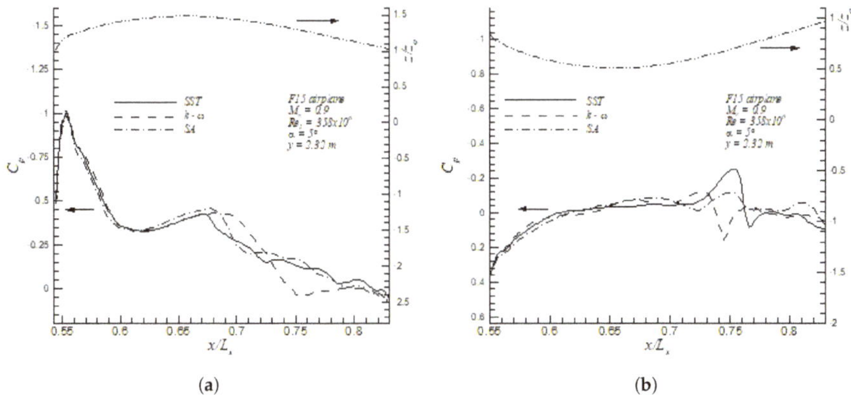

(a)

(b)

Figure 18. Pressure coefficients of F15 in the upper (**a**) and lower (**b**) surfaces at $y/L_y = 0.35$.

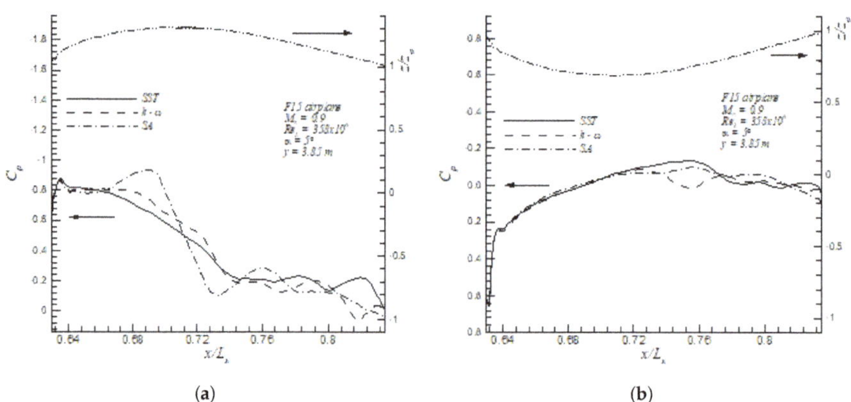

Figure 19. Pressure coefficients of F15 in the upper (**a**) and lower (**b**) surfaces at $y/L_y = 0.59$.

8. Conclusions

An assessment of three popular turbulent models was performed in 3D aerodynamic cases. The FLITE3D flow solver, in conjugation with the Spalart–Allmaras, the Wilcox k-ω and the Menter SST turbulence models, was applied to the supersonic flat plate, RAE2822 airfoil, ONERA M6 wing, and F15 aircraft.

The Menter SST model exhibited a short developing section in the supesonic flat plate downstream of the edge. The transonic RAE2822 airfoil possessed a strong shock–boundary layer interaction with an induced separation. The ONERA M6 wing exhibited a very complex flow phenomena such as double shocks, strong APG and streamline curvature effects. In addition, the computed global lift and drag coefficients (C_L and C_D) over the F15 aircraft by all turbulence models were very consistent; however, local pressure coefficients by the Menter SST model were in better agreement with experimental values in [33].

Generally speaking, numerical results have been very consistent for the three turbulence models. For attached flows or middle separated flows, there was not a clear superiority of the two-equation turbulence models over the one-equation Spalart–Allmaras model. However, the contrary was observed in separated flows with recirculation zones. Particularly, the Menter SST model showed the best compromise between accurately describing the physics of the flow and numerical stability. The Shear Stress Transport (SST) formulation might be the reason for this. Future investigation will focus on unsteady flow simulations. This could be more appropriate in realistically capturing the complex turbulent structures governed by shock-boundary layer interactions and highly-separated flows; in particular, at the transonic regime.

Funding: This material was based upon work supported by the Air Force Office of Scientific Research (AFOSR) under award number FA9550-17-1-0051.

Acknowledgments: The author acknowledges Oubay Hassan and Kenneth Morgan for valuable insight. Christian Lagares is acknowledged for important discussion.

Conflicts of Interest: The author declare no conflict of interest.

References

1.	Pope, S.B. *Turbulent Flows*; Cambridge University Press: Cambridge, UK, August 2000.
2.	Moin, P.; Mahesh, K. Direct Numerical Simulation: A Tool in Turbulence Research. *Annu. Rev. Fluid Mech.* **1998**, *30*, 539–578. [CrossRef]

3. Meneveau, C.; Katz, J. Scale-invariance and turbulence models for large-eddy simulation. *Annu. Rev. Fluid Mech.* **2000**, *32*, 1–32. [CrossRef]

4. Frohlich, J.; von Terzi, D. Hybrid LES/RANS methods for the simulation of turbulent flows. *Prog. Aerosp. Sci.* **2008**, *44*, 349–377. [CrossRef]

5. Catalano, P.; Amato, M. An evaluation of RANS turbulence modelling for aerodynamic applications. *Aerosp. Sci. Technol.* **2003**, *7*, 493–509. [CrossRef]

6. Knight, D.; Yan, H.; Panaras, A.G.; Zheltovodov, A. Advances in CFD prediction of shockwave turbulent boundary layer interactions. *Prog. Aerosp. Sci.* **2003**, *39*, 121–184. [CrossRef]

7. Manzari, M.T.; Hassan, O.; Morgan, K.; Weatherill, N.P. Turbulent flow computations on 3D unstructured grids. *Finite Elem. Anal. Des.* **1998**, *30*, 353–363. [CrossRef]

8. Sørensen, K.A.; Hassan, O.; Morgan, K.; Weatherill, N.P. A multigrid accelerated hybrid unstructured mesh method for 3D compressible turbulent flow. *Comput. Mech.* **2003**, *31*, 101–114. [CrossRef]

9. Peiró, J.; Peraire, J.; Morgan, K. *FELISA System Reference Manual. Part 1—Basic Theory*; Swansea Report C/R/821/94; University of Wales: Wales, UK, 1994.

10. Peraire, J.; Morgan, K.; Peiró, J. Unstructured finite element mesh generation and adaptive procedures for CFD. In *Applications of Mesh Generation to Complex 3D Configurations*; AGARD: Paris, France, 1990.

11. Hassan, O.; Morgan, K.; Probert, E.J.; Peraire, J. Unstructured tetrahedral mesh generation for three-dimensional viscous flows. *Int. J. Numer. Methods Eng.* **1996**, *39*, 549–567. [CrossRef]

12. Weatherill, N.P.; Hassan, O. Efficient three–dimensional Delaunay triangulation with automatic boundary point creation and imposed boundary constraints. *Int. J. Numer. Methods Eng.* **1994**, *37*, 2005–2039. [CrossRef]

13. Jameson, A.; Schmidt, W.; Turkel, E. Numerical simulation of the Euler equations by finite volume methods using Runge–Kutta timestepping schemes. In Proceedings of the 14th Fluid and Plasma Dynamics Conference, Palo Alto, CA, USA, 13–25 June 1981; AIAA Paper; pp. 81–1259.

14. Harten, A.; Lax, P.D.; van Leer, B. On upstreaming differencing and Godunov-type schemes for hyperbolic conservation laws. *SIAM Rev.* **1983**, *25*, 35. [CrossRef]

15. Morgan, K.; Peraire, J.; Peiro, J.; Hassan, O. The computation of 3-dimensional flows using unstructured grids. *Comput. Methods Appl. Mech. Eng.* **1991**, *87*, 335–352. [CrossRef]

16. Sørensen, K.A. A Multigrid Accelerated Procedure for the Solution of Compressible Fluid Flows on Unstructured Hybrid Meshes. Ph.D. Thesis, University of Wales, Swansea, Wales, 2002.

17. Wilcox, D.C. *Turbulence Modeling for CFD*; DWC Industries, Inc.: La Canada, CA, USA, 2006.

18. Spalart, P.R.; Allmaras, S.R. A one equation turbulence model for aerodynamics flows. In Proceedings of the 30th Aerospace Science Meeting and Exhibit, Reno, NV, USA, 6–9 January 1992; AIAA Paper 92-0439.

19. Menter, F.R. Review of the shear-stress transport turbulence model experience from an industrial perspective. *Int. J. Comput. Fluid Dyn.* **2009**, *23*, 305–316. [CrossRef]

20. Geuzaine, P. An Implicit Upwind Finite Volume Method for Compressible Turbulent Flows on Unstructured Meshes. Ph.D. Thesis, Université de Liège: Liège, Belgium, 1999.

21. Spalart, P.R.; Rumsey, C.L. Effective Inflow Conditions for Turbulence Models in Aerodynamic Calculations. *AIAA J.* **2007**, *45*, 2544–2553. [CrossRef]

22. White, F.M. *Viscous Fluid Flow*; McGraw Hill: New York, NY, USA, 1974.

23. Schlichting, H. *Boundary Layer Theory*; McGraw Hill: New York, NY, USA, 1968.

24. Araya, G.; Jansen, K. Compressibility effect on spatially-developing turbulent boundary layers via DNS. In Proceedings of the 4th Thermal and Fluids Engineering Conference (TFEC2019), Las Vegas, NV, USA, 14–17 April 2019.

25. Cook, P.; McDonald, M.; Firmin, M. *Aerofoil RAE2822 Pressure Distributions and Boundary Layer and Wake Measurements*; Report AR-138; AGARD: Paris, France, 1979.

26. Hirschel, E.H. *Finite Approximations in Fluid Mechanics II: DFG Priority Research Program, Results 1986–1988*; Notes on Numerical Fluid Mechanics; Vieweg: Braunschweig, Germany; Wiesbaden, Germany, 1989; Volume 25.

27. Swanson, R.C.; Rossow, C.C. An efficient solver for the RANS equations and a one-equation turbulence model. *Comput. Fluids* **2011**, *42*, 13–25. [CrossRef]

28. Schmitt, V.; Charpin, F. *Pressure Distributions of the ONERA M6 Wing at Transonic Mach Numbers*; Report AR-138; AGARD: Paris, France, 1979.

29. LeMoigne, A.; Qin, N. Variable-fidelity aerodynamic optimization for turbulent flows using a discrete adjoint formulation. *AIAA J.* **2004**, *42*, 1281–1192.

30. Nielsen, E.J.; Anderson, W.K. Recent improvements in aerodynamic sesign optimization on unstructured meshes. *AIAA J.* **2002**, *40*, 1155–1163. [CrossRef]

31. Huang, J.C.; Lin, H.; Yang, J.Y. Implicit preconditioned WENO scheme for steady viscous flow computation. *J. Comput. Phys.* **2009**, *228*, 420–438. [CrossRef]

32. Jakirlić, S.; Eisfeld, B.; Jester-Zurker, R.; Kroll, N. Near-wall, Reynolds-stress model calculations of transonic flow configurations relevant to aircraft aerodynamics. *Int. J. Heat Fluid Flow* **2007**, *28*, 602–615. [CrossRef]

33. Webb, L.; Varda, D.; Whitmore, S. *Flight and Wind-Tunnel Comparisons of the Inlet/Airframe Interaction of the F-15 Airplane*; Technical Paper 2374; NASA: Washington, DC, USA, 1984.

Article

Baropycnal Work: A Mechanism for Energy Transfer across Scales

Aarne Lees [1,2] and Hussein Aluie [1,2,*]

[1] Department of Mechanical Engineering, University of Rochester, Rochester, NY 14627, USA;
aarnelees@gmail.com

[2] Laboratory for Laser Energetics, University of Rochester, Rochester, NY 14627, USA

* Correspondence: hussein@rochester.edu

Received: 5 April 2019; Accepted: 13 May 2019; Published: 18 May 2019

Abstract: The role of baroclinicity, which arises from the misalignment of pressure and density gradients, is well-known in the vorticity equation, yet its role in the kinetic energy budget has never been obvious. Here, we show that baroclinicity appears naturally in the kinetic energy budget after carrying out the appropriate scale decomposition. Strain generation by pressure and density gradients, both barotropic and baroclinic, also results from our analysis. These two processes underlie the recently identified mechanism of "baropycnal work", which can transfer energy across scales in variable density flows. As such, baropycnal work is markedly distinct from pressure-dilatation into which the former is implicitly lumped in Large Eddy Simulations. We provide numerical evidence from 1024^3 direct numerical simulations of compressible turbulence. The data shows excellent pointwise agreement between baropycnal work and the nonlinear model we derive, supporting our interpretation of how it operates.

Keywords: multiscale; energy transfer; cascade; turbulence; variable density flow; baroclinic vortex generation

---

1. Introduction

Energy transfer across length scales is one of the defining characteristics of turbulent flows, the subject of which fits well under the "Multiscale Turbulent Transport" theme of this special issue in *Fluids*. In constant density canonical turbulence, the only pathway for transferring energy across scales is *deformation work* [1,2], which we represent below by Π. This is often referred to as the *turbulence production* term in the turbulent kinetic energy (TKE) budget within the Reynolds averaging decomposition [3] or the *spectral flux* within the Fourier decomposition [4–6]. Deformation work gives rise to the cascade in incompressible turbulence, which is largely believed to operate by vortex stretching in 3-dimensions [1,7–10], an idea which may be traced back to Taylor [11,12].

Recent studies [13–18] have shown that in the presence of density variations, such as in compressible turbulence, there exists another pathway across scales called "baropycnal work", represented by Λ below. In the traditional formulation of the compressible Large Eddy Simulation (LES) equations [19–22], which are essentially a coarse-graining decomposition of scales, Λ is almost always implicitly lumped with $P\nabla\cdot\mathbf{u}$, where P is pressure and $\mathbf{u}$ is velocity and treated as a large scale (resolved) pressure-dilatation which does not require modeling. Reference [15] argued that baropycnal work, Λ, is more similar in nature to deformation work, Π, in that it involves large-scales interacting with small-scales thereby allowing it to transfer energy *across* scales. As such, it is fundamentally distinct from pressure dilatation which involves only large-scales and cannot transfer energy directly across scales.

In this work, we shall investigate the mechanisms by which baropycnal work transfers energy across scales. The main result is embodied in Equation (16) below, which shows that Λ transfers energy by two processes:

(I)   Barotropic and baroclinic generation of strain, **S**, from gradients of pressure and density, ρ:
$$(\text{const.})\,\ell^2\rho^{-1}\left[\boldsymbol{\nabla}P\boldsymbol{\cdot}\mathbf{S}\boldsymbol{\cdot}\boldsymbol{\nabla}\rho\right] = (\text{const.})\,\ell^2\rho^{-1}\left[\left(\boldsymbol{\nabla}\rho\left(\boldsymbol{\nabla}P\right)^T\right):\mathbf{S}\right],$$

(II)   Baroclinic generation of vorticity, $\boldsymbol{\omega}$:
$$(\text{const.})\,\ell^2\rho^{-1}\left(\boldsymbol{\nabla}\rho\times\boldsymbol{\nabla}P\right)\boldsymbol{\cdot}\boldsymbol{\omega},$$

where the dyadic product $\boldsymbol{\nabla}\rho\left(\boldsymbol{\nabla}P\right)^T$ is a tensor. Length scale ℓ is that at which density and pressure gradients are evaluated as we make clear in Equations (16)–(18) below. To our knowledge, these results are the first to show how baroclinicity, $\boldsymbol{\nabla}\rho\times\boldsymbol{\nabla}P$, appears in the kinetic energy budget. Baroclinicity is often analyzed within the vorticity budget but its role in the kinetic energy budget has never been obvious. We shall show here that baroclinicity appears naturally in the kinetic energy budget after performing the appropriate scale decomposition. Strain generation by pressure and density gradients (both barotropic and baroclinic) also results from our analysis, highlighting its potential significance which is often overlooked in the literature. The processes of strain and vortex generation show baropycnal work Λ to be markedly distinct from the process of pressure dilatation, further supporting the argument against lumping the two terms.

There is a diverse array of applications in this research subject. Density variability can arise in high Mach number flows, but it is also pertinent in the limit of low Mach numbers along contact discontinuities such as in multi-species or multi-phase flows [23]. Variable density (VD) flows are relevant in a wide range of systems, such as in molecular clouds in the interstellar medium [24–26]), in inertial confinement fusion [27–29], in high-speed flight and combustion [30,31], and in air-sea interaction in geophysical flows [32–35].

The outline of the paper is as follows. In Section 2, we shall use coarse-graining to decompose scales and identify baropycnal work, Λ. In Section 3, we use scale locality to approximate Λ with a nonlinear model, which in turn shows how Λ is due to a combination of strain generation and baroclinic vorticity generation. In Section 4, we describe our direct numerical simulations (DNS) used in Section 5 to present evidence that indeed Λ and its nonlinear model exhibit excellent agreement. This justifies our analysis of Λ via its nonlinear model, similar to what was done in Reference [7] in their analysis of Π. We conclude with Section 6.

2. Multi-Scale Dynamics

To analyze the dynamics of different scales in a compressible flow, we use the coarse-graining approach, which has proven to be a natural and versatile framework to understand and model scale interactions (e.g., Reference [36,37]). The approach is standard in partial differential equations and distribution theory (e.g., see References [38,39]). It became common in Large Eddy Simulation (LES) modeling of turbulence thanks to the original work of Leonard [40] and the later work of Germano [41]. Eyink [37,42–44] subsequently developed the formalism mathematically to analyze the fundamental physics of scale coupling in turbulence.

Coarse-graining has been used in many fluid dynamics applications, ranging from DNS of incompressible turbulence (e.g., References [45–48]), to 2D laboratory flows (e.g., References [49–55]), to experiments of turbulent jets [56] and flows through a grid [57], through a duct [58], in a water channel [59] and in turbomachinery (e.g., References [60,61]). Moreover, the framework has been extended to geophysical flows [62–64], magnetohydrodynamics [65,66] and compressible turbulence (e.g., Reference [14]), and most recently, as a framework to extract the spectrum in a flow [67].

For any field $\mathbf{a}(\mathbf{x})$, a coarse-grained or (low-pass) filtered field, which contains modes at scales $> \ell$, is defined in n-dimensions as

$$\overline{\mathbf{a}}_\ell(\mathbf{x}) = \int d^n\mathbf{r}\, G_\ell(\mathbf{r})\,\mathbf{a}(\mathbf{x}+\mathbf{r}), \tag{1}$$

where $G(\mathbf{r})$ is a normalized convolution kernel and $G_\ell(\mathbf{r}) = \ell^{-n} G(\mathbf{r}/\ell)$ is a dilated version of the kernel having its main support over a region of diameter ℓ. The scale decomposition in (1) is essentially a partitioning of scales in the system into large ($\gtrsim \ell$), captured by $\overline{\mathbf{a}}_\ell$ and small ($\lesssim \ell$), captured by the residual $\mathbf{a}'_\ell = \mathbf{a} - \overline{\mathbf{a}}_\ell$. In the remainder of this paper, we shall omit subscript ℓ from variables if there is no risk for confusion.

Variable Density Flows

In incompressible turbulence, our understanding of the scale dynamics of kinetic energy centers on analyzing $|\overline{\mathbf{u}}_\ell|^2/2$. In variable density turbulence, scale decomposition is not as straightforward due to the density field $\rho(\mathbf{x})$. Several definitions of "large-scale" kinetic energy have been used in the literature, corresponding to different scale-decompositions as discussed in Reference [15]. These include $\overline{\rho}_\ell|\overline{\mathbf{u}}_\ell|^2/2$, which has been used in several studies (e.g., References [68–71]) and $|\overline{(\sqrt{\rho}\mathbf{u})}_\ell|^2/2$, which has also been used extensively in compressible turbulence studies (e.g., References [17,72–74]). A "length-scale" within these different decompositions corresponds to different flow variables, each of which can yield quantities with units of energy. However, as demonstrated in Reference [75], such decompositions can violate the so-called *inviscid criterion*, yielding difficulties with disentangling viscous from inertial dynamics in turbulent flows. The inviscid criterion stipulates that a scale decomposition should guarantee a negligible contribution from viscous terms in the evolution equation of the large length-scales. It was shown mathematically in Reference [15] and demonstrated numerically in Reference [75] that a Hesselberg-Favre decomposition, introduced by Hesselberg [76] but often associated with Favre [19,77], $|\overline{\rho\mathbf{u}}_\ell|^2/2\overline{\rho}_\ell$, satisfies the inviscid criterion, which allows for properly disentangling the dynamical ranges of scales. We will use the common notation

$$\widetilde{\mathbf{u}}_\ell(\mathbf{x}) = \overline{\rho\mathbf{u}}_\ell/\overline{\rho}_\ell \ , \tag{2}$$

which yields $\overline{\rho}_\ell|\widetilde{\mathbf{u}}_\ell|^2/2$ for kinetic energy at scales larger than ℓ. The budget for the large-scale KE can be easily derived [15] from the momentum Equation (20) below:

$$\partial_t \overline{\rho}_\ell \frac{|\widetilde{\mathbf{u}}_\ell|^2}{2} + \boldsymbol{\nabla}\!\cdot\!\mathbf{J}_\ell = -\Pi_\ell - \Lambda_\ell + \overline{P}_\ell\boldsymbol{\nabla}\!\cdot\!\overline{\mathbf{u}}_\ell - D_\ell + \epsilon_\ell^{inj}, \tag{3}$$

where $\mathbf{J}_\ell(\mathbf{x})$ is space transport of large-scale kinetic energy, $-\overline{P}_\ell\boldsymbol{\nabla}\!\cdot\!\overline{\mathbf{u}}_\ell$ is large-scale pressure dilatation, $D_\ell(\mathbf{x})$ is viscous dissipation acting on scales $> \ell$ and $\epsilon_\ell^{inj}(\mathbf{x})$ is the energy injected due to external stirring. These terms are defined in Equations (16)–(18) of Reference [15]. The $\Pi_\ell(\mathbf{x})$ and $\Lambda_\ell(\mathbf{x})$ terms account for the transfer of energy *across* scale ℓ and are defined as

$$\Pi_\ell(\mathbf{x}) \;\; = \;\; -\overline{\rho}\,\partial_j\widetilde{u}_i\,\widetilde{\tau}(u_i, u_j) \tag{4}$$

$$\Lambda_\ell(\mathbf{x}) \;\; = \;\; \frac{1}{\overline{\rho}}\partial_j\overline{P}\,\overline{\tau}(\rho, u_j), \tag{5}$$

where

$$\overline{\tau}_\ell(f, g) \equiv \overline{(fg)}_\ell - \overline{f}_\ell\overline{g}_\ell \tag{6}$$

is a 2nd-order *generalized central moment* of fields $f(\mathbf{x}), g(\mathbf{x})$ (see Reference [41]).

The first flux term, Π_ℓ, is similar to its incompressible counterpart and is often called deformation work. The second flux term, Λ_ℓ, was identified in References [13,15] and called "baropycnal work". It is inherently due to the presence of a variable density and vanishes in the incompressible limit. Recent work by Eyink and Drivas [18,78] identified a third possible pathway for energy transfer, which they called "pressure-dilatation defect" and arises when the joint limits of $\kappa, \mu \to 0$ and $\ell \to 0$ do not commute. Eyink and Drivas [18] showed that the pressure-dilatation defect mechanism transfers energy downscale in 1D normal shocks. In this paper, we shall focus on understanding the mechanisms by which baropycnal work transfers energy across scales.

3. The Mechanism of Baropycnal Work

Our investigation of the mechanism behind baropycnal work is inspired by the work of Borue & Orszag [7], where they used the nonlinear model of the energy flux Π_ℓ to show that it operates, on average, by vortex stretching (see also Equation (32) in Reference [8]).

Our derivation of the nonlinear model of Λ_ℓ will follow that in Reference [79], which is somewhat different from the standard derivation of nonlinear models [7,56,80]. We utilize the property of scale-locality [37] (specifically, ultraviolet locality) of the subscale mass flux which was proved to hold in variable density flows by Reference [13] under weak assumptions. Specifically, for our present purposes, we require that the spectra of density and velocity decay faster than k^{-1} in wavenumber. In other words, density and velocity should have finite second-order moments, $\langle \rho^2 \rangle < \infty$ and $\langle |\mathbf{u}|^2 \rangle < \infty$, in the limit of infinite Reynolds number. Here, $\langle \ldots \rangle$ is a space average. Ultraviolet scale locality implies that contributions to the subscale mass flux $\overline{\tau}_\ell(\rho, \mathbf{u})$ at scale ℓ from smaller scales $\delta \ll \ell$ are negligible [13]:

$$|\overline{\tau}_\ell(\rho'_\delta, \mathbf{u}'_\delta)| \ll |\overline{\tau}_\ell(\rho, \mathbf{u})| \tag{7}$$

By assuming the validity of Equation (7) in the limit $\delta \to \ell$, we can justify the approximation

$$\overline{\tau}_\ell(\rho, \mathbf{u}) \approx \overline{\tau}_\ell(\overline{\rho}_\ell, \overline{\mathbf{u}}_\ell), \tag{8}$$

which neglects any contribution from scales $< \ell$ to the subscale mass flux. Using the usual definition of an increment:

$$\delta f(\mathbf{x}; \mathbf{r}) = f(\mathbf{x} + \mathbf{r}) - f(\mathbf{x}), \tag{9}$$

the subscale mass flux term can be rewritten exactly in terms of $\delta\rho$ and $\delta\mathbf{u}$ [13,37]:

$$\overline{\tau}(\overline{\rho}_\ell, \overline{\mathbf{u}}_\ell) = \langle \delta\overline{\rho}_\ell \, \delta\overline{\mathbf{u}}_\ell \rangle_\ell - \langle \delta\overline{\rho}_\ell \rangle_\ell \langle \delta\overline{\mathbf{u}}_\ell \rangle_\ell \tag{10}$$

Equation (10) is exact, where

$$\left\langle \delta\overline{f}_\ell(\mathbf{x}; \mathbf{r}) \right\rangle_\ell = \int d\mathbf{r} \, G_\ell(\mathbf{r}) \delta\overline{f}_\ell(\mathbf{x}; \mathbf{r}) \tag{11}$$

is a local average around $\mathbf{x}$ over all separations $\mathbf{r}$ weighted by the kernel G_ℓ. A spatially localized kernel effectively limits the average to separations $|\mathbf{r}| \lesssim \ell/2$. Since a filtered field $\overline{f}_\ell(\mathbf{x})$ is smooth, we can Taylor expand its increments around $\mathbf{x}$

$$\delta\overline{f}(\mathbf{x}; \mathbf{r}) = \overline{f}(\mathbf{x} + \mathbf{r}) - \overline{f}(\mathbf{x}) \approx \mathbf{r} \cdot \nabla \overline{f}(\mathbf{x}) + \ldots \tag{12}$$

where we neglect higher order terms.

Substituting the first term in the Taylor expansion of each of $\delta\overline{\rho}$ and $\delta\overline{\mathbf{u}}$ into Equation (10) gives

$$\begin{aligned}
\overline{\tau}(\overline{\rho}, \overline{\mathbf{u}}_i) &= (\partial_k \overline{\rho})(\partial_m \overline{u}_i) \left[\langle r_k r_m \rangle_\ell - \langle r_k \rangle_\ell \langle r_m \rangle_\ell \right] \\
&= (\partial_k \overline{\rho})(\partial_m \overline{u}_i) \left[\frac{1}{3} \delta_{km} \ell^2 \int d^3\mathbf{r} \, G(\mathbf{r}) \, |\mathbf{r}|^2 \right] \\
&= \frac{1}{3} \ell^2 C_2 \, \partial_k \overline{\rho} \, \partial_k \overline{u}_i
\end{aligned} \tag{13}$$

This is the nonlinear model of the subscale mass flux $\overline{\tau}_\ell(\rho, u_i)$. In deriving the second line, we used the symmtery of the kernel such that $\langle r_k \rangle_\ell = 0$. In the final expression, $C_2 = \int d^3\mathbf{r} \, G(\mathbf{r}) \, |\mathbf{r}|^2$ depends solely on the shape of kernel G and, in particular, is independent of scale ℓ.

We now have a nonlinear model of Λ_ℓ that is only a function of filtered fields, which are resolved in LES simulations. We can use this model to gain insight into the mechanism by which Λ transfers energy across scales. The velocity gradient tensor $\partial_k \overline{u}_j$ can be decomposed into symmetric and antisymmetric parts

$$\partial_i \overline{u}_j = \overline{S}_{ij} + \overline{\Omega}_{ij}, \tag{14}$$

with

$$\begin{aligned}
\overline{S}_{ij} &= \frac{1}{2} \left(\partial_i \overline{u}_j + \partial_j \overline{u}_i \right) \\
\overline{\Omega}_{ij} &= \frac{1}{2} \left(\partial_i \overline{u}_j - \partial_j \overline{u}_i \right) = \frac{1}{2} \epsilon_{ijk} \overline{\omega}_k,
\end{aligned} \tag{15}$$

where $\omega = \boldsymbol{\nabla} \times \mathbf{u}$ is vorticity and ϵ_{ijk} is the Levi-Civita symbol. Therefore, Λ at any scale ℓ can be approximated by a nonlinear model, Λ_m, everywhere in space:

$$\begin{aligned}
\Lambda(\mathbf{x}) \approx \Lambda_\mathrm{m}(\mathbf{x}) &= \frac{1}{3} C_2 \, \ell^2 \, \frac{1}{\overline{\rho}} \left(\partial_j \overline{P} \, \partial_k \overline{\rho} \, \partial_k \overline{u}_j \right) \\
&= \frac{1}{3} C_2 \, \ell^2 \, \frac{1}{\overline{\rho}} \left[\boldsymbol{\nabla} \overline{P} \cdot \overline{\mathbf{S}} \cdot \boldsymbol{\nabla} \overline{\rho} + \frac{1}{2} \overline{\omega} \cdot \left(\boldsymbol{\nabla} \overline{\rho} \times \boldsymbol{\nabla} \overline{P} \right) \right] \\
&= \Lambda_{SR} + \Lambda_{BC}
\end{aligned} \tag{16}$$

where

$$\Lambda_{SR} = \frac{1}{3} C_2 \, \ell^2 \, \frac{1}{\overline{\rho}} \left[\boldsymbol{\nabla} \overline{P} \cdot \overline{\mathbf{S}} \cdot \boldsymbol{\nabla} \overline{\rho} \right] \tag{17}$$

is the strain generation process of baropycnal work and

$$\Lambda_{BC} = \frac{1}{3} C_2 \, \ell^2 \, \frac{1}{\overline{\rho}} \left[\frac{1}{2} \overline{\omega} \cdot \left(\boldsymbol{\nabla} \overline{\rho} \times \boldsymbol{\nabla} \overline{P} \right) \right] \tag{18}$$

is its baroclinic vorticity generation process. Equation (16) is the main result of this paper. In the following sections, we will provide numerical support showing excellent pointwise agreement between Λ and its nonlinear model Λ_m.

To illustrate how strain generation by baropycnal work takes place, consider an unstably stratified flow configuration in which $\boldsymbol{\nabla} \overline{P}$ and $\boldsymbol{\nabla} \overline{\rho}$ in Λ_{SR} are anti-aligned ($\boldsymbol{\nabla} \overline{\rho} \cdot \boldsymbol{\nabla} \overline{P} < 0$) as illustrated in Figure 1 of Reference [15] and both are parallel to a contracting eigenvector of $\overline{\mathbf{S}}$ (associated with a negative eigenvalue of $\overline{\mathbf{S}}$). Remember that the strain $\overline{\mathbf{S}}$ in our nonlinear model arises from $\overline{\tau}_\ell(\rho, \mathbf{u})$ which represents scales smaller than ℓ. In such a configuration, the contraction (and therefore strain) is enhanced leading to the generation of kinetic energy in the form of straining motion at scales smaller than ℓ ($\Lambda_\ell > 0$ in Equation (3)). The ultimate source of kinetic energy being transferred by Λ to motions at scales $< \ell$ is potential energy due to the large-scale pressure gradient, $\boldsymbol{\nabla} \overline{P}$.

The baroclinic component, Λ_{BC}, demonstrates how baroclinicity, $\boldsymbol{\nabla} \rho \times \boldsymbol{\nabla} P$, plays a role in the energetics across scales. The importance of baroclinicity is well known [81,82] but it has always been analyzed within the vorticity budget. Its contribution to the energy budget has never been clear. Baroclinicity in the kinetic energy budget arises naturally from our scale decomposition and the identification of Λ as a scale-transfer mechanism. The need for a scale decomposition in order for Λ and, as a result, baroclinic energy transfer, to appear in the kinetic energy budget should not be surprising. This is similar to the scale transfer term Π, which does not appear in the budget without disentangling scales due to energy conservation. In the same vein, the appearance of baroclinicity in the vorticity equation can be interpreted as being a consequence of an effective scale decomposition performed by the curl operator $\boldsymbol{\nabla} \times$, which a high-pass filter.

As mentioned in the introduction, in the compressible LES literature, Λ is almost always lumped with pressure-dilatation, $\overline{P} \boldsymbol{\nabla} \cdot \overline{\mathbf{u}}$ in the form of $\overline{P} \boldsymbol{\nabla} \cdot \tilde{\mathbf{u}}$ [19–22], thereby completely missing the physical processes inherent in baropycnal work. Our analysis here supports the argument in References [13,15] to separate Λ from pressure-dilatation. In those studies, it was reasoned that Λ and $\overline{P} \boldsymbol{\nabla} \cdot \overline{\mathbf{u}}$ are fundamentally different; the former involves interactions between the large scale pressure gradient with subscale fluctuations, allowing the transfer energy across scales, whereas the latter is solely due to large-scale fields and cannot participate in the transfer of energy *across* scales.

4. Simulations

To provide empirical support to our nonlinear model of Λ, we carry out a suite of DNS of forced compressible turbulence in a periodic box of size 2π on which we perform *a priori* tests of our derived model against simulation data. The DNS solve the fully compressible Navier Stokes equations:

$$\partial_t \rho \quad + \partial_j(\rho u_j) = 0 \tag{19}$$

$$\partial_t(\rho u_i) \quad + \partial_j(\rho u_i u_j) = -\partial_i P + \partial_j \sigma_{ij} + \rho F_i \tag{20}$$

$$\partial_t(\rho E) \quad + \partial_j(\rho E u_j) = -\partial_j(P u_j) + \partial_j[2\mu\, u_i(S_{ij} - \frac{1}{d}S_{kk}\delta_{ij})] - \partial_j q_j + \rho u_i F_i - \mathcal{RL} \tag{21}$$

Here, $\mathbf{u}$ is velocity, ρ is density, $E = |\mathbf{u}|^2/2 + e$ is total energy per unit mass, where e is specific internal energy, P is thermodynamic pressure, μ is dynamic viscosity, $\mathbf{q} = -\kappa \nabla T$ is the heat flux with a thermal conductivity κ and temperature T. Both dynamic viscosity and thermal conductivity are spatially variable, where $\mu(\mathbf{x}) = \mu_0(T(\mathbf{x})/T_0)^{0.76}$. Thermal conductivity is set to satisfy a Prandtl number $Pr = c_p \mu/\kappa = 0.7$, where $c_p = R\gamma/(\gamma - 1)$ is the specific heat with specific gas constant R and $\gamma = 5/3$. We use the ideal gas equation of state, $P = \rho RT$. We stir the flow using an external acceleration field F_i and $\mathcal{RL}$ represents radiation losses from internal energy. $S_{ij} = (\partial_j u_i + \partial_i u_j)/2$ is the symmetric strain tensor and σ_{ij} is the the deviatoric (traceless) viscous stress

$$\sigma_{ij} = 2\mu(S_{ij} - \frac{1}{3}S_{kk}\delta_{ij}) \tag{22}$$

We solve the above equations using the pseudo-spectral method with 2/3rd dealiasing. We advance in time using the 4th-order Runge-Kutta scheme with a variable time step.

The acceleration $\mathbf{F}$ we use is similar to that used in Reference [83]. In Fourier space, the acceleration is defined as

$$\widehat{F}_i(\mathbf{k}) = \widehat{f}_j(\mathbf{k}) P_{ij}^\zeta(\mathbf{k}), \tag{23}$$

where the complex vector $\widehat{\mathbf{f}}$ is constructed from independent Ornstein-Uhlenbeck stochastic processes [84] and the projection operator $P_{ij}^\zeta(\mathbf{k}) = \zeta\delta_{ij} + (1 - 2\zeta)\frac{k_i k_j}{|\mathbf{k}|^2}$ allows to control the ratio of solenoidal ($\nabla \cdot \mathbf{F} = 0$) and dilatational ($\nabla \times \mathbf{F} = 0$) components of the forcing using the parameter ζ. When $\zeta = 0$, the forcing is purely dilatational and when $\zeta = 1$, the forcing is purely solenoidal. The acceleration is constrained to low wavenumbers $|\mathbf{k}| < k_F$.

For the internal energy loss term $\mathcal{RL}$, we have tested two schemes: a spatially varying radiative loss, $\mathcal{RL} = \rho \mathbf{u} \cdot \mathbf{F}$ and another that is independent of space, $\mathcal{RL} = \langle \rho \mathbf{u} \cdot \mathbf{F} \rangle$. The two schemes yield indistinguishable results. Including an internal energy loss term is similar to what was done in other studies [85,86] to allow total energy to remain stationary. After an initial transient, mean kinetic energy reaches a statistically stationary state as shown in Figure 1.

Table 1 summarizes the simulations we ran for this study and various metrics characterizing the importance of compressibility effects in each run. The turbulent Mach number is $M_t = \langle u_i u_i \rangle^{1/2} / \langle c \rangle$ and the Taylor Reynolds number is $Re_\lambda = \langle (u_i u_i)/3 \rangle^{1/2} \lambda / \langle \mu/\rho \rangle$. Here $c = \sqrt{\gamma p/\rho}$ is the sound speed and $\lambda = \langle u_i u_i \rangle^{1/2} / \left\langle u_{i,i}^2 \right\rangle^{1/2}$ is the Taylor microscale. The compressibility metrics in Table 1 show the relative importance of dilatational versus solenoidal velocity modes. We use the Helmholtz decomposition, $\mathbf{u} = \mathbf{u}^d + \mathbf{u}^s$ to obtain the dilatational ($\nabla \times \mathbf{u}^d = 0$) and solenoidal ($\nabla \cdot \mathbf{u}^s = 0$) components of the velocity field. The dilatational kinetic energy is $K^d = \left\langle \rho u_i^d u_i^d /2 \right\rangle$ and the solenoidal kinetic energy is $K^s = \langle \rho u_i^s u_i^s /2 \rangle$. Their ratio K^d/K^s yields a measure of compressibility at large scales.

It is well-known [83,87] that the dilatational kinetic energy K^d becomes significant when forced directly using **F** with a small ζ. This holds, even though the low-ζ runs have a lower Mach number than high-ζ runs. The ratio $(\boldsymbol{\nabla}\cdot\mathbf{u})_{\mathrm{rms}}/(\boldsymbol{\nabla}\times\mathbf{u})_{\mathrm{rms}}$ yields a measure of compressibility at small scales.

Table 1. Comparison of compressibility metrics and cascade terms at different grid resolution. Low-ζ corresponds to high compressibility in the external forcing. The spatially averaged cascade terms $\langle\Lambda_\ell\rangle$ and $\langle\Pi_\ell\rangle$ are calculated using the sharp-spectral cutoff filter with $k_\ell = 2\pi/\ell = 6$. $\frac{\Delta x}{\eta}$ is the ratio of grid size to the Kolmogorov length.

Run	N	ζ	M_t	Re_λ	$\dfrac{(\boldsymbol{\nabla}\cdot\mathbf{u})_{\mathrm{rms}}}{(\boldsymbol{\nabla}\times\mathbf{u})_{\mathrm{rms}}}$	$\dfrac{K^d}{K^s}$	$\dfrac{\Delta x}{\eta}$	$\langle\Pi_\ell\rangle$	$\langle\Lambda_\ell\rangle$
1	1024	0.01	0.23	65	0.50	0.74	0.23	8.9×10^{-3}	-5.6×10^{-3}
2	512	0.01	0.22	33	0.46	0.51	0.28	6.8×10^{-3}	-4.8×10^{-3}
3	512	0.6	0.33	206	0.05	0.02	1.78	2.3×10^{-2}	-9.6×10^{-5}
4	256	0.01	0.21	18	0.54	0.56	0.25	3.6×10^{-3}	-3.1×10^{-3}
5	256	0.6	0.42	150	0.04	0.01	2.1	5.0×10^{-2}	-6.0×10^{-4}
6	256	1.0	0.46	175	0.03	0.003	2.2	5.0×10^{-2}	-3.8×10^{-4}
7	128	0.01	0.20	10	0.65	0.80	0.24	8.0×10^{-4}	-7.5×10^{-4}
8	128	0.6	0.50	105	0.05	0.01	2.3	4.0×10^{-2}	-4.0×10^{-4}
9	128	1.0	0.40	95	0.03	0.01	2.0	2.5×10^{-2}	-2.0×10^{-4}

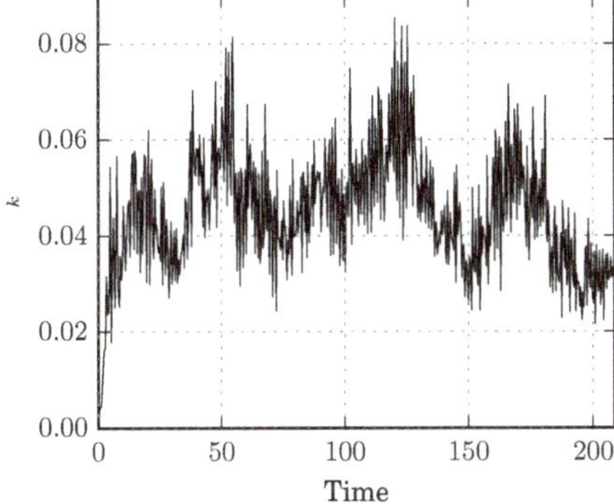

Figure 1. Time-series of average kinetic energy showing a statistically steady state.

The last two columns in Table 1 summarize the effect of ζ on the relative importance of Π and Λ. While the deformation work Π is significant in all cases, baropycnal work Λ, which arises only in variable density flows, is greatly affected by the type of forcing used. At the Reynolds numbers we simulate, we find that Λ becomes important only for low ζ when the dilatational modes are directly forced. Even at relatively high Mach numbers, Λ remains small for high ζ. We caution, however, that these observations might be Reynolds number dependent. Moreover, Λ has been shown to dominate in non-dilatational low Mach number variable density flows [88,89].

Figure 2 shows typical visualizations of the flows arising from low-ζ and high-ζ forcing. The stark qualitative difference shows the significance of dilatational forcing on the flow [83,90], at least in limited resolution simulations. It has been argued [85,91,92] that at sufficiently high Reynolds numbers, flows forced dilatationally will produce sufficient vortical motion to resemble those forced solenoidally. Since we are primarily interested in the Λ term here, unless stated otherwise, plots that follow will be from low-ζ simulations where Λ is significant.

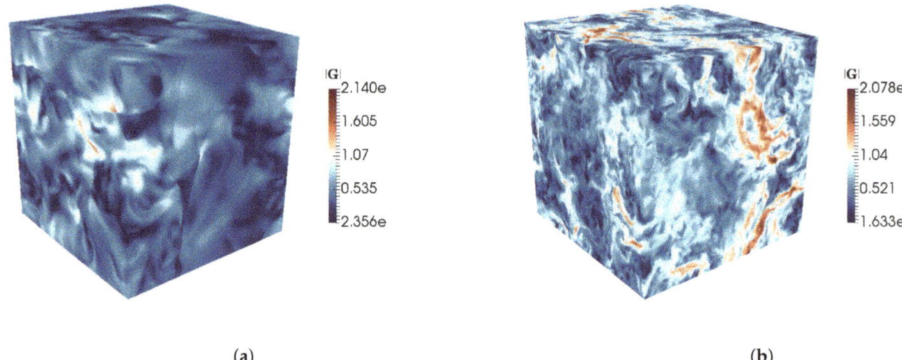

(a)   (b)

Figure 2. Momentum magnitude $|\rho\mathbf{u}|$ from (**a**) Run 1 ($\zeta = 0.01$) and (**b**) Run 3 ($\zeta = 0.6$).

Figure 3 shows the velocity spectra in the case of highly compressive (low-ζ) forcing. At intermediate scales, the spectrum of u^d seems to follow a power law close to k^{-2}, which is expected for the dilatational component [17,25,87]. It is well-known (e.g., References [17,25]) that obtaining a clear power-law scaling of the solenoidal velocity is challenging when forcing dilatationally, even at our 1024^3 resolution.

Figure 4 shows the cascade terms Π_ℓ and Λ_ℓ averaged over the domain as a function of the filter wavenumber k. As is the case in 3D isotropic incompressible turbulence, Π_ℓ is positive for all wavenumbers, transferring kinetic energy from large to small scales. On the other hand, Λ_ℓ is negative, effectively reducing the total amount of energy transferred across scales. This is consistent with previous studies which measured Λ_ℓ in homogeneous isotropic compressible turbulence [17]. Across a shock, the pressure and density gradients have the same direction and are aligned with the contracting strain eigenvector, leading to negative baropycnal work [15,18], thereby reducing the intensity of the cascade. Using the terminology of [2], this is a "bi-directional cascade." The situation is different in buoyancy driven (unstably stratified) flows, where pressure and density gradients are in opposite directions leading to positive baropycnal work [15]. Within the framework of Reynolds-averaged Navier-Stokes (RANS), it has been shown that for variable density flows, such as turbulence generated by the Rayleigh-Taylor instability, the RANS equivalent of Λ is the largest contributor to the kinetic energy cascade [88]. We shall present our results on Λ in buoyancy driven flows in forthcoming work [89].

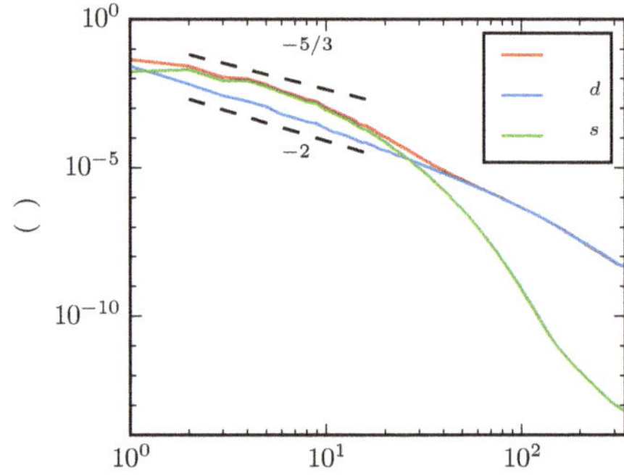

Figure 3. Spectra of velocity (u) and its dilatational and solenoidal (u^d and u^s, respectively) components from Run 1. The reference dashed black lines have slopes of $-5/3$ and -2.

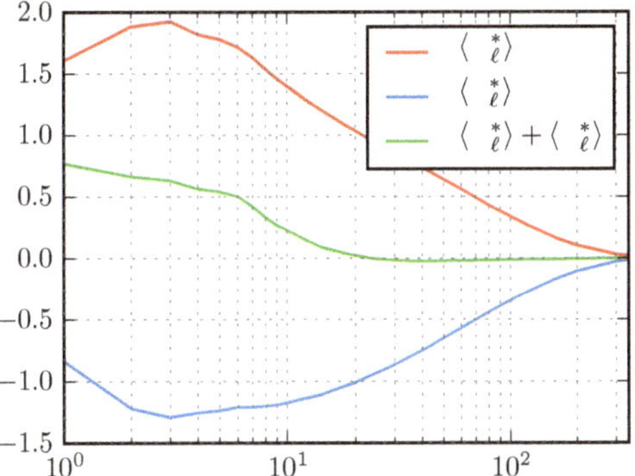

Figure 4. Flux terms Π_ℓ and Λ_ℓ from Run 1, as well as their sum averaged over space and time, as a function of the filtering wavenumber $k = 2\pi/\ell$. Filtering here uses the sharp-spectral filter kernel. The star superscript indicates normalization by the effective kinetic energy injection, $\varepsilon^{eff} = \varepsilon^{inj} + \langle p \boldsymbol{\nabla} \cdot \mathbf{u} \rangle$.

5. Numerical Results

To quantify the pointwise agreement between baropycnal work Λ and its nonlinear model Λ_m in our DNS, we measure the correlation coefficient:

$$R_c = \frac{\langle \Lambda_m \Lambda \rangle - \langle \Lambda_m \rangle \langle \Lambda \rangle}{\left[\left(\langle \Lambda_m^2 \rangle - \langle \Lambda_m \rangle^2 \right) \left(\langle \Lambda^2 \rangle - \langle \Lambda \rangle^2 \right) \right]^{1/2}}. \tag{24}$$

We also analyze the joint probability density function (PDF) in Figures 5–7 and visualize Λ and Λ_m in x-space in Figure 8.

In our study, we use the filters defined in Table 2. Both the box and Gaussian filters are positive in physical space, which is important to guarantee physical realizability of filtered quantities [46]. On the other hand, the sharp spectral filter is not sign definite in x-space, which limits its utility in analyzing scale process in physical space.

Table 2. The types of filters used in calculating Λ and Λ_m. The Heaviside function $H(x) = 1$ for $x \geq 0$ and $H(x) = 0$ for $x < 0$. The correlation coefficient R_c is shown at two scales $k_\ell = 2\pi/\ell$.

Filter Type	Kernel	$R_c \vert k_\ell = 8$	$R_c \vert k_\ell = 16$
Box	$G_\ell(\mathbf{x}) = \prod_{i=1}^{3} \frac{1}{\ell} H\left(\frac{\ell}{2} - \vert x_i \vert \right)$	0.93	0.94
Gaussian	$G_\ell(\mathbf{x}) = \frac{1}{\ell^3} \left(\frac{6}{\pi} \right)^{3/2} e^{-\frac{6\vert \mathbf{x} \vert^2}{\ell^2}}$	0.97	0.97
Sharp spectral	$\hat{G}_\ell(\mathbf{k}) = \prod_{i=1}^{3} H\left(\frac{2\pi}{\ell} - \vert k_i \vert \right)$	0.27	0.28

Our results indicate an excellent agreement between baropycnal work and its nonlinear model. Using either the Gaussian or Box filters, the correlation coefficients are very high, $R_c > 0.9$, for all the length scales we analyzed. The sharp spectral filter, on the other hand, yields poor agreement. This is not surprising since the sharp spectral filter can yield negative filtered densities [15,75] and physically unrealizable subscale stresses [46] due to its non-positivity in x-space.

Figures 5–7, using the box, Gaussian and sharp spectral filters, respectively, plot $\Lambda(\mathbf{x})$ and $\Lambda_{\mathrm{m}}(\mathbf{x})$ along a line in the domain to show the typical agreement between the two quantities. Also shown are the joint PDFs, which exhibit excellent linear agreement when using either the box or Gaussian kernels but not the sharp spectral filter. Instantaneous visualizations in Figure 8 of $\Lambda(\mathbf{x})$ and $\Lambda_{\mathrm{m}}(\mathbf{x})$ are consistent with the excellent statistical agreement, showing an almost perfect pointwise correlation. We note that in our dilatationally forced flows, most of the contribution to Λ_{m} is from its straining component, Λ_{SR}, with a negligible contribution from Λ_{BC} (see Equation (16)). This is due to the shocks which contribute mostly to Λ_{SR}. In flows dominated by baroclinicity, such as in the Rayleigh-Taylor instability, a significant contribution to Λ comes from Λ_{BC}, as will be shown in forthcoming work [89].

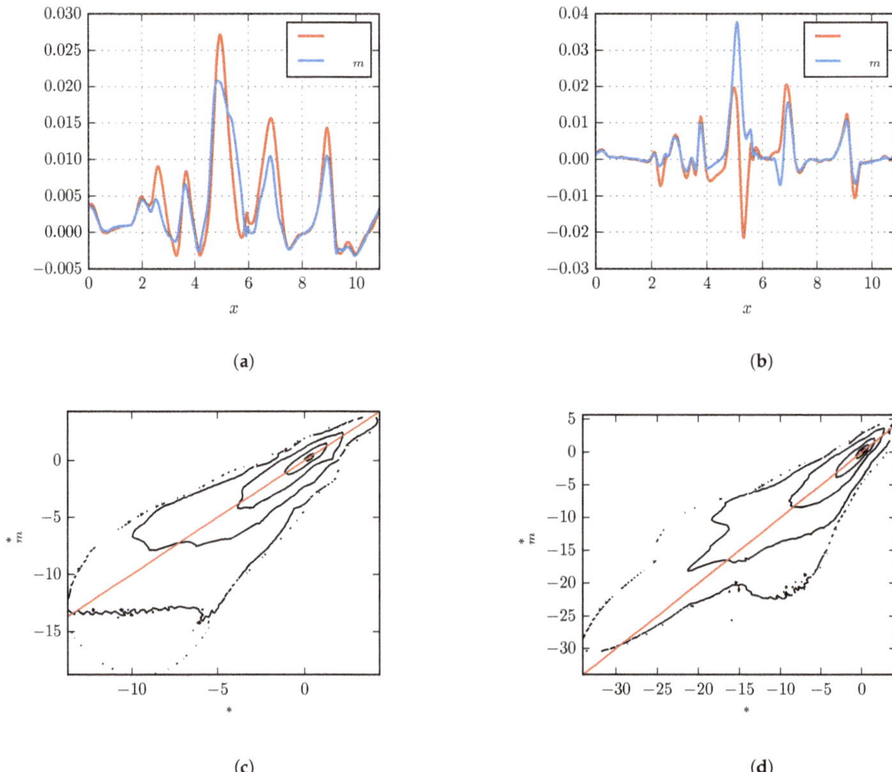

Figure 5. Correlation between baropycnal work and its nonlinear model from Run 1 using the box kernel with filter scale $k_\ell = 8$ in (**a**,**c**) and $k_\ell = 16$ in (**b**,**d**). Top two panels plot Λ and Λ_{m} along a diagonal line through the domain from a single snapshot. Lower two panels show time-averaged isocontours of the logarithm of the joint PDF between Λ and Λ_{m}, where star superscripts indicate that means have been subtracted and the values are normalized by their variance. Straight-red lines are $y = x$. The correlation coefficients are $R_c = 0.93$ at filter scale $k_\ell = 8$ and $R_c = 0.94$ at $k_\ell = 16$. All four panels indicate excellent correlation between Λ and Λ_{m}.

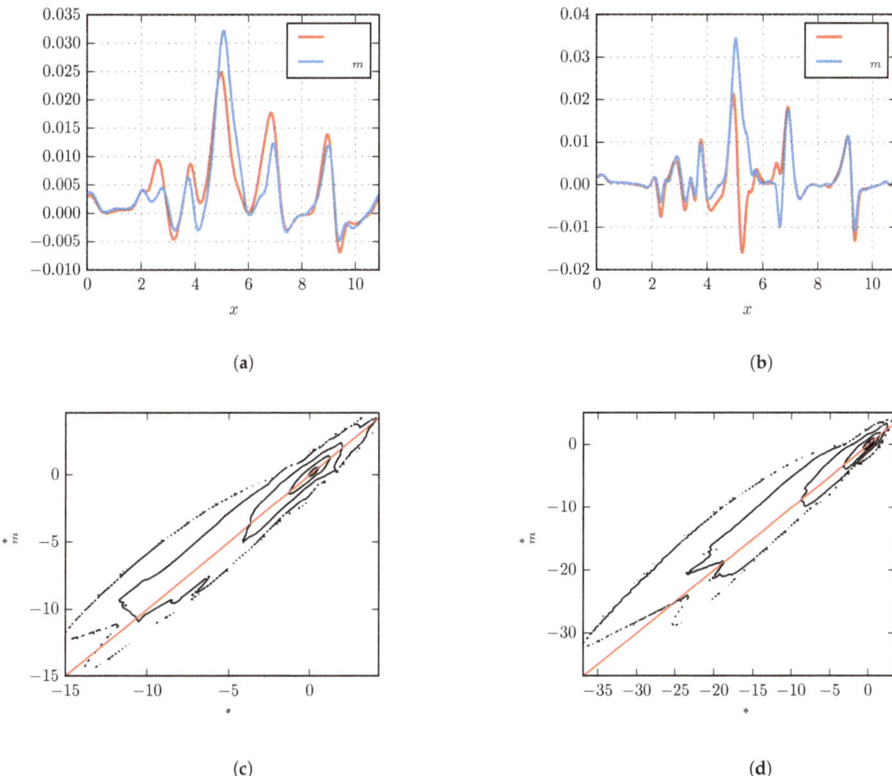

Figure 6. Same as in Figure 5 but using the Gaussian filter. The correlation coefficients are $R_c = 0.97$ at filter scale $k_\ell = 8$ and $R_c = 0.97$ at $k_\ell = 16$. All four panels indicate excellent correlation between Λ and Λ_{m}.

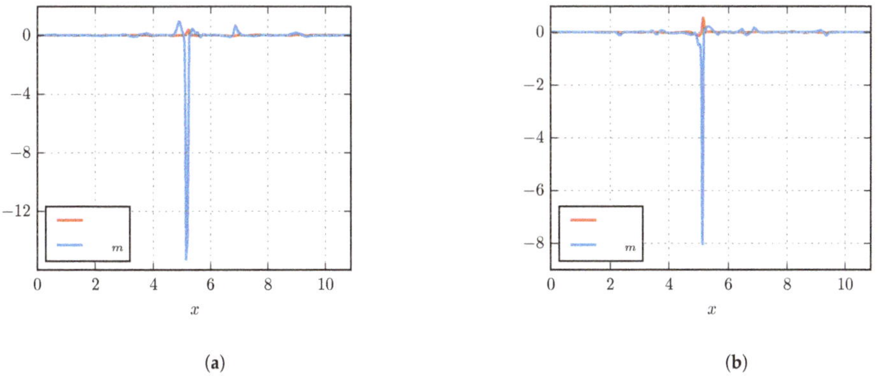

Figure 7. *Cont.*

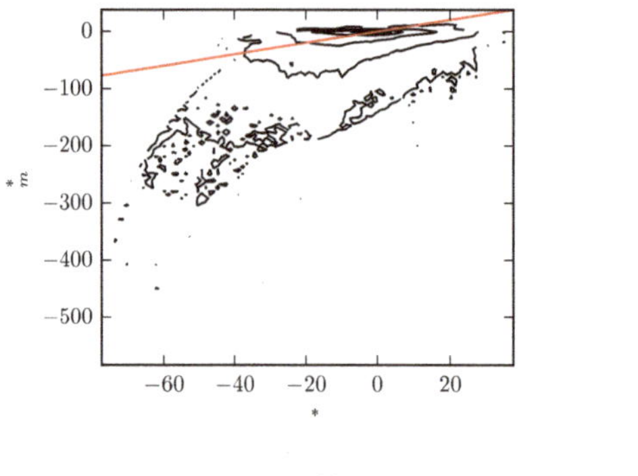
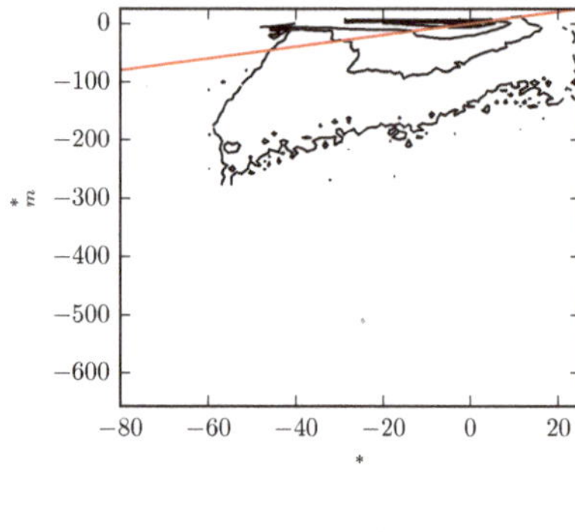

(c)   (d)

Figure 7. Same as in Figure 5 but using the sharp-spectral filter. The correlation coefficients are $R_c = 0.27$ at filter scale $k_\ell = 8$ and $R_c = 0.28$ at $k_\ell = 16$. The correlation between Λ and $\Lambda_{\rm m}$ is poor when using a sharp-spectral filter due to its nonpositivity, which can yield negative filtered densities [15] and physically unrealizable stresses [46].

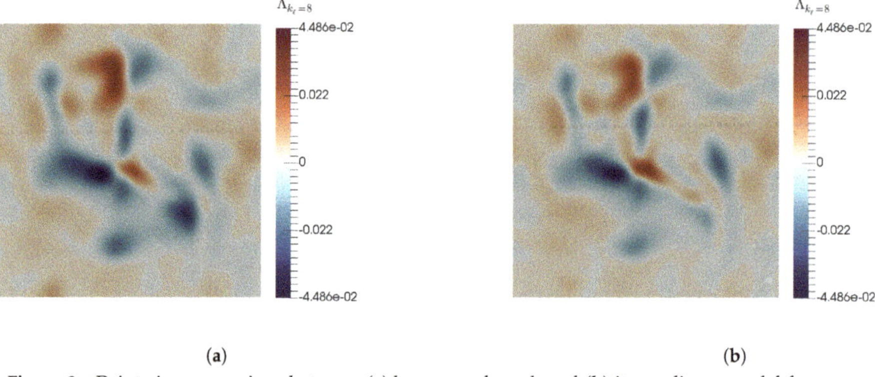

(a)   (b)

Figure 8. Pointwise comparison between (**a**) baropycnal work and (**b**) its nonlinear model from a 2D slice of the 3D domain in Run 1, at one instant in time. A Gaussian kernel at scale $k_\ell = 8$ is used. The visualizations show excellent pointwise correlation.

6. Summary

Past work [13,15,18] has identified baropycnal work, Λ, as a process capable of transferring kinetic energy across scales in addition to deformation work, Π. This paper aimed at elucidating the physical mechanism by which Λ operates.

Using scale-locality [13] and a multiscale gradient expansion [79], we derived a nonlinear model, $\Lambda_{\rm m}$, of baropycnal work. Using DNS, we showed excellent agreement between Λ and $\Lambda_{\rm m}$ everywhere in space and at any time, giving further empirical justification for our analysis of Λ via its model $\Lambda_{\rm m}$.

We found that baropycnal work operates by the baroclinic generation of vorticity and also by strain generation due to pressure and density gradients, both barotropic and baroclinic. While the role of pressure and density gradients in generating vorticity is well recognized, their role in strain generation has been less emphasized in the literature.

As far as we know, this is the first direct demonstration of how baroclinicity enters the kinetic energy budget, which arises naturally from our scale decomposition and the identification of Λ as a scale-transfer mechanism. Baroclinicity is often analyzed within the vorticity budget but its role in the energetics has never been obvious. The need for a scale decomposition in order for Λ and, as a result, baroclinic energy transfer, to appear in the kinetic energy budget is similar to the scale transfer term Π, which only appears in the budget after decomposing scales due to energy conservation. In the same vein, the appearance of baroclinicity in the vorticity equation can be interpreted as being a consequence of an effective scale decomposition performed by the curl operator $\nabla \times$, which is a high-pass filter. Our findings here support the argument in References [13,15] to separate Λ from pressure-dilatation, $\overline{P}\nabla\cdot\overline{\mathbf{u}}$ in compressible LES, where the two terms are often lumped together in the form of $\overline{P}\nabla\cdot\widetilde{\mathbf{u}}$.

In forthcoming work, we shall present further evidence of the excellent agreement between Λ and Λ_m using low Mach number buoyancy driven flows with significant density variability [89].

Author Contributions: Conceptualization, H.A.; methodology, H.A.; software, A.L. and H.A.; validation, A.L.; formal analysis, A.L. and H.A.; investigation, A.L.; data curation, A.L.; writing—original draft preparation, A.L.; writing—review and editing, H.A. and A.L.; visualization, A.L.; supervision, H.A.; project administration, H.A.; funding acquisition, H.A.

Funding: This research was funded by DOE FES grant number DE-SC0014318. A.L. and H.A. were also supported DOE NNSA award DE-NA0003856. H.A. was also supported by NASA grant 80NSSC18K0772 and DOE grant DE-SC0019329. This report was prepared as an account of work sponsored by an agency of the U.S. Government. Neither the U.S. Government nor any agency thereof, nor any of their employees, makes any warranty, express or implied, or assumes any legal liability or responsibility for the accuracy, complete- ness, or usefulness of any information, apparatus, product, or process disclosed, or represents that its use would not infringe privately owned rights. Reference herein to any specific commercial product, process, or service by trade name, trademark, manufacturer, or otherwise does not necessarily constitute or imply its endorsement, recommendation, or favoring by the U.S. Government or any agency thereof. The views and opinions of authors expressed herein do not necessarily state or reflect those of the U.S. Government or any agency thereof.

Acknowledgments: Computing time was provided by the National Energy Research Scientific Computing Center (NERSC) under Contract No. DE-AC02-05CH11231.

Conflicts of Interest: The authors declare no conflict of interest.

References

1. Tennekes, H.; Lumley, J.L. *A First Course in Turbulence*; The MIT Press: Cambridge, MA, USA, 1972.
2. Alexakis, A.; Biferale, L. Cascades and transitions in turbulent flows. *Phys. Rep.* **2018**, *767–769*, 1–101. [CrossRef]
3. Pope, S.B. *Turbulent Flows*; Cambridge University Press: New York, NY, USA, 2000.
4. Frisch, U. *Turbulence. The Legacy of A. N. Kolmogorov*; Cambridge University Press: Cambridge, UK, 1995.
5. Alexakis, A.; Mininni, P.D.; Pouquet, A. Imprint of large-scale flows on turbulence. *Phys. Rev. Lett.* **2005**, *95*, 264503. [CrossRef] [PubMed]
6. Mininni, P.D.; Alexakis, A.; Pouquet, A. Large-scale flow effects, energy transfer, and self-similarity on turbulence. *Phys. Rev. E* **2006**, *74*, 016303. [CrossRef]
7. Borue, V.; Orszag, S.A. Local energy flux and subgrid-scale statistics in three-dimensional turbulence. *J. Fluid Mech.* **1998**, *366*, 1–31. [CrossRef]
8. Eyink, G.L. Cascade of circulations in fluid turbulence. *Phys. Rev. E* **2006**, *74*, 066302. [CrossRef] [PubMed]
9. Eyink, G.L. Dissipative anomalies in singular Euler flows. *Phys. D Nonlinear Phenom.* **2008**, *237*, 1956–1968. [CrossRef]
10. Xu, H.; Pumir, A.; Bodenschatz, E. The pirouette effect in turbulent flows. *Nat. Phys.* **2011**, *7*, 709–712. [CrossRef]
11. Taylor, G.I.; Green, A.E. Mechanism of the production of small eddies from large ones. *Proc. R. Soc. Lond. Ser. A-Math. Phys. Sci.* **1937**, *158*, 499–521.
12. Taylor, G.I. Production and dissipation of vorticity in a turbulent fluid. *Proc. R. Soc. Lond. Ser. A-Math. Phys. Sci.* **1938**, *164*, 15–23.
13. Aluie, H. Compressible Turbulence: The Cascade and its Locality. *Phys. Rev. Lett.* **2011**, *106*, 174502. [CrossRef]
14. Aluie, H.; Li, S.; Li, H. Conservative Cascade of Kinetic Energy in Compressible Turbulence. *Astrophys. J. Lett.* **2012**, *751*, L29. [CrossRef]
15. Aluie, H. Scale decomposition in compressible turbulence. *Phys. D Nonlinear Phenom.* **2013**, *247*, 54–65. [CrossRef]

16. Kritsuk, A.G.; Wagner, R.; Norman, M.L. Energy cascade and scaling in supersonic isothermal turbulence. *J. Fluid Mech.* **2013**, *729*, R1. [CrossRef]

17. Wang, J.; Yang, Y.; Shi, Y.; Xiao, Z.; He, X.T.; Chen, S. Cascade of Kinetic Energy in Three-Dimensional Compressible Turbulence. *Phys. Rev. Lett.* **2013**, *110*, 214505. [CrossRef] [PubMed]

18. Eyink, G.L.; Drivas, T.D. Cascades and Dissipative Anomalies in Compressible Fluid Turbulence. *Phys. Rev. X* **2018**, *8*, 011022. [CrossRef]

19. Favre, A. Statistical equations of turbulent gases. In *Problems of Hydrodynamic and Continuum Mechanics*; SIAM: Philadelphia, PA, USA, 1969; pp. 231–266.

20. Lele, S.K. Compressibility effects on turbulence. *Annu. Rev. Fluid Mech.* **1994**, *26*, 211–254. [CrossRef]

21. Garnier, E.; Adams, N.; Sagaut, P. *Large Eddy Simulation for Compressible Flows*; Springer: Amsterland, The Netherlands, 2009.

22. O'Brien, J.; Urzay, J.; Ihme, M.; Moin, P.; Saghafian, A. Subgrid-scale backscatter in reacting and inert supersonic hydrogen–air turbulent mixing layers. *J. Fluid Mech.* **2014**, *743*, 554–584. [CrossRef]

23. Mukherjee, S.; Zarghami, A.; Haringa, C.; van As, K.; Kenjereš, S.; Van den Akker, H.E. Simulating liquid droplets: A quantitative assessment of lattice Boltzmann and Volume of Fluid methods. *Int. J. Heat Fluid Flow* **2018**, *70*, 59–78. [CrossRef]

24. Kritsuk, A.G.; Norman, M.L.; Padoan, P.; Wagner, R. The statistics of supersonic isothermal turbulence. *Astrophys. J.* **2007**, *665*, 416–431. [CrossRef]

25. Federrath, C.; Roman-Duval, J.; Klessen, R.S.; Schmidt, W.; Mac Low, M.M. Comparing the statistics of interstellar turbulence in simulations and observations. *Astron. Astrophys.* **2010**, *512*, A81. [CrossRef]

26. Pan, L.; Padoan, P.; Haugbølle, T.; Nordlund, Å. Supernova Driving. II. Compressive Ratio In Molecular-Cloud Turbulence. *Astrophys. J. Lett.* **2016**, *825*, 30. [CrossRef]

27. Yan, R.; Betti, R.; Sanz, J.; Aluie, H.; Liu, B.; Frank, A. Three-dimensional single-mode nonlinear ablative Rayleigh-Taylor instability. *Phys. Plasmas* **2016**, *23*, 022701. [CrossRef]

28. Zhang, H.; Betti, R.; Gopalaswamy, V.; Yan, R.; Aluie, H. Nonlinear excitation of the ablative Rayleigh-Taylor instability for all wave numbers. *Phys. Rev. E* **2018**, *97*, 011203. [CrossRef] [PubMed]

29. Zhang, H.; Betti, R.; Yan, R.; Zhao, D.; Shvarts, D.; Aluie, H. Self-Similar Multimode Bubble-Front Evolution of the Ablative Rayleigh-Taylor Instability in Two and Three Dimensions. *Phys. Rev. Lett.* **2018**, *121*, 185002. [CrossRef] [PubMed]

30. Larsson, J.; Laurence, S.; Bermejo-Moreno, I.; Bodart, J.; Karl, S.; Vicquelin, R. Incipient thermal choking and stable shock-train formation in the heat-release region of a scramjet combustor. Part II: Large eddy simulations. *Combust. Flame* **2015**, *162*, 907–920. [CrossRef]

31. Urzay, J. Supersonic Combustion in Air-Breathing Propulsion Systems for Hypersonic Flight. *Annu. Rev. Fluid Mech.* **2018**, *50*, 593–627. [CrossRef]

32. Emanuel, K.A. An air-sea interaction theory for tropical cyclones. Part I: Steady-state maintenance. *J. Atmos. Sci.* **1986**, *43*, 585–605. [CrossRef]

33. Bourassa, M.A.; Rodriguez, E.; Chelton, D. Winds and currents mission: Ability to observe mesoscale AIR/SEA coupling. In Proceedings of the 2016 IEEE International Geoscience and Remote Sensing Symposium (IGARSS), Beijing, China, 10–15 July 2016; IEEE: Piscataway, NJ, USA, 2016; pp. 7392–7395.

34. Deike, L.; Melville, W.K.; Popinet, S. Air entrainment and bubble statistics in breaking waves. *J. Fluid Mech.* **2016**, *801*, 91–129. [CrossRef]

35. Renault, L.; McWilliams, J.C.; Masson, S. Satellite observations of imprint of oceanic current on wind stress by air-sea coupling. *Sci. Rep.* **2017**, *7*, 17747. [CrossRef]

36. Meneveau, C.; Katz, J. Scale-Invariance and Turbulence Models for Large-Eddy Simulation. *Annu. Rev. Fluid Mech.* **2000**, *32*, 1–32. [CrossRef]

37. Eyink, G.L. Locality of turbulent cascades. *Physica D* **2005**, *207*, 91–116. [CrossRef]

38. Strichartz, R.S. *A Guide to Distribution Theory and Fourier Transforms*; World Scientific Publishing Company: Singapore, 2003.

39. Evans, L.C. *Partial Differential Equations*; American Mathematical Society: Providence, RI, USA 2010.

40. Leonard, A. Energy Cascade in Large-Eddy Simulations of Turbulent Fluid Flows. *Adv. Geophys.* **1974**, *18*, A237.

41. Germano, M. Turbulence: The filtering approach. *J. Fluid Mech.* **1992**, *238*, 325–336. [CrossRef]

42. Eyink, G.L. Local energy flux and the refined similarity hypothesis. *J. Stat. Phys.* **1995**, *78*, 335–351. [CrossRef]

43. Eyink, G.L. Exact Results on Scaling Exponents in the 2D Enstrophy Cascade. *Phys. Rev. Lett.* **1995**, *74*, 3800–3803. [CrossRef] [PubMed]

44. Eyink, G.; Aluie, H. Localness of energy cascade in hydrodynamic turbulence. I. Smooth coarse graining. *Phys. Fluids* **2009**, *21*, 115107. [CrossRef]

45. Piomelli, U.; Cabot, W.H.; Moin, P.; Lee, S. Subgrid-scale backscatter in turbulent and transitional flows. *Phys. Fluids A Fluid Dyn.* **1991**, *3*, 1766–1771. [CrossRef]

46. Vreman, B.; Geurts, B.; Kuerten, H. Realizability conditions for the turbulent stress tensor in large-eddy simulation. *J. Fluid Mech.* **1994**, *278*, 351–362. [CrossRef]

47. Aluie, H.; Eyink, G. Localness of energy cascade in hydrodynamic turbulence. II. Sharp spectral filter. *Phys. Fluids* **2009**, *21*, 115108. [CrossRef]

48. Buzzicotti, M.; Linkmann, M.; Aluie, H.; Biferale, L.; Brasseur, J.; Meneveau, C. Effect of filter type on the statistics of energy transfer between resolved and subfilter scales from a-priori analysis of direct numerical simulations of isotropic turbulence. *J. Turbul.* **2018**, *19*, 167–197. [CrossRef]

49. Rivera, M.K.; Daniel, W.B.; Chen, S.Y.; Ecke, R.E. Energy and Enstrophy Transfer in Decaying Two-Dimensional Turbulence. *Phys. Rev. Lett.* **2003**, *90*, 104502. [CrossRef] [PubMed]

50. Chen, S.; Ecke, R.E.; Eyink, G.L.; Wang, X.; Xiao, Z. Physical Mechanism of the Two-Dimensional Enstrophy Cascade. *Phys. Rev. Lett.* **2003**, *91*, 214501. [CrossRef] [PubMed]

51. Chen, J.; Meneveau, C.; Katz, J. Scale interactions of turbulence subjected to a straining relaxation destraining cycle. *J. Fluid Mech.* **2006**, *562*, 123–150. [CrossRef]

52. Kelley, D.H.; Ouellette, N.T. Spatiotemporal persistence of spectral fluxes in two-dimensional weak turbulence. *Phys. Fluids* **2011**, *23*, 5101. [CrossRef]

53. Rivera, M.K.; Aluie, H.; Ecke, R.E. The direct enstrophy cascade of two-dimensional soap film flows. *Phys. Fluids* **2014**, *26*, 055105. [CrossRef]

54. Liao, Y.; Ouellette, N.T. Long-range ordering of turbulent stresses in two-dimensional flow. *Phys. Rev. E* **2015**, *91*, 063004. [CrossRef]

55. Fang, L.; Ouellette, N.T. Advection and the Efficiency of Spectral Energy Transfer in Two-Dimensional Turbulence. *Phys. Rev. Lett.* **2016**, *117*, 104501. [CrossRef]

56. Liu, S.; Meneveau, C.; Katz, J. On the properties of similarity subgrid-scale models as deduced from measurements in a turbulent jet. *J. Fluid Mech.* **1994**, *275*, 83–119. [CrossRef]

57. Meneveau, C. Statistics of Turbulence Subgrid-Scale Stresses: Necessary Conditions and Experimental Tests. *Phys. Fluids* **1994**, *6*, 815–833. [CrossRef]

58. Tao, B.; Katz, J.; Meneveau, C. Statistical geometry of subgrid-scale stresses determined from holographic particle image velocimetry measurements. *J. Fluid Mech.* **2002**, *457*, 35–78. [CrossRef]

59. Bai, K.; Meneveau, C.; Katz, J. Experimental study of spectral energy fluxes in turbulence generated by a fractal, tree-like object. *Phys. Fluids* **2013**, *25*, 110810. [CrossRef]

60. Chow, Y.C.; Uzol, O.; Katz, J.; Meneveau, C. Decomposition of the spatially filtered and ensemble averaged kinetic energy, the associated fluxes and scaling trends in a rotor wake. *Phys. Fluids* **2005**, *17*, 085102. [CrossRef]

61. Akbari, G.; Montazerin, N. On the role of anisotropic turbomachinery flow structures in inter-scale turbulence energy flux as deduced from SPIV measurements. *J. Turbul.* **2013**, *14*, 44–70. [CrossRef]

62. Aluie, H.; Kurien, S. Joint downscale fluxes of energy and potential enstrophy in rotating stratified Boussinesq flows. *EPL Europhys. Lett.* **2011**, *96*, 44006. [CrossRef]

63. Aluie, H.; Hecht, M.; Vallis, G.K. Mapping the Energy Cascade in the North Atlantic Ocean: The Coarse-Graining Approach. *J. Phys. Oceanogr.* **2018**, *48*, 225–244. [CrossRef]

64. Buzzicotti, M.; Aluie, H.; Biferale, L.; Linkmann, M. Energy transfer in turbulence under rotation. *Phys. Rev. Fluids* **2018**, *3*, 291. [CrossRef]

65. Aluie, H.; Eyink, G. Scale Locality of Magnetohydrodynamic Turbulence. *Phys. Rev. Lett.* **2010**, *104*, 081101. [CrossRef]

66. Aluie, H. Coarse-grained incompressible magnetohydrodynamics: Analyzing the turbulent cascades. *New J. Phys.* **2017**, *19*, 025008. [CrossRef]

67. Sadek, M.; Aluie, H. Extracting the spectrum of a flow by spatial filtering. *Phys. Rev. Fluids* **2018**, *3*, 124610. [CrossRef]

68. Chassaing, P. An alternative formulation of the equations of turbulent motion for a fluid of variable density. *J. Mec. Theor. Appl.* **1985**, *4*, 375–389.

69. Bodony, D.J.; Lele, S.K. On using large-eddy simulation for the prediction of noise from cold and heated turbulent jets. *Phys. Fluids* **2005**, *17*, 085103. [CrossRef]

70. Burton, G.C. Study of ultrahigh Atwood-number Rayleigh–Taylor mixing dynamics using the nonlinear large-eddy simulation method. *Phys. Fluids* **2011**, *23*, 045106. [CrossRef]

71. Karimi, M.; Girimaji, S.S. Influence of orientation on the evolution of small perturbations in compressible shear layers with inflection points. *Phys. Rev. E* **2017**, *95*, 033112. [CrossRef] [PubMed]

72. Kida, S.; Orszag, S.A. Energy and spectral dynamics in forced compressible turbulence. *J. Sci. Comput.* **1990**, *5*, 85–125. [CrossRef]

73. Cook, A.W.; Zhou, Y. Energy transfer in Rayleigh-Taylor instability. *Phys. Rev. E* **2002**, *66*, 026312. [CrossRef] [PubMed]

74. Grete, P.; O'Shea, B.W.; Beckwith, K.; Schmidt, W.; Christlieb, A. Energy transfer in compressible magnetohydrodynamic turbulence. *Phys. Plasmas* **2017**, *24*, 092311. [CrossRef]

75. Zhao, D.; Aluie, H. Inviscid criterion for decomposing scales . *Phys. Rev. Fluids* **2018**, *3*, 054603. [CrossRef]

76. Hesselberg, T. Die Gesetze der ausgeglichenen atmosphärischen Bewegungen. *Beiträge Zur Phys. Der Freien Atmosphäre* **1926**, *12*, 141–160.

77. Favre, A.; Gaviglio, J.; Dumas, R. Further space-time correlations of velocity in a turbulent boundary layer. *J. Fluid Mech.* **1958**, *3*, 344–356. [CrossRef]

78. Drivas, T.D.; Eyink, G.L. An Onsager singularity theorem for turbulent solutions of compressible Euler equations. *Commun. Math. Phys.* **2018**, *359*, 733–763. [CrossRef]

79. Eyink, G.L. Multi-scale gradient expansion of the turbulent stress tensor. *J. Fluid Mech.* **2006**, *549*, 159–190. [CrossRef]

80. Bardina, J.; Ferziger, J.H.; Reynolds, W.C. Improved subgrid-scale models for large-eddy simulation. In Proceedings of the 13th Fluid and Plasma Dynamics Conference, Snowmass, CO, USA, 14–16 July 1980.

81. Sharp, D.H. An overview of Rayleigh-Taylor instability. *Phys. D Nonlinear Phenom.* **1984**, *12*, 3–18. [CrossRef]

82. Kundu, P.K.; Cohen, I.M. *Fluid Mechanics*; Academic Press: Oxford, UK, 2008.

83. Federrath, C.; Klessen, R.S.; Schmidt, W. The Density Probability Distribution in Compressible Isothermal Turbulence: Solenoidal versus Compressive Forcing. *Astrophys. J. Lett.* **2008**, *688*, L79. [CrossRef]

84. Eswaran, V.; Pope, S. An examination of forcing in direct numerical simulations of turbulence. *Comput. Fluids* **1988**, *16*, 257–278. [CrossRef]

85. Jagannathan, S.; Donzis, D.A. Reynolds and Mach number scaling in solenoidally-forced compressible turbulence using high-resolution direct numerical simulations. *J. Fluid Mech.* **2016**, *789*, 669–707. [CrossRef]

86. Wang, J.; Wang, L.P.; Xiao, Z.; Shi, Y.; Chen, S. A hybrid numerical simulation of isotropic compressible turbulence. *J. Comput. Phys.* **2010**, *229*, 5257–5279. [CrossRef]

87. Petersen, M.R.; Livescu, D. Forcing for statistically stationary compressible isotropic turbulence. *Phys. Fluids* **2010**, *22*, 116101. [CrossRef]

88. Livescu, D.; Ristorcelli, J.R.; Gore, R.A.; Dean, S.H.; Cabot, W.H.; Cook, A.W. High-Reynolds number Rayleigh-Taylor turbulence. *J. Turbul.* **2009**, *10*, 1–32. [CrossRef]

89. Zhao, D.; Betti, R.; Aluie, H. The cascade in compressible Rayleigh-Taylor turbulence. **2019**, in preparation.

90. Kritsuk, A.G.; Ustyugov, S.D.; Norman, M.L.; Padoan, P. Self-organization in Turbulent Molecular Clouds: Compressional Versus Solenoidal Modes. In *Numerical Modeling of Space Plasma Flows, Astronum-2009*; Pogorelov, N.V., Audit, E., Zank, G.P., Eds.; Astronomical Society of the Pacific: San Francisco, CA, USA, 2010; Volume 429, p. 15.

91. Sarkar, S.; Erlebacher, G.; Hussaini, M.Y.; Kreiss, H.O. The analysis and modelling of dilatational terms in compressible turbulence. *J. Fluid Mech.* **1991**, *227*, 473–493. [CrossRef]
92. Shivamoggi, B.K. Equilibrium statistical mechanics of compressible isotropic turbulence. *Europhys. Lett.* **1997**, *38*, 657–662. [CrossRef]

 *fluids*

Article

Computation of Kinematic and Magnetic α-Effect and Eddy Diffusivity Tensors by Padé Approximation

Sílvio M.A. Gama [1,*], Roman Chertovskih [2] and Vladislav Zheligovsky [3]

[1] Centro de Matemática (Faculdade de Ciências) da Universidade do Porto, Rua do Campo Alegre 687, 4169-007 Porto, Portugal

[2] Research Center for Systems and Technologies, Faculty of Engineering, University of Porto, Rua Dr. Roberto Frias, s/n, 4200-465 Porto, Portugal; roman@fe.up.pt

[3] Institute of Earthquake Prediction Theory and Mathematical Geophysics, Russian Ac. Sci., 84/32 Profsoyuznaya St, 117997 Moscow, Russian; vlad@mitp.ru

* Correspondence: smgama@fc.up.pt

Received: 13 May 2019; Accepted: 7 June 2019; Published: 14 June 2019

Abstract: We present examples of Padé approximations of the α-effect and eddy viscosity/diffusivity tensors in various flows. Expressions for the tensors derived in the framework of the standard multiscale formalism are employed. Algebraically, the simplest case is that of a two-dimensional parity-invariant six-fold rotation-symmetric flow where eddy viscosity is negative, indicating intervals of large-scale instability of the flow. Turning to the kinematic dynamo problem for three-dimensional flows of an incompressible fluid, we explore the application of Padé approximants for the computation of tensors of magnetic α-effect and, for parity-invariant flows, of magnetic eddy diffusivity. We construct Padé approximants of the tensors expanded in power series in the inverse molecular diffusivity $1/\eta$ around $1/\eta = 0$. This yields the values of the dominant growth rate to satisfactory accuracy for η, several dozen times smaller than the threshold, above which the power series is convergent. We do computations in Fortran in the standard "double" (real*8) and extended "quadruple" (real*16) precision, and perform symbolic calculations in Mathematica.

Keywords: incompressible fluid; magnetic mode; alpha-effect; eddy diffusivity; eddy viscosity; Padé approximant

---

1. Introduction

Power series expansion of analytic functions is perhaps the most powerful tool of numerical analysis. Let us just note that most algorithms for the numerical integration of ordinary differential equations, such as the Runge–Kutta methods, rely on Taylor series expansions for derivation. The truncated series—i.e., polynomials—are easy to compute, and thus provide an important basic algorithm for the evaluation of many analytic functions.

However, there are two caveats. One is related to the finite precision of computations, stemming from the hardware architecture and employed in the overwhelming majority of computer codes. A well-known example of the resultant failure of a computational procedure is the straightforward attempt to compute an exponent of a real large negative number by this technique [1]. Mathematically, this does not present any difficulty—the large individual terms in the Taylor expansion around zero are guaranteed to mutually cancel out and yield the final result, which is less than unity. For finite-precision computations, however, the cancellation is no longer guaranteed, and the initial growth of individual terms can result in the ultimate loss of accuracy.

The other stems from the finiteness of the radius of convergence of most power series encountered in computational practice. A complementary technique is then needed to analytically continue a function defined by the power series outside its circle of convergence. This can be achieved by constructing the so-called Padé approximants [2–4], i.e., an implementation of the continuation in the form of the ratio of two polynomials. Let us cite the words of appraisal in [5]: "Padé approximation has the uncanny knack of picking the function you had in mind from among all the possibilities. Except when it doesn't! That is the downside of Padé approximation: it is uncontrolled. There is, in general, no way to tell how accurate it is, or how far out in x it can usefully be extended. It is a powerful, but in the end still mysterious, technique".

A no less mysterious notion is that of eddy diffusivity [6], also known as eddy (or turbulent) viscosity when fluid viscosity, the source of diffusion in hydrodynamics, is considered. Like the magnetic α-effect and anisotropic kinetic alpha (AKA)-effect, eddy diffusivities are often encountered in magnetohydrodynamics when the generation of large-scale magnetic fields by flows of electrically conductive fluids is considered. At first sight, it is in direct contradiction with the second principle of thermodynamics. Of course, in fact no physical laws are violated. Both notions just describe the mean influence of the small scales on large-scale structures.

According to the modern paradigm, cosmic magnetic fields (such as the solar or geomagnetic field) exist due to the dynamo processes in the moving, electrically conductive medium (such as melted rocks in the outer Earth's core) [7]. The generating flows are typically turbulent and feature a vast hierarchy of spatial and temporal scales. Small-scale fluctuations of the flow (called "cyclonic events" by Parker) give rise to small fluctuations in the magnetic field. The interaction of the small-scale components of the magnetic field and the flow velocity produces a mean electromotive force (e.m.f.) that may have a non-zero component parallel to the mean magnetic field, and this can be beneficial for magnetic field generation [8]. The part of the mean e.m.f. linear in the mean field gives rise to the so-called magnetic α-effect. If the flow is parity-invariant, the α-effect disappears and the impact of yet smaller spatial scales becomes apparent; the mean e.m.f. is then a linear combination of the first-order spatial derivatives of the mean field, giving rise to eddy (turbulent) diffusivity. These physical ideas are treated under various simplifying assumptions in mean-field electrodynamics [9,10], and relying only on the first principles, by asymptotic methods of homogenization of elliptic operators in magnetohydrodynamic multiscale stability theory [11–17]. Weakly nonlinear stability problems are also amenable to these methods [18].

The analysis of equations makes it evident that eddy viscosity/diffusivity acquires unusual properties because it acts on mean fields only, i.e., essentially an open physical system is considered. In fact, in this class of MHD systems, the inverse energy cascade is important, with energy proliferating from small scales toward large ones; the source of energy for the developing large-scale magnetic, hydrodynamic, or combined MHD perturbation is the forcing applied to maintain the perturbed (also sometimes called basic) fluid flow.

Evaluation of the α-effect and eddy diffusivity tensors involves solving the so-called auxiliary problems, which are linear problems for the respective elliptic operators of linearization. In the large-scale dynamo problem, computing the magnetic α-effect tensor requires considering three such problems; the number increases to 12 when the tensor of eddy diffusivity is sought (unless auxiliary problems for the adjoint operator come into play, decreasing the number of auxiliary problems to be treated to just 6, see, e.g., [19,20]). Interesting results (e.g., instability to large-scale perturbation or dynamos) are typically obtained for relatively small molecular viscosity or magnetic diffusivity. Consequently, high spatial resolution is needed when solving the auxiliary problems, which makes the problems computationally intensive. However, the tensors can be easily expanded in the respective Reynolds number (i.e., in the inverse viscosity or diffusivity, provided the size of the flow periodicity box and the flow velocity are order unity) when it is small, i.e., for large viscosities and diffusivities. When the parameter tends to the critical value for the onset of the small-scale instability (i.e., in the dynamo context, to the value for which the generation of small-scale magnetic fields starts), the tensors usually exhibit singular behavior [20–23] of a simple pole type, bounding from above the radius of convergence of the series. This suggests trying Padé approximants for computing the tensors and the respective large-scale magnetic field/instability growth rates.

Here, we report numerical experiments exploring these ideas. The paper is organized as follows. In Section 2, we apply the Padé approximants techniques for evaluation of the eddy viscosity in a two-dimensional flow with two symmetries, in whose presence the eddy viscosity tensor reduces to a scalar. In view of the first caveat discussed in the beginning of this introduction, we have chosen to perform the calculations in precise arithmetics allowing an arbitrary number of correct digits; for this purpose, we have used the programming language Mathematica, giving an opportunity to make symbolic computations. In Section 3, we revert to the standard "double precision" computations (real*8, in Fortran speak) of the magnetic α-effect tensor, using the "quadruple precision" (real*16) computations for comparison. In Section 4, we again employ the symbolic capabilities of Mathematica to evaluate the magnetic eddy diffusivity tensor. Our findings are summarized in Section 5.

2. Calculation of Eddy Viscosity

No truly two-dimensional flows exist in nature, but they mimick properties of natural objects, such as the atmosphere or ocean [24–26]. We analyze the eddy viscosity tensor, $\varepsilon_{ijk\ell}$ [11], of a two-dimensional flow of incompressible fluid that has two symmetries: parity invariance (S1) and six-fold rotation symmetry (S2).

Since an S1-symmetric flow has a center of symmetry, it cannot possess the large-scale anisotropic kinetic α-effect [27]. This is important because in the presence of the AKA effect, the large-scale dynamics are essentially dispersive and non-diffusive, concealing the impact of the eddy viscosity. The symmetry S2 implies the isotropy of fourth-order tensors (see, for instance, [28]), in particular, $\varepsilon_{ijk\ell} = \nu_E \delta_{ij}\delta_{k\ell}$, where the scalar ν_E is the (standard) eddy viscosity and δ_{mn} is the Kronecker symbol. Although the assumption that a flow features the two symmetries is mathematically convenient, it may be non-realistic when considering natural or engineering problems [29].

Two-dimensional flows endowed with the symmetries S1 and S2 can be constructed as follows. A space-periodic flow is assumed, the periodicity cell being the rectangle

$$[0, L_1] \times [0, L_2] \ni \mathbf{x} = (x_1, x_2), \qquad L_1 = \sqrt{3}L_2 = 2\pi.$$

Its stream-function $\Psi(t, \mathbf{x})$ is then a sum of Fourier modes whose wave vectors are $p(2,0) + q(1, \sqrt{3})$, where p and q are integers. Any two such modes that have wave vectors mutually related by rotations by $\pi/3$ must both be involved in the sum with the same real amplitude.

We begin this section by recalling the expression for scalar eddy viscosity in terms of the solutions to two auxiliary problems and the analytical framework for evaluating the expansion of the eddy viscosity tensor in powers of the inverse of the molecular viscosity. We carry on by recalling the standard terminology and definitions of Padé approximants to a series. Finally, we discuss our results and conclusions.

2.1. Eddy Viscosities and Multiscale Techniques

For parity-invariant and six-fold rotation-symmetric flows of an incompressible fluid, eddy viscosities were calculated in [11] by multiscale techniques. They can be expressed in terms of solutions to two auxiliary problems, which can be solved analytically only in special cases [11] (e.g., if the flow depends only on a single spatial coordinate).

Briefly, the eddy viscosity in an isotropically forced two-dimensional flow is calculated as follows [11,30]. In terms of the stream-function $\Psi(t, \mathbf{x})$, the two-dimensional forced Navier–Stokes equation for incompressible fluid takes the form

$$\partial_t \nabla^2 \Psi + J(\nabla^2 \Psi, \Psi) = \nu \nabla^2 \nabla^2 \Psi + \partial f_1 / \partial x_2 - \partial f_2 / \partial x_1 .$$

Its solution, Ψ, defines a basic flow. Here, $J(g_1, g_2) = (\partial g_1 / \partial x_1)(\partial g_2 / \partial x_2) - (\partial g_1 / \partial x_2)(\partial g_2 / \partial x_1)$ is the Jacobian determinant of functions $g_1(x_1, x_2)$ and $g_2(x_1, x_2)$, the operator $\nabla^2 = \partial^2 / \partial x_1^2 + \partial^2 / \partial x_2^2$ is the Laplacian, ν the kinematic molecular viscosity, and $\mathbf{f} = (f_1, f_2)$ the external force. (In our numerical examples, the flow norm and the size of the small-scale periodicity cell are order unity; thus, the inverse molecular viscosity can be regarded as the local Reynolds number, which is the key dimensionless parameter of the problem.) Now assume that the basic flow possesses the symmetries $S1$ and $S2$. Then the (scalar) eddy viscosity coefficient, $\nu_E = \nu_E(\Psi, \nu)$, which depends only on the basic flow and molecular viscosity, is [30]

$$\nu_E(\Psi, \nu) = \nu - \left\langle \left(Q + 2\frac{\partial S}{\partial x_1} \right) \frac{\partial \Psi}{\partial x_2} \right\rangle. \tag{1}$$

Here, the angle brackets denote the average over the periodicity cell,

$$\langle g(x_1, x_2) \rangle = \frac{1}{L_1 L_2} \int_{x_1=0}^{L_1} \int_{x_2=0}^{L_2} g(x_1, x_2) \, dx_2 \, dx_1,$$

the scalar functions $Q(x_1, x_2)$ and $S(x_1, x_2)$ are solutions to two auxiliary problems

$$\mathfrak{G}Q = \frac{\partial \nabla^2 \Psi}{\partial x_2}, \qquad \mathfrak{G}S = Q\frac{\partial \nabla^2 \Psi}{\partial x_2} + 2J\left(\Psi, \frac{\partial Q}{\partial x_1}\right) - \nabla^2 Q\frac{\partial \Psi}{\partial x_2} + 4\nu\frac{\partial \nabla^2 Q}{\partial x_1}, \tag{2}$$

and $\mathfrak{G}$ denotes the linearization of the Navier–Stokes equation around Ψ,

$$\mathfrak{G} : \psi \mapsto J(\nabla^2 \psi, \Psi) + J(\nabla^2 \Psi, \psi) - \nu \nabla^2 \nabla^2 \psi.$$

We restrict it to zero-mean functions of the same space periodicity as the basic flow. In this functional space, we can define the inverse Laplace operator, which we denote as ∇^{-2}. A field G from this space can be expressed as a Fourier series

$$G(x_1, x_2) = \sum_{p,q \in \mathbb{Z}} \widehat{G}_{p,q} \exp(2\pi i(px_1/L_1 + qx_2/L_2)), \tag{3}$$

where $\langle G \rangle = \widehat{G}_{0,0} = 0$. The equation $\nabla^2 F = G \iff F = \nabla^{-2}G$ can now be readily solved: given $\langle F \rangle = 0$, we find

$$\nabla^{-2}G = -(2\pi)^{-2} \sum_{p,q \in \mathbb{Z}\backslash\{0\}} \frac{\widehat{G}_{p,q}}{(p/L_1)^2 + (q/L_2)^2} \exp(2\pi i(px_1/L_1 + qx_2/L_2)). \tag{4}$$

The existence of a deterministic time-independent space-periodic flow, which has an isotropic negative eddy viscosity when the molecular viscosity is below a critical value, was established in [31]. The so-called decorated hexagonal flow (DHF), on which we will also focus here, is

$$\begin{aligned}
\Psi(x_1, x_2) = [\ &-\cos 2x_1 - \cos(x_1 + \sqrt{3}x_2) - \cos(x_1 - \sqrt{3}x_2) \\
&+ \cos(4x_1 + 2\sqrt{3}x_2) + \cos(5x_1 - \sqrt{3}x_2) + \cos(x_1 - 3\sqrt{3}x_2) \\
&- \cos 4x_1 - \cos(2x_1 + 2\sqrt{3}x_2) - \cos(2x_1 - 2\sqrt{3}x_2) \\
&+ \cos(4x_1 - 2\sqrt{3}x_2) + \cos(5x_1 + \sqrt{3}x_2) + \cos(x_1 + 3\sqrt{3}x_2)]/2.
\end{aligned} \tag{5}$$

The phenomenon of negative eddy viscosity is quite common among two-dimensional divergenceless space-periodic basic flows with the symmetries $S1$ and $S2$: about $1/3$ of such flows feature negative eddy viscosity for sufficiently low molecular viscosity [30]. The auxiliary problems (2) can be solved either numerically by spectral methods or by expanding in powers of ν^{-1} to high orders and afterwards extending analytically (relying on their meromorphy [30]) beyond the disk of convergence. We examine the latter method, enabling us to perform all calculations exactly.

2.2. Eddy Viscosity Expansion in Powers of ν^{-1}

Let us expand (1):

$$\nu_E(\Psi, \nu) = \nu + \sum_{n=1}^{\infty} \nu_E^{(n)}(\Psi)\nu^{-n}. \tag{6}$$

To calculate $\nu_E^{(n)}$, we expand the solutions Q and S to (2) in Maclaurin series in ν^{-1}:

$$Q = \sum_{n=1}^{\infty} Q^{(n)}\nu^{-n}, \qquad S = \sum_{n=1}^{\infty} S^{(n)}\nu^{-n}.$$

Substituting the series into (1) and integrating by parts yields

$$\nu_E^{(n)}(\Psi) = \left\langle -Q^{(n)}\frac{\partial\Psi}{\partial x_2} + 2S^{(n)}\frac{\partial^2\Psi}{\partial x_1 \partial x_2} \right\rangle. \tag{7}$$

Here, $Q^{(1)} = -\nabla^{-2}(\partial\Psi/\partial x_2)$ and $S^{(1)} = -4\nabla^{-2}(\partial Q^{(1)}/\partial x_1)$, and the subsequent terms satisfy the recurrence relations:

$$Q^{(n)} = \mathcal{B}Q^{(n-1)}, \tag{8}$$

$$S^{(n)} = \mathcal{B}S^{(n-1)} - \nabla^{-2}\nabla^{-2}\left[Q^{(n-1)}\frac{\partial\nabla^2\Psi}{\partial x_2} + 2J\left(\Psi, \frac{\partial Q^{(n-1)}}{\partial x_1}\right) - \nabla^2 Q^{(n-1)}\frac{\partial\Psi}{\partial x_2} + 4\frac{\partial\nabla^2 Q^{(n)}}{\partial x_1}\right], \tag{9}$$

where the operator $\mathcal{B}$ is defined as

$$\mathcal{B} : f \mapsto \nabla^{-2}\nabla^{-2}\left[J(\nabla^2 f, \Psi) + J(\nabla^2\Psi, f)\right].$$

Since here we consider parity-invariant flows, their stream-functions being even, i.e., $\Psi(-\mathbf{x}) = \Psi(\mathbf{x})$, the series (6) involves only odd powers of ν^{-1} [30], i.e., $\nu_E^{(n)} = 0$ for all even n.

By definition, the $[L/M]$ Padé approximant to a series, whose first $m \geq L + M + 1$ terms are known, is the ratio of a polynomial of degree $\leq L$ to a polynomial of degree $\leq M$, such that the first $L + M + 1$ terms of the expansion of the ratio coincide with the respective terms of the series.

We use Padé approximants to reconstruct the dependence of the eddy viscosity on the inverse molecular viscosity employing the expansion (6). The Padé approximation techniques can also be naturally applied for exploring the poles of the eddy viscosity. A pole on the real axis can usually be linked to the onset of linear instability to small-scale perturbations, or sometimes (if it appears again on decreasing $\nu > 0$) to its cessation.

2.3. Results of Calculations

The performance of present-day computers and efficiency of symbolic programming software gives an opportunity to calculate the coefficients (7) using recurrence relations (8)-(9) and construct the Padé approximants exactly. All calculations reported in this section are performed with full precision by Mathematica [32]. Table 1 shows some of the first twenty non-zero coefficients $\nu_E^{(n)}$ ($n = 1, 3, 5, ..., 39$) for the DHF. One of the reasons for performing full precision symbolic computations with Mathematica has been a hope of discovering useful relations between the coefficients of the approximants. Unfortunately, none are visible in the data in Table 1.

We truncate the series (6) at orders up to 39, even terms missing. Roots of their Padé approximants quickly stabilize near $\nu = \nu^\star \approx 0.58$ (see Figure 1), indicating a transition to negative eddy viscosity at lower molecular viscosities. The root $\nu^\star$ is simple; a sharpened estimate is $1/\nu^\star = 1.72144 \pm 10^{-5}$.

Table 1. The first 39 non-zero coefficients of (6) for the decorated hexagonal flow (DHF, prime decomposition where presented). The first 5 significant figures of these coefficients are given in [31]. "$a \lll p \ggg c$" denotes a natural number containing p decimal digits between a and b.

n	Coefficient $\nu_{\mathrm{E}}^{(n)}$ (Exact Rational Number)
1	$\dfrac{3}{2^2}$
3	$-\dfrac{3\times5\times11\times1931\times80491}{2^9\times7^4\times13^2\times19^2}$
5	$-\dfrac{3\times53^2\times222967\times199451798303381365128830607922192539}{2^{19}\times5^2\times7^9\times13^6\times19^7\times31^3\times37^2\times43^3\times61^2}$
7	$\dfrac{3^3\times23\times17401\times11608063\times57039665830751679518604082987471 0499\times595146062519802577066082838776447096016784218965582671080441286999}{2^{25}\times5^2\times7^{17}\times13^{10}\times19^{11}\times31^5\times37^7\times43^5\times61^6\times67^3\times73^3\times79^3\times97^3\times103^2\times109^2}$
9	$-\dfrac{9606359879 \lll 188 \ggg 5777697637}{2^{33}\times5^2\times7^{24}\times11^2\times13^{14}\times19^{15}\times31^7\times37^{11}\times43^7\times61^{10}\times67^5\times73^5\times79^5\times97^5\times103^3\times109^6\times127^3\times139^3\times151^3\times157^2\times163^2}$
11	$-\dfrac{7129561983 \lll 324 \ggg 7108258721}{8493879641 \lll 326 \ggg 4312960000}$
$\vdots$	$\vdots$
39	$-\dfrac{1648936106 \lll 9785 \ggg 2091564067}{3775138782 \lll 9788 \ggg 0000000000}$

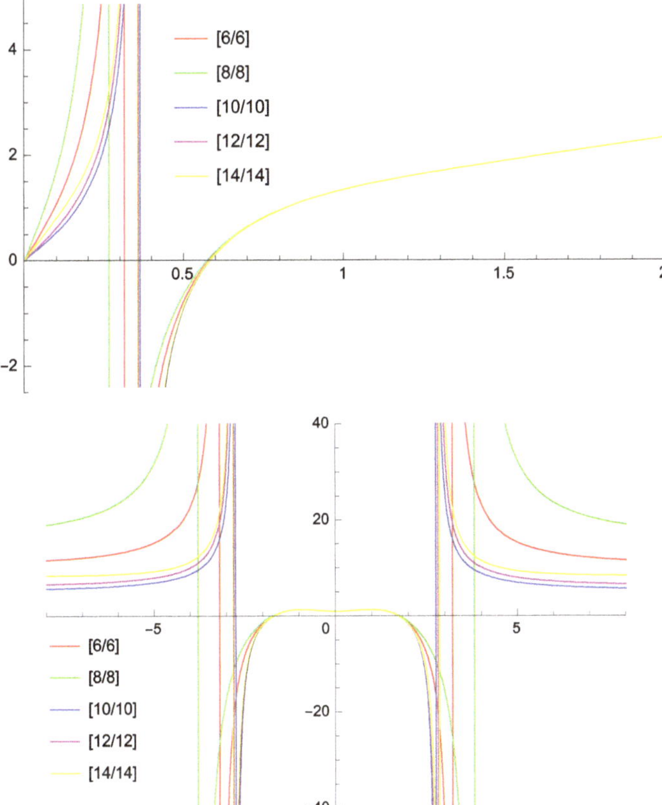

Figure 1. Padé approximants (vertical axes) $[L/L]_{\nu_{\mathrm{E}}}(\nu)$ (top) and $[L/L]_{\nu_{\mathrm{E}}/\nu}(\nu^{-1})$ (bottom) for $6 \leq L \leq 14$ step 2. Horizontal axes: ν (top), ν^{-1} (bottom). In the bottom panel, we extend ν to negative values on purpose to highlight that the Maclaurin expansion of $\nu_{\mathrm{E}}(\nu^{-1})/\nu$ is an even function of ν^{-1}.

These results indicate that Padé approximants are a reliable alternative to other methods for the calculation of the point of the onset of large-scale instability in this type of flow. Not only do they furnish stable estimates (provided the approximated function is meromorphic and the series is long enough), but they also serve as precursors to future work. An illustration is Figure 1 (right), where the [14/14] approximant (as many others) exhibits a singular behavior near $1/\nu \approx 2.81$ apparently related to the onset of linear instability to small-scale perturbations. The non-monotonicity of the eddy viscosity as a function of ν can be regarded as a manifestation of the complexity of the two-dimensional turbulent flow.

3. Computation of the Magnetic α-Effect Tensor

As we have seen in the previous section, the use of Padé expansions for reconstructing the dependence of eddy viscosity on the molecular one is possible and does not require very high degrees of the polynomials involved, at least for moderate (not very small) molecular viscosities. However, for 3D fields, arbitrary-precision symbolic calculations cannot be regarded as a practical realization of the approximation algorithm. Here, we construct Padé approximants for evaluation of the magnetic α-effect tensor using arithmetics of floating point numbers of the conventional double (real*8) and the extended quadruple (real*16) precision. The problem now at hand is to construct approximations applicable for small magnetic molecular diffusivities.

3.1. The Multiscale Formalism Revealing the Magnetic α-Effect

We review here the multiscale expansions [17] describing the kinematic generation of large-scale magnetic fields by small-scale zero-mean space-periodic steady flows. Our magnetic modes depend on two three-dimensional spatial variables, the fast, $\mathbf{x}$, and slow, $\mathbf{X} = \varepsilon\mathbf{x}$, one (the flow $\mathbf{v}$ depends exclusively on $\mathbf{x}$). A magnetic mode $\mathbf{b}$ is an eigenfield of the magnetic induction operator $\mathcal{L}$:

$$\mathcal{L}\mathbf{b} = \lambda\mathbf{b}, \tag{10}$$

$$\mathcal{L} : \mathbf{b} \mapsto \eta\nabla^2\mathbf{b} + \nabla \times (\mathbf{v} \times \mathbf{b}). \tag{11}$$

Here, η is the magnetic molecular diffusivity and Re λ the growth rate of the mode $\mathbf{b}$. Both the mode and flow are solenoidal.

The scale ratio ε is a small parameter of the problem, in which the magnetic mode $\mathbf{b}$ and its growth rate are expanded:

$$\mathbf{b} = \sum_{n=0}^{\infty} \mathbf{b}_n(\mathbf{X}, \mathbf{x})\,\varepsilon^n, \tag{12}$$

$$\lambda = \sum_{n=0}^{\infty} \lambda_n\varepsilon^n. \tag{13}$$

By substituting the expansions into (10) and the solenoidality conditions, we derive a hierarchy of equations for the coefficients in (12) and (13).

As in the previous section, we denote by angle brackets the mean over the periodicity cell $\mathbb{T}^3$ in the fast variables and by the braces, the fluctuating part:

$$\langle\mathbf{f}\rangle = (2\pi)^{-3} \int_{\mathbb{T}^3} \mathbf{f}(\mathbf{X}, \mathbf{x})\,d\mathbf{x} = \sum_{k=1}^{3} \langle\mathbf{f}\rangle_k\,\mathbf{e}_k, \qquad \{\mathbf{f}\} = \mathbf{f} - \langle\mathbf{f}\rangle.$$

Here, $\mathbf{e}_k$ are unit Cartesian coordinate vectors.

The relevant solution to the first (order ε^0) equation $\mathfrak{L}\mathbf{b}_0 = \lambda_0\mathbf{b}_0$ is $\lambda_0 = 0$ and a linear combination $\mathbf{b}_0 = \sum_{k=1}^{3} \langle\mathbf{b}_0\rangle_k \mathbf{s}_k$ of small-scale solenoidal neutral magnetic modes $\mathbf{s}_k(\mathbf{x})$ that are solutions to the three so-called auxiliary problems of type I:

$$\mathfrak{L}\mathbf{s}_k = 0, \qquad \langle\mathbf{s}_k\rangle = \mathbf{e}_k. \tag{14}$$

Averaging the second (order ε^1) equation,

$$\mathfrak{L}\mathbf{b}_1 + 2\eta(\nabla \cdot \nabla_{\mathbf{X}})\mathbf{b}_0 + \nabla_{\mathbf{X}} \times (\mathbf{v} \times \mathbf{b}_0) = \lambda_1\mathbf{b}_0, \tag{15}$$

we obtain an eigenvalue problem

$$\nabla_{\mathbf{X}} \times (\mathfrak{A} \langle\mathbf{b}_0\rangle) = \lambda_1 \langle\mathbf{b}_0\rangle, \qquad \nabla_{\mathbf{X}} \cdot \langle\mathbf{b}_0\rangle = 0 \tag{16}$$

(the subscript $\mathbf{X}$ denotes differentiation in the slow variables). Here, $\mathfrak{A}$ is the tensor of the magnetic α-effect. The kth column of this 3×3 matrix is

$$\mathfrak{A}_k = \langle\mathbf{v} \times \mathbf{s}_k\rangle, \tag{17}$$

in agreement with Parker's [8] concept of the interaction of fine structures of the flow, $\mathbf{v}$, and magnetic field, $\sum_{k=1}^{3} \langle\mathbf{b}_0\rangle_k \{\mathbf{s}_k\}$, giving rise to a mean e.m.f., $\mathfrak{A} \langle\mathbf{b}_0\rangle$, linear in the large-scale magnetic field $\langle\mathbf{b}_0\rangle$. For space-periodic mean fields

$$\langle\mathbf{b}_0\rangle = \mathbf{B}e^{i\mathbf{q}\cdot\mathbf{X}}, \qquad \mathbf{B} \cdot \mathbf{q} = 0, \tag{18}$$

where $\mathbf{q}$ is an arbitrary unit vector, straightforward algebra [20] yields solutions to the eigenvalue problem (16):

$$\lambda_{1\pm}(\mathbf{q}) = \frac{i}{2}\left((\mathfrak{A}_3^2 - \mathfrak{A}_2^3)q_1 + (\mathfrak{A}_1^3 - \mathfrak{A}_3^1)q_2 + (\mathfrak{A}_2^1 - \mathfrak{A}_1^2)q_3\right) \pm \sqrt{a}, \qquad a = \mathbf{q} \cdot (\det{}^{s}\mathfrak{A}){}^{s}\mathfrak{A}^{-1}\mathbf{q}. \tag{19}$$

Here, ${}^{s}\mathfrak{A}_k^l = (\mathfrak{A}_k^l + \mathfrak{A}_l^k)/2$ are entries of the symmetrized α-tensor ${}^{s}\mathfrak{A} = (\mathfrak{A} + \mathfrak{A}^*)/2$.

For $a \leq 0$, the α-effect just sustains harmonic oscillations of the mean magnetic field in the slow time $T_1 = \varepsilon t$. When $a > 0$, the slow-time growth rate $\mathrm{Re}\,\lambda_1(\mathbf{q}) = \sqrt{a}$ of the large-scale magnetic mode depends only on the symmetrized tensor ${}^{s}\mathfrak{A}$, whose eigenvalues α_i are real. In the Cartesian coordinate system, whose axes coincide with eigenvectors of ${}^{s}\mathfrak{A}$, (19) takes the form

$$a = \alpha_1\alpha_2(q_3')^2 + \alpha_2\alpha_3(q_1')^2 + \alpha_1\alpha_3(q_2')^2,$$

where q_i' are components of $\mathbf{q}$ in this basis. Thus,

$$\gamma_\alpha \equiv \max_{|\mathbf{q}|=1} \mathrm{Re}\,\lambda_{1\pm}(\mathbf{q}) = \sqrt{\max(\alpha_1\alpha_2, \alpha_2\alpha_3, \alpha_1\alpha_3)} \tag{20}$$

is the maximum slow-time growth rate of large-scale magnetic modes generated by the α-effect. While the entries of the α-effect tensor, $\mathfrak{A}_l^k$, are smooth functions of η, the graph of γ_α (20) has cusps at the points η, where $\alpha_1 < \alpha_2 = 0 < \alpha_3$ [20] (see, e.g., two such cusps in Figure 2).

When $a \neq 0$ and the kernel of the magnetic induction operator $\mathfrak{L}$ does not involve small-scale zero-mean modes (generically both conditions are satisfied), all terms in the expansions (12) and (13) can be determined from the hierarchy of equations obtained by substituting the series into the eigenvalue Equation (10). If the symmetrized tensor ${}^{s}\mathfrak{A}$ is positively or negatively defined (and if the spatial periodicity of the eigenfunction is compatible with that of the flow), then the series (12) and (13) are summable [16] for sufficiently small ε and constitute an analytical in the ε eigensolution for the large-scale magnetic induction operator; a unique ε-parameterized branch of the eigenvalues (13) originates from any simple eigenvalue λ_1 of the α-effect operator.

3.2. Padé Approximation

The modes $\mathbf{s}_k$ (and consequently, elements of the magnetic α-effect tensor) are functions of molecular eddy diffusivity η, meromorphic in this parameter. (By contrast, the slow-time growth rates of modes generated by the α-effect are not, because the square root present in (19) gives rise to branch points.) A power series expansion of $\mathbf{s}_k$ in η^{-1} for large η can be constructed like in the hydrodynamic problem considered in the previous section. We divide (14) by η to obtain

$$\nabla^2 \mathbf{s}_k = -\eta^{-1} \nabla \times (\mathbf{v} \times \mathbf{s}_k).$$

Consequently, the coefficients in the expansion

$$\mathbf{s}_k = \sum_{n=0}^{\infty} \mathbf{s}_k^{(n)} \eta^{-n} \tag{21}$$

satisfy the recurrence relations

$$\mathbf{s}_k^{(0)} = \mathbf{e}_k, \qquad \mathbf{s}_k^{(n)} = -\nabla^{-2} \left(\nabla \times (\mathbf{v} \times \mathbf{s}_k^{(n-1)}) \right) \quad \text{for } n \geq 1 \tag{22}$$

(cf. (5.14)–(5.15) in [13]). Here, ∇^{-2} denotes the inverse Laplacian in the fast variables acting in the functional space of zero-mean fields. (Actually, we consider the problem for a flow, whose r.m.s. velocity is unity; the size of the periodicity box also being order unity, the inverse molecular diffusivity η^{-1} can be regarded, like in the hydrodynamic case, to be equal to the local magnetic Reynolds number, the dimensionless parameter characterizing the mathematical properties of the problem. (21) can thus be understood as an expansion in a small Reynolds number.) These recurrence relations can be used to compute the coefficients in (21) by pseudospectral methods. It must be noted that mathematically, they are perfectly suitable for numerical work: indeed, the presence of the inverse Laplacian in the second relation (22) suggests that on increasing the number n of the coefficient $\mathbf{s}_k^{(n)}$, they become smoother and their energy spectrum decay is steeper. This is in sharp contrast, for instance, with the recurrence relations for the Lagrangian time-Taylor coefficients in the expansions of solutions to the Euler equation for incompressible fluid flow [33].

It is straightforward to determine the radius of convergence of the series (21), ρ, regarded as a function of $1/\eta$. The recurrence relations (22) involve the operator

$$\mathfrak{M} : \mathbf{s} \mapsto -\nabla^{-2} (\nabla \times (\mathbf{v} \times \mathbf{s})) \tag{23}$$

acting in the functional space of solenoidal zero-mean space-periodic fields. Since it is compact, its spectrum consists of a countable set of eigenvalues μ_i, tending to zero. Consequently, $\rho \geq 1/\max_i |\mu_i|$; generically the equality holds, but $\rho > 1/\max_i |\mu_i|$ if the expansion of $\mathbf{s}_k^{(1)}$ in the basis of eigenfunctions of the operator $\mathfrak{M}$ does not involve eigenfunctions associated with any eigenvalue μ_i such that $|\mu_i| = \max_i |\mu_i|$. Therefore, generically the radii of convergence of the series for all the three $\mathbf{s}_k$ are the same.

Clearly, the radius of convergence of the ensuing series for the α-effect tensor (17),

$$\mathfrak{A}_k = \sum_{n=1}^{\infty} \mathfrak{A}_k^{(n)} \eta^{-n}, \qquad \mathfrak{A}_k^{(n)} = \left\langle \mathbf{v} \times \mathbf{s}_k^{(n)} \right\rangle, \tag{24}$$

is not smaller than that of the series (21) for $\mathbf{s}_k$. Let us note a symmetry property of (24). Denote by the superscript minus objects pertinent to the reverse flow $-\mathbf{v}$:

$$\mathfrak{L}^- : \mathbf{b} \mapsto \eta \nabla^2 \mathbf{b} - \nabla \times (\mathbf{v} \times \mathbf{b}), \qquad \mathfrak{L}^- \mathbf{s}_k^- = 0, \qquad \left\langle \mathbf{s}_k^- \right\rangle = \mathbf{e}_k, \qquad \mathfrak{A}_k^- = \left\langle -\mathbf{v} \times \mathbf{s}_k^- \right\rangle. \tag{25}$$

The α-effect tensor $\mathfrak{A}^-$ for the reverse flow $-\mathbf{v}$ is obtained from the tensor $\mathfrak{A}$ for $\mathbf{v}$ by transposition, i.e., $(\mathfrak{A}_k^-)_l = (\mathfrak{A}_l)_k$ [20]. Since the recurrence relations (22) are linear in $\mathbf{v}$,

$$\mathbf{s}_k^{(n)} = (-1)^n (\mathbf{s}_k^-)^{(n)}, \tag{26}$$

where $(\mathbf{s}_k^-)^{(n)}$ denote the coefficients in the expansion of $\mathbf{s}_k^-$ in power series (21). This implies

$$(\mathfrak{A}_l)_k^{(n)} = (\mathfrak{A}_k^-)_l^{(n)} = \left\langle -\mathbf{v} \times (\mathbf{s}_k^-)^{(n)} \right\rangle_l = (-1)^{n+1} \left\langle \mathbf{v} \times \mathbf{s}_k^{(n)} \right\rangle_l = (-1)^{n+1} (\mathfrak{A}_k)_l^{(n)}.$$

Therefore, the coefficients in the series (24) are symmetric matrices for odd n and antisymmetric ones for even n. In other words, the symmetrized α-effect tensor $^s\mathfrak{A}$ involved in the computation of the discriminant a in (19) is expanded in odd powers of η^{-1}, and of the common imaginary part of $\lambda_{1\pm}(\mathbf{q})$ in even powers; the latter expansion is not needed when computing just the growth rates.

3.3. Numerical Results

Here, we construct Padé approximants for the α-effect tensor (24) for a sample solenoidal flow and compare the maximum growth rate values γ_α (20) obtained for the approximated tensor to those computed directly at individual values of η by spectral methods.

For this purpose, a sample solenoidal flow has been synthesized as a Fourier series with pseudo-random coefficients, corrected to make it solenoidal and zero-mean. It involves Fourier harmonics for wave numbers not exceeding 10. The coefficients are scaled so that the energy spectrum decays exponentially by 10 orders of magnitude and the r.m.s. velocity is unity.

Solutions to auxiliary problems have been computed by the code in [34] employing standard pseudo-spectral methods. For $\eta > 0.05$, the problem was preconditioned by the operator $(-\nabla^2)^{-1/2}$, readily available in the Fourier space. The resolution of 64^3 Fourier harmonics was used. Energy spectra of the neutral modes $\mathbf{s}_k$ decay for this flow by at least 9 orders of magnitude for the smallest considered $\eta = 0.035$. The Lebesgue space L_2 norms of $\mathbf{s}_k^{(n)}$ from $n = 0$ to $n = 128$ decay by 38 orders of magnitude, and hence the power series (21) and (24) converge for $\eta \gtrsim 0.50$.

3.3.1. Approximation by the Algorithm *I*

We have tried two approaches for the construction of Padé approximants of the entries of the α-effect tensor. Here we discuss the results obtained by the algorithm proposed in [35].

Padé approximant $[M/L]_f$ of a function $f(y) = \sum_{m=0}^{\infty} f^{(n)} y^n$ is the ratio of two polynomials of degrees M (numerator) and L (denominator), whose $M + L + 1$ first Taylor expansion coefficients coincide with those of f. For the sake of argument, let us assume $M \geq L$. Then $L + 1$ coefficients of the denominator $d(y) = \sum_{n=0}^{L+1} d_n y^n$ satisfy the linear system of equations of the form

$$\begin{bmatrix} f^{(M+1)} & f^{(M)} & \dots & f^{(M+2-L)} & f^{(M+1-L)} \\ f^{(M+2)} & f^{(M+1)} & \dots & f^{(M+3-L)} & f^{(M+2-L)} \\ & & \dots & & \\ f^{(M+L-1)} & f^{(M+L-2)} & \dots & f^{(M+2)} & f^{(M+1)} \\ f^{(M+L)} & f^{(M+L-1)} & \dots & f^{(M+1)} & f^{(M)} \end{bmatrix} \begin{bmatrix} d_0 \\ d_1 \\ \dots \\ d_{L-1} \\ d_L \end{bmatrix} = 0. \tag{27}$$

(finding the coefficients of the numerator upon solving (27) is straightforward; see [3,4] for details).

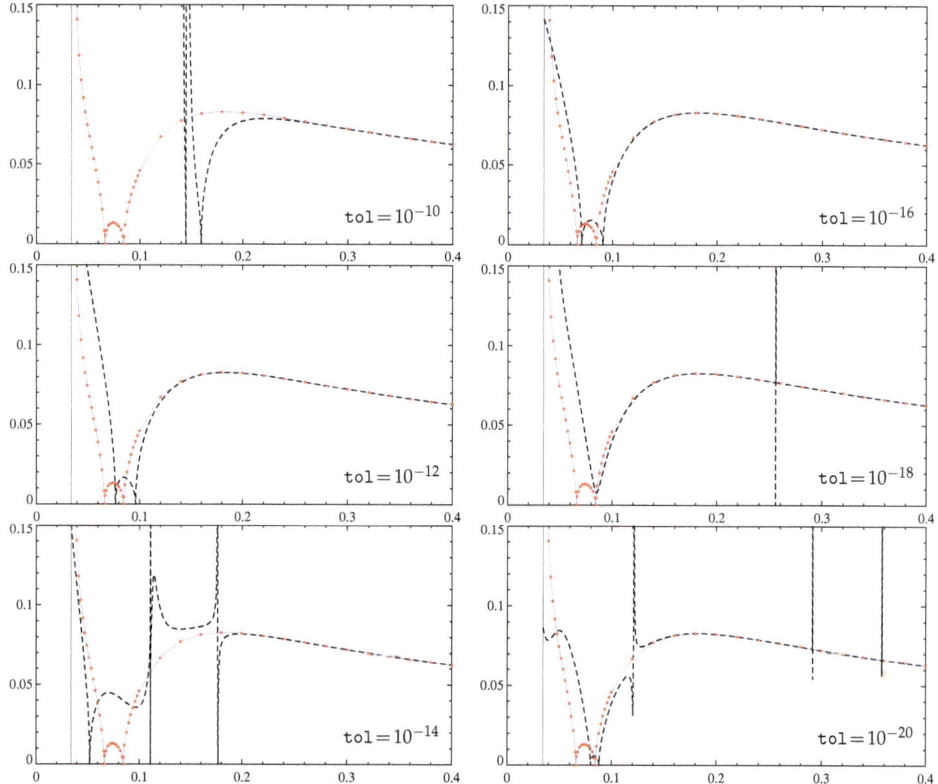

Figure 2. Maximum slow-time growth rates γ_α (20) (vertical axis) of large-scale magnetic modes generated by the α-effect as a function of the molecular diffusivity η (horizontal axis), computed using the α-effect tensor, Padé-approximated by the algorithm in [35] for a varying tolerance `tol`. Thin solid line: the dependence determined by computation of γ_α at individual η values (red solid circles) by spectral methods (resolution 64^3 Fourier harmonics), thick dashed line: the approximate dependence.

In our case, the coefficients $\mathbf{s}^{(n)}$ are obtained by iteratively applying the operator $\mathfrak{M}$ (23). It is compact, and its eigenvalues, except for a finite number of them, are below unity in absolute value. The respective spectral components decay during the iterations according to the power law and sooner or later reduce in magnitude below the accuracy of computations. Consequently, for large L the Toeplitz matrix of size $L \times (L+1)$ in the l.h.s. of (27) becomes numerically degenerate, i.e., its rank effectively falls below L. To construct an approximant under such adverse numerical conditions, it was proposed in [35] to compute the singular value decomposition of this matrix and to regard its effective rank as equal to the number of singular values whose absolute value exceeds the given relative tolerance `tol` (i.e., is not smaller than $\mathrm{tol}\|(f^{(1)}, f^{(2)}, ..., f^{(M+L-1)}, f^{(M+L)})\|$, where $\|\cdot\|$ is the standard Lebesgue space L_2 norm), decreasing the degrees of the polynomials involved in the Padé approximation, M and L, by the number of the "missing" dimensions. To counter noise due to rounding errors, $\mathrm{tol} = 10^{-14}$ was often used in [35].

Beyond the poor spectral properties of the system of equations for the Padé coefficients, there exist two other reasons for amplification of the numerical noise originally due to round-off errors:

- Pseudospectral methods used in the computation of space-periodic solutions $\mathbf{s}_k$ to the auxiliary problems (14) and their coefficients $\mathbf{s}_k^{(n)}$ (22) involve fast Fourier transforms. These algorithms are very efficient. However, they operate by computing various linear combinations of the Fourier coefficients. Typically, at least for moderate molecular diffusivities, the energy spectra of these fields decay fast. In a sum of a large coefficient with a small (in absolute values) one, a significant part of the accuracy of the smaller coefficient is lost.

- Insufficiency of the spatial resolution can result in significant numerical errors. We may note that while increasing the resolution improves solutions, it aggravates the FFT accuracy problems.

We have tried the algorithm in [35] for a set of tol values ranging between 10^{-10} and 10^{-20} using the MATLAB procedure provided by the authors of [35]. We have computed 65 first coefficients of the power series expansions of the symmetrized α-effect tensor entries $\left({}^s\mathfrak{A}_k\right)_l^{(n)}$ (which involve only odd powers of $1/\eta$; see Section 3.2) up to order η^{-129} terms with the spatial resolution of 64^3 Fourier harmonics. The MATLAB procedure has been requested to construct the [63/64] Padé approximants for each entry. The results are shown in Figure 2. We observe that the approximations of the maximum growth rates γ_α are relatively accurate for tol $= 10^{-12}$ and 10^{-16} for $\eta \gtrsim 0.1$. This bound is roughly 5 times smaller than the minimal η for which the power series for ${}^s\mathfrak{A}_k^l$ converge. Table 2 sheds light on the reasons why the gain is unsatisfactory (for $\eta \geq 0.1$ spectral computations of the α-effect, growth rates for individual η's are efficient): since the rank decreases, when small in absolute value, singular values are discarded, and the algorithm ends up with very moderate orders $[2L/2L - 1]$.

Table 2. Order parameter L of the Padé approximants $[2L - 1/2L]$ (ratios of polynomials in $1/\eta$) constructed by the algorithm in [35] for six independent entries of the symmetrized α-effect tensor ${}^s\mathfrak{A}$.

tol	${}^s\mathfrak{A}_1^1$	${}^s\mathfrak{A}_1^2$	${}^s\mathfrak{A}_1^3$	${}^s\mathfrak{A}_2^2$	${}^s\mathfrak{A}_2^3$	${}^s\mathfrak{A}_3^3$
10^{-10}	5	5	6	5	5	5
10^{-12}	6	6	6	6	6	6
10^{-14}	7	7	8	7	7	7
10^{-16}	8	8	9	8	8	8
10^{-18}	9	8	10	10	9	9
10^{-20}	10	10	11	10	10	10

The four remaining panels in Figure 2 reveal the presence of the so-called Froissart doublets in the approximants of some entries. Froissart doublets is a factor of the form $(1/\eta - a)/(1/\eta - \tilde{a})$ in the approximant, where the two constants a and $\tilde{a}$ are close but distinct. Such a factor implies a singular behavior of the approximant for $1/\eta$ close to $\tilde{a}$, not altering its behavior much at distances from $\tilde{a}$ significantly larger than $|a - \tilde{a}|$. Often, such factors are artifacts emerging due to noise in the data. Almost vertical segments of the plots are signatures of the Froissart doublets (see Figure 2). They extend to both positive and negative infinity in the graphs of the approximants, but since (20) are nonlinear functions of the tensor entries, the respective segments of graphs of γ_α may be bounded from below and/or above. (Their detection has proven unexpectedly difficult; in order to reliably show their full range in the vertical direction, we have plotted the approximation step 10^{-9} along the abscissa.) We thus see that although the algorithm in [35] is supposed to be robust, it is prone to yielding approximants involving Froissart doublets.

3.3.2. Approximation by the Algorithm *II*

Since we have failed to obtain satisfactory results with the use of the algorithm *I* in [35], we have also tested the algorithm *II* in [5]. Quoting from [5], although the equations for the coefficients of the approximant involve a matrix in the Toeplitz form, "experience shows that the equations are frequently close to singular, so that one should not solve them by the methods" relying on this form, "but rather by full LU decomposition. Additionally, it is a good idea to refine the solution by iterative improvement (routine mprove in §2.5)". This is implemented in their pade procedure. We have used it with one alteration: the routine mprove stops when the discrepancy increases; instead, it has been allowed to make up to 1000 improvement iterations, permitting the discrepancy to temporarily grow and storing the minimum-discrepancy solution obtained in the course of these iterations (however, it has often been forced to stop before the allowed number of iterations has been performed, the iterative process blowing up with an overflow).

Approximate maximum slow-time growth rates (20) of large-scale magnetic modes generated by the α-effect have been computed for the same flow again using $[2L - 1/2L]$ Padé approximants for the entries ${}^s\mathfrak{A}_l^k(\eta)$. They are compared in Figure 3 for increasing orders L with the actual maximum growth rates obtained by direct computation of the fields $\mathbf{s}_k$ using spectral methods for individual molecular diffusivities η.

Four Padé approximants $[2L - 1/2L]$ of the entries of the α-effect tensor have been constructed for each considered L, using the resolution of 64^3 or 512^3 Fourier harmonics and running our code with the double (real*8) or extended quadruple (real*16) precision of the floating-point number arithmetics. In computers built around Intel and compatible processors, the former is standard, and the latter is not supported by hardware but is software-emulated; however, many compilers do not require modifying the Fortran source code to use it, all of the floating-point data and computations can be readily promoted to the real*16 precision by using the appropriate compiler option such as $-r16$. Higher precision and resolution has been expected to improve the accuracy of the coefficients of the approximants, to augment the orders of the approximants beyond those produced by the algorithm in [35], and to increase the η interval, where the growth rate values determined for the approximated α-effect tensor are close to the actual growth rates.

In Figure 3, we show the resultant approximations of γ_α for $L = 4, 8, 14, 20$, and 28 to 31. The four graphs for $L = 4$ are visually indistinguishable, and we show only one of them; the same holds true for $L = 8$. For $L = 14$ and 20, the plots of the approximated γ_α, computed with quadruple precision for the two spatial resolutions, also visually coincide (Figure 3d,f,h,j), but this is wrong for the respective double precision approximations. For higher L, all four plots are visually distinct. The quadruple precision approximations are plagued much less by the Froissart doublets (and never involve multiple occurrences of the doublets) than the double precision ones. Three high-L quadruple precision approximations are reasonably accurate for $L = 28$ and the 512^3 resolution for $\eta \geq 0.055$ (Figure 3n), for $L = 29$ and the 64^3 resolution for $\eta \geq 0.05$ (Figure 3p), and for $L = 30$ and the 64^3 resolution for $\eta \geq 0.05$ (Figure 3t). Thus, the left end of the interval of validity of Padé approximations has decreased roughly twice compared to that obtained by the algorithm in [35]. All other quadruple-precision approximations for $L \geq 29$ (Figure 3r,v,x,z) can also give reasonable accuracy for $\eta \gtrsim 0.06 \div 0.07$ upon removal of Froissart doublets from the affected approximants of ${}^s\mathfrak{A}_l^k$.

These results suggest that Padé approximants are useful for representing the functional dependence of the slow-time growth rates (20) of large-scale magnetic modes generated by the α-effect for fairly low magnetic molecular diffusivities. However, for the construction of Padé approximants, accurate enough to serve small η, the quadruple precision arithmetics must be used, and hence run times become comparable to those of direct computation of the growth rates at individual ηs (note that real*16 computations are typically ten times slower than real*8 ones). Therefore, another strategy is perhaps also sensible: in computations for an individual η, to use relatively low-order, not very precise Padé approximants for neutral modes $\mathbf{s}_k$ (constructed, for instance, at each point in space or for each Fourier harmonics within the employed resolution) as the initial data for further refinement by the usual iterative methods.

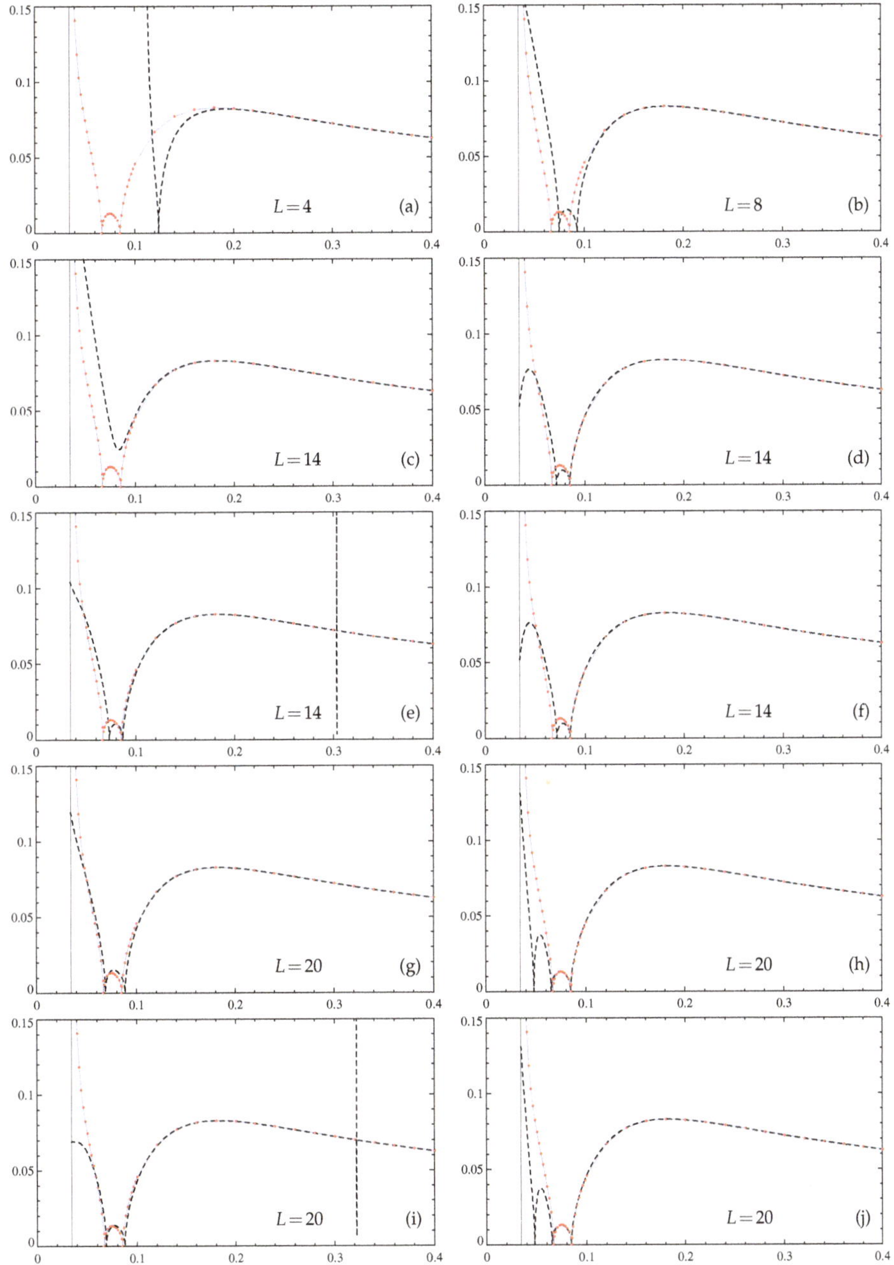

Figure 3. *Cont.*

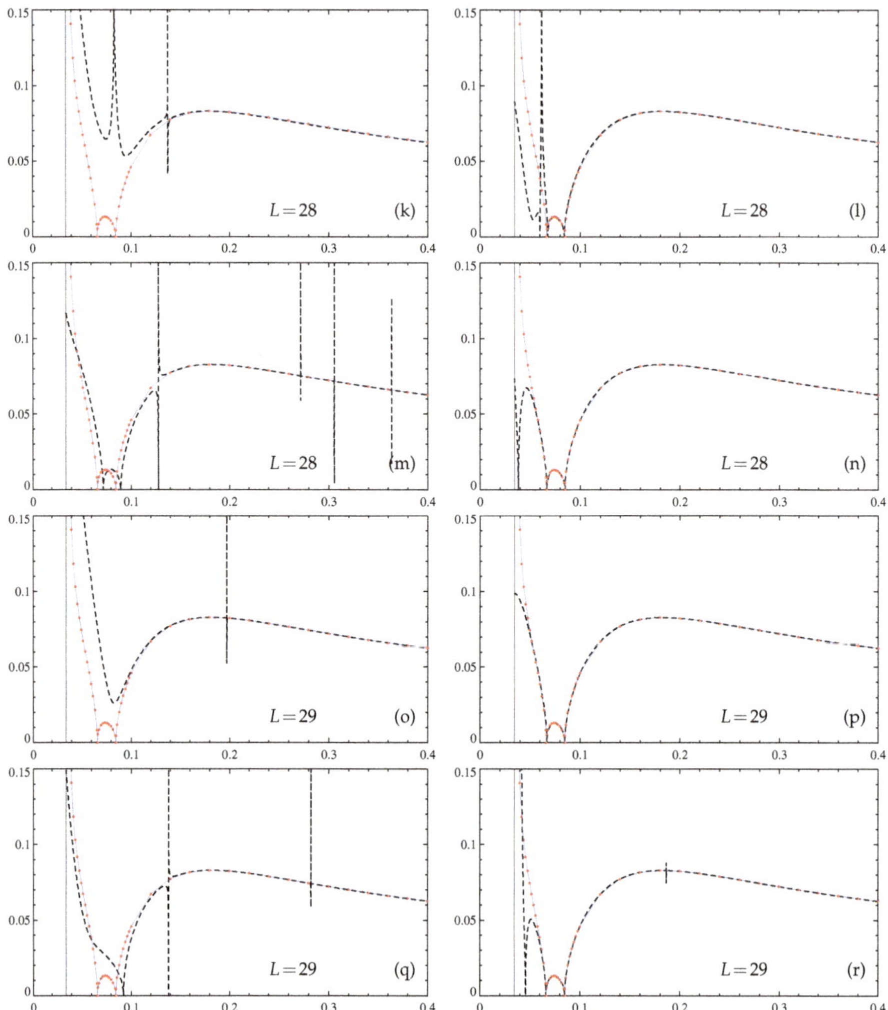

Figure 3. *Cont.*

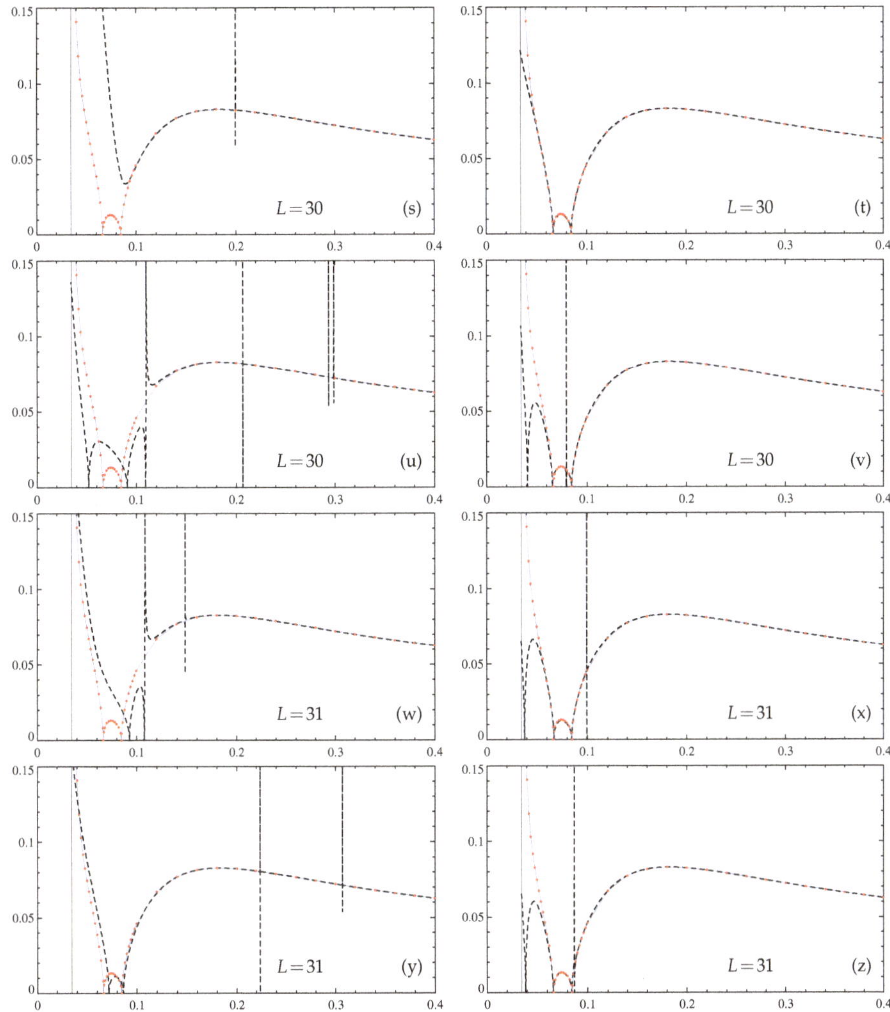

Figure 3. Approximate dependencies of the maximum slow-time growth rates γ_α (20) (vertical axis) of large-scale magnetic modes generated by the α-effect on molecular diffusivity η (horizontal axis). Padé approximants $[2L-1/2L]$ of α-effect tensor entries are constructed by the algorithm in [5] for the specified L. Resolution of 64^3 (**c,d,g,h,k,l,o,p,s,t,w,x**) and 512^3 (**a,b,e,f,i,j,m,n,q,r,u,v,y,z**) Fourier harmonics, computations with the double (real*8, left panels except (**a**)) and quadruple (real*16, right panels and (**a**)) precision. Thin solid line: the dependence revealed by computation of γ_α at individual η values (red solid circles) by spectral methods (resolution 64^3 Fourier harmonics), wide dashed line: Padé approximants.

4. Computation of the Magnetic Eddy Diffusivity Tensor

As discussed in Section 2, an important class are parity-invariant flows. This symmetry is compatible with the equations of fluid dynamics (the Navier–Stokes or Euler equations) provided the forcing has the same property. For such a flow, the domain of the operator of magnetic induction $\mathfrak{L}$ splits into the subspaces of parity-invariant fields (such that $\mathbf{f}(-\mathbf{x}) = -\mathbf{f}(\mathbf{x})$) and of parity-anti-invariant ones (such that $\mathbf{f}(-\mathbf{x}) = \mathbf{f}(\mathbf{x})$). Solutions to the auxiliary problems (14), $\mathbf{s}_k(\mathbf{x})$, are therefore parity-anti-invariant. This implies $\mathfrak{A} = 0$ (see (17)), i.e., no α-effect acts in such flows, and hence $\lambda_1 = 0$.

4.1. The Multiscale Formalism Revealing the Magnetic Eddy Diffusivity

By (15),

$$\mathbf{b}_1 = \sum_{k=1}^{3} \sum_{m=1}^{3} \frac{\partial \langle \mathbf{b}_0 \rangle_k}{\partial X_m} \mathbf{g}_{mk} \,,$$

where the small-scale zero-mean (non-solenoidal!) fields $\mathbf{g}_{mk}(\mathbf{x})$ solve nine auxiliary problems of type II:

$$\mathfrak{L}\mathbf{g}_{mk} = -2\eta \frac{\partial \mathbf{s}_k}{\partial x_m} - \mathbf{e}_m \times (\mathbf{v} \times \mathbf{s}_k) \,. \tag{28}$$

For parity-invariant $\mathbf{v}$, $\mathbf{g}_{mk}(\mathbf{x})$ are also parity-invariant. Moreover, $\mathbf{b}_n$ are parity-anti-invariant for all even n and parity-invariant for odd n [17] in the expansion (12), and no odd powers of ε enter the series (13) for the eigenvalue λ.

Averaging the third (order ε^2) equation in the hierarchy yields

$$\eta \nabla_\mathbf{X}^2 \langle \mathbf{b}_0 \rangle + \nabla_\mathbf{X} \times \sum_{k=1}^{3} \sum_{m=1}^{3} \mathfrak{D}_{mk} \frac{\partial \langle \mathbf{b}_0 \rangle_k}{\partial X_m} = \lambda_2 \langle \mathbf{b}_0 \rangle \,, \tag{29}$$

where

$$\mathfrak{D}_{mk} = \langle \mathbf{v} \times \mathbf{g}_{mk} \rangle \tag{30}$$

is the so-called tensor of magnetic eddy diffusivity correction. Again assuming that the mean field $\langle \mathbf{b}_0 \rangle$ is a Fourier harmonic (18), we find [20]

$$\lambda_{2\pm}(\mathbf{q}) = -\eta - \frac{1}{2} \sum_{j,l,n} (D_n^l - D_l^n) q_j \pm \sqrt{d} \,, \tag{31}$$

$$d = \sum_{j,l,n} \left(((^s\!D_n^l)^2 - {}^s\!D_l^l \, {}^s\!D_n^n) q_j^2 - 2q_j q_n (^s\!D_n^l \, {}^s\!D_j^l - {}^s\!D_l^l \, {}^s\!D_j^n) \right) \,, \tag{32}$$

where both sums are over even permutations of indices 1, 2, and 3 (i.e., (j, l, n) are combinations (1,2,3), (2,3,1), and (3,1,2)), and it is denoted

$$D_n^l = \sum_m \mathfrak{D}_{mn}^l q_m \,, \qquad {}^s\!D_n^l = (D_n^l + D_l^n)/2 \,. \tag{33}$$

The minimum

$$\eta_{\text{eddy}} \equiv \min_{|\mathbf{q}|=1} \left(-\text{Re}\, \lambda_{2\pm}(\mathbf{q}) \right) \tag{34}$$

is called the minimum magnetic eddy diffusivity. When it is negative, the interaction of the fluctuating small-scale velocity and magnetic field is capable of generating large-scale magnetic fields. For this reason, this quantity is of prime interest in large-scale dynamo theory.

Expression (30) can be transformed into

$$\mathfrak{D}_{mk}^l = \left\langle \mathbf{Z}_l \cdot \left(2\eta \frac{\partial \mathbf{s}_k}{\partial x_m} + \mathbf{e}_m \times (\mathbf{v} \times \mathbf{s}_k) \right) \right\rangle \,, \tag{35}$$

where $\mathbf{Z}_l$ are zero-mean solutions to three auxiliary problems for the adjoint operator:

$$\mathfrak{L}^* \mathbf{Z}_l = \mathbf{v} \times \mathbf{e}_l \,, \tag{36}$$

and $\mathfrak{L}^* : \mathbf{z} \mapsto \eta \nabla^2 \mathbf{z} - \{\mathbf{v} \times (\nabla \times \mathbf{z})\}$ is the operator adjoint to $\mathfrak{L}$ acting in the space of zero-mean space-periodic fields.

4.2. Padé Approximation

Relation (30) suggests constructing expansions of the solutions to the auxiliary problems in the inverse molecular diffusivity, (21), and

$$\mathbf{g}_{mk} = \sum_{n=1}^{\infty} \mathbf{g}_{mk}^{(n)} \eta^{-n}. \tag{37}$$

Dividing (28) by η yields

$$\nabla^2 \mathbf{g}_{mk} = -\eta^{-1} \nabla \times (\mathbf{v} \times \mathbf{g}_{mk}) - 2 \frac{\partial \mathbf{s}_k}{\partial x_m} - \eta^{-1} \mathbf{e}_m \times (\mathbf{v} \times \mathbf{s}_k),$$

whereby

$$\mathbf{g}_{mk}^{(1)} = -\nabla^{-2} \left(2 \frac{\partial \mathbf{s}_k^{(1)}}{\partial x_m} + \mathbf{e}_m \times (\mathbf{v} \times \mathbf{e}_k) \right), \tag{38}$$

$$\mathbf{g}_{mk}^{(n)} = -\nabla^{-2} \left(\nabla \times (\mathbf{v} \times \mathbf{g}_{mk}^{(n-1)}) + 2 \frac{\partial \mathbf{s}_k^{(n)}}{\partial x_m} + \mathbf{e}_m \times (\mathbf{v} \times \mathbf{s}_k^{(n-1)}) \right) \quad \text{for } n > 1. \tag{39}$$

Clearly, the flow being parity-invariant, all $\mathbf{s}_k^{(n)}$ are parity-anti-invariant and all $\mathbf{g}_{mk}^{(n)}$, parity-invariant. By (30) and (37),

$$\mathfrak{D}_{mk} = \sum_{n=1}^{\infty} \mathfrak{D}_{mk}^{(n)} \eta^{-n}, \qquad \mathfrak{D}_{mk}^{(n)} = \left\langle \mathbf{v} \times \mathbf{g}_{mk}^{(n)} \right\rangle. \tag{40}$$

Thus, algorithm I consists of the following steps:

- find the fields $\mathbf{s}_k^{(n)}$ employing (22);
- find the fields $\mathbf{g}_{mk}^{(n)}$ employing (38) and (39);
- calculate the coefficients $\mathfrak{D}_{mk}^{(n)}$ employing (40).

Expressions (40) reveal symmetry properties of coefficients in the eddy diffusivity tensor expansion (similar to those of the coefficients of the α-effect tensor expansion). Eddy diffusivity tensor for the reverse flow $-\mathbf{v}$ is related to that of the flow $\mathbf{v}$ by the relations $(\mathfrak{D}_{ml}^-)_k = -\mathfrak{D}_{mk}^l$ [19]. By (38) and (39), $\mathbf{g}_{mk}^{(n)} = (-1)^n (\mathbf{g}_{mk}^-)^{(n)}$. Thus, identities, analogous to those used in the case of the α-effect tensor, reveal that for each fixed m, the coefficients $(\mathfrak{D}_{mk}^l)^{(n)}$ in the series (40) are symmetric 3×3 matrices for even n, and antisymmetric ones for odd n. This implies that the symmetrized matrix $^s\mathbf{D}$ (33), determining the discriminant d (32), is expanded in even powers of $1/\eta$, and the antisymmetric one $\mathbf{D} - {}^s\mathbf{D}$, determining the common part of $\lambda_{2\pm}(\mathbf{q})$ (31), in odd powers.

It is simple to show that, like in the case of α-effect dynamos, for a given flow $\mathbf{v}$, the radius of convergence of all of the series (37) and (40) (regarded as functions of $1/\eta$) is generically equal to $1/\max_i |\mu_i|$, where μ_i are eigenvalues of the compact operator $\mathfrak{M}$ (23); convergence of the series is guaranteed for $\eta > \max_i |\mu_i|$.

An alternative form of the eddy diffusivity tensor can be exploited. Comparing (14) and (36), we find

$$\mathbf{e}_l + \nabla \times \mathbf{Z}_l = \mathbf{s}_l^- \quad \Rightarrow \quad \mathbf{Z}_l = \eta^{-1} \nabla^{-2} (\mathbf{v} \times \mathbf{s}_l^-), \tag{41}$$

where the superscript minus denotes objects pertinent to the reverse flow $-\mathbf{v}$ (see (25)). Using (41) to eliminate $\mathbf{s}_k$ in (35) yields [19]

$$\mathfrak{D}^l_{mk} = \eta \left\langle \mathbf{Z}_l \cdot \left(2\,\nabla \times \frac{\partial \mathbf{Z}_k^-}{\partial x_m} - \mathbf{e}_m \times \nabla^2 \mathbf{Z}_k^- \right) \right\rangle. \tag{42}$$

By (41), all $\mathbf{Z}_l$ are parity-invariant, and hence (36) is equivalent to

$$\nabla^2 \mathbf{Z}_l = \eta^{-1}(\mathbf{v} \times (\mathbf{e}_l + \nabla \times \mathbf{Z}_l)),$$

which implies a power series expansion

$$\mathbf{Z}_l = \sum_{n=1}^{\infty} \mathbf{Z}_l^{(n)} \eta^{-n}, \tag{43}$$

where the coefficients satisfy recurrence relations

$$\mathbf{Z}_l^{(1)} = \nabla^{-2}(\mathbf{v} \times \mathbf{e}_l), \qquad \mathbf{Z}_l^{(n)} = \nabla^{-2}(\mathbf{v} \times (\nabla \times \mathbf{Z}_l^{(n-1)})) \quad \text{for } n > 1. \tag{44}$$

By linearity of (44) in $\mathbf{v}$, the coefficients of such an expansion for the reverse flow are linked:

$$(\mathbf{Z}_l^-)^{(n)} = (-1)^n \mathbf{Z}_l^{(n)}.$$

This implies algorithm II for calculation of $(\mathfrak{D}^l_{mk})^{(n)}$ based on (42):
- find the fields $\mathbf{Z}_l^{(n)}$ applying (44);
- for $n \geq 1$ calculate

$$(\mathfrak{D}^l_{mk})^{(n)} = \sum_{j=1}^{n} (-1)^j \left\langle \mathbf{Z}_l^{(n+1-j)} \cdot \left(2\,\nabla \times \frac{\partial \mathbf{Z}_k^{(j)}}{\partial x_m} - \mathbf{e}_m \times \nabla^2 \mathbf{Z}_k^{(j)} \right) \right\rangle. \tag{45}$$

Using (41), it is straightforward, albeit tedious, to transform (42) into

$$\mathfrak{D}^l_{mk} = \eta \left\langle \{\mathbf{s}_l^-\} \times \{\mathbf{s}_k\} - \{\mathbf{s}_k\} \nabla \cdot \mathbf{Z}_l + \{\mathbf{s}_l^-\} \nabla \cdot \mathbf{Z}_k^- \right\rangle_m. \tag{46}$$

Relations (41) imply

$$\nabla \times \mathbf{Z}_l^{(n)} = (-1)^n \mathbf{s}_l^{(n)}, \qquad \mathbf{Z}_l^{(n)} = (-1)^{n-1} \nabla^{-2}(\mathbf{v} \times \mathbf{s}_l^{(n-1)}) \quad \text{for } n \geq 1. \tag{47}$$

Thus, the coefficients in the expansion (40) have the entries

$$(\mathfrak{D}^l_{mk})^{(n)} = \sum_{j=1}^{n} \left\langle (-1)^j \mathbf{s}_l^{(j)} \times \mathbf{s}_k^{(n+1-j)} - \mathbf{s}_k^{(n+1-j)} \nabla \cdot \mathbf{Z}_l^{(j)} - (-1)^n \mathbf{s}_l^{(n+1-j)} \nabla \cdot \mathbf{Z}_k^{(j)} \right\rangle_m. \tag{48}$$

Algorithm III consists of the following steps:

- determine coefficients in the expansion of the neutral magnetic modes $\mathbf{s}_k$ applying recurrence relations (22);
- in the course of these calculations, determine coefficients in the expansion of $\nabla \cdot \mathbf{Z}_l$ using (47);
- calculate the coefficients $(\mathfrak{D}^l_{mk})^{(n)}$ applying (48).

If the flow $\mathbf{v}$ involves a small number of Fourier harmonics, it is unclear a priori which of the three algorithms is more efficient. Algorithm I involves the computation of two sets of coefficients, for $\mathbf{s}_k$ and $\mathbf{g}_{mk}$, algorithm II, only of the set of coefficients for $\mathbf{Z}_l$, and algorithm III is an intermediate case involving two sets of coefficients, for $\mathbf{s}_k$ and $\nabla \cdot \mathbf{Z}_l$. However, the calculation of the n-term sums (45) and (48) in algorithms II and III, respectively, is computer-intensive.

4.3. Numerical Results

For numerical experimentation, we have applied two types of flows: the so-called cosine flows introduced in [20], and their curls, considered in [21]. They are of interest in that the latter have a point-wise zero vorticity (kinematic) helicity, and the former have a point-wise zero velocity helicity, but nevertheless they are capable of both small- and large-scale magnetic field generation (see ibid.). Involving a small number of trigonometric functions, they are particularly useful for calculating the Taylor series coefficients (40).

The cosine flows are defined as follows:

$$
\begin{aligned}
v_1 &= \beta n (b_1 \sin(\mathbf{a} \cdot \mathbf{x}) + a_1 \sin(\mathbf{b} \cdot \mathbf{x})) \cos nx_3, \\
v_2 &= \beta n (b_2 \sin(\mathbf{a} \cdot \mathbf{x}) + a_2 \sin(\mathbf{b} \cdot \mathbf{x})) \cos nx_3, \\
v_3 &= -\beta (\mathbf{a} \cdot \mathbf{b})(\cos(\mathbf{a} \cdot \mathbf{x}) + \cos(\mathbf{b} \cdot \mathbf{x})) \sin nx_3.
\end{aligned}
\tag{49}
$$

Here, $\mathbf{a} = (a_1, a_2, 0)$ and $\mathbf{b} = (b_1, b_2, 0)$ are constant horizontal vectors, and

$$
\beta = 2(n^2(|\mathbf{a}|^2 + |\mathbf{b}|^2) + 2(\mathbf{a} \cdot \mathbf{b})^2)^{-1/2},
$$

so that the r.m.s. flow velocity is unity.

Because of many symmetries of the cosine flows, all entries of the eddy diffusivity tensor $\mathfrak{D}$ vanish, except for five pairs (see [20]):

$$
\mathfrak{D}^2_{31} = -\mathfrak{D}^1_{32}, \quad \mathfrak{D}^3_{12} = -\mathfrak{D}^2_{13}, \quad \mathfrak{D}^1_{23} = -\mathfrak{D}^3_{21}, \quad \mathfrak{D}^3_{22} = -\mathfrak{D}^2_{23}, \quad \mathfrak{D}^1_{13} = -\mathfrak{D}^3_{11}.
$$

Consequently, the minimum eddy diffusivity (34) takes a simple form:

$$
\eta_{\text{eddy}} = \eta - \max \left(\mathfrak{D}^2_{31}, \frac{1}{2} \left(\mathfrak{D}^3_{12} + \mathfrak{D}^1_{23} + \sqrt{(\mathfrak{D}^3_{12} - \mathfrak{D}^1_{23})^2 + (\mathfrak{D}^3_{22} + \mathfrak{D}^1_{13})^2} \right) \right).
\tag{50}
$$

We have considered the particular sample flow for

$$
\mathbf{a} = (1, 0, 0), \qquad \mathbf{b} = (1, 1, 0), \qquad n = 1
\tag{51}
$$

and used Mathematica again to implement algorithm I: we have calculated exactly the coefficients of the expansions (21) and (37) of solutions to the auxiliary problems (14) and (28) using the recurrence relations (22) and (38) and (39), and of the coefficients $\mathfrak{D}^{(n)}_{mk}$ of the series (40) up to order η^{-49}. The precise coefficients $\mathbf{s}_k^{(49)}$ and $\mathbf{g}_{mk}^{(49)}$ require about 2 Gbytes of memory for storage (in the ASCII form). For the flow (49), the coefficients $\mathfrak{D}^{(n)}_{mk}$ turn out to be rational; in the ASCII form, the vectors $\mathfrak{D}^{(49)}_{mk}$ occupy 10 to 20 Kbytes of memory.

Figure 4 shows the sequence of the ratios $|(\mathfrak{D}^l_{mk})^{(2n-1)} / (\mathfrak{D}^l_{mk})^{(2n+1)}|^{1/2}$ for five independent entries. The limit of this sequence for $n \to \infty$ is equal to the radius of convergence of the series (40) (regarded as a function of $1/\eta$) for the respective entry (note that due to the antisymmetry in l and k, the entries involve only odd powers of $1/\eta$). The figure demonstrates that the series for the five entries have the same radius of convergence and converge for $\eta \gtrsim 0.5$.

Because precise Mathematica calculations require considerable computer resources, we have not considered high-order Padé approximants. The amount of calculations reduces if meromorphic functions

$$
q_1 = \mathfrak{D}^2_{31}, \qquad q_2 = (\mathfrak{D}^3_{12} + \mathfrak{D}^1_{23})/2, \qquad q_3 = ((\mathfrak{D}^3_{12} - \mathfrak{D}^1_{23})^2 + (\mathfrak{D}^3_{22} + \mathfrak{D}^1_{13})^2)/4
\tag{52}
$$

are approximated only, in terms of which (see (50))

$$\eta_{\text{eddy}} = \eta - \max(q_1, q_2 + \sqrt{q_3}).$$

(53)

The poor quality of the resultant approximation (see Figure 5a) is due to the presence of two Froissart doublets in the approximants of q_2 and q_3 (Figure 5b). Gaps are present in the plot in Figure 5a, where the approximant of q_3 becomes negative due to its singular behavior, and thus the square root in (53) cannot be extracted. Upon factoring the doublets out (which is simple in Mathematica) in the two plagued approximants, the quality of the approximation becomes very similar to that obtained by using [18/18] approximants of q_i (see Figure 5c). Increasing the orders to [24/24] does not significantly improve the approximated η_{eddy} (see Figure 5d). A better approximation is obtained if the elements of the eddy diffusivity tensor $\mathfrak{D}$ are Padé-approximated individually (see Figure 6); this gives reasonably accurate values of η_{eddy} for $\eta \gtrsim 0.02$, which is roughly 25 times larger than the minimum η, for which the power series in $1/\eta$ for the fields $\mathbf{s}_k$ and $\mathbf{g}_{mk}$, as well as the elements of the tensor $\mathfrak{D}$, are convergent.

Upon shifting by a quarter of the period in the vertical coordinate x_3, the curl of (49) takes the form

$$v_1 = \beta \left(((\mathbf{a} \cdot \mathbf{b})a_2 + n^2 b_2) \sin(\mathbf{a} \cdot \mathbf{x}) + ((\mathbf{a} \cdot \mathbf{b})b_2 + n^2 a_2) \sin(\mathbf{b} \cdot \mathbf{x}) \right) \cos nx_3,$$

$$v_2 = -\beta \left(((\mathbf{a} \cdot \mathbf{b})a_1 + n^2 b_1) \sin(\mathbf{a} \cdot \mathbf{x}) + ((\mathbf{a} \cdot \mathbf{b})b_1 + n^2 a_1) \sin(\mathbf{b} \cdot \mathbf{x}) \right) \cos nx_3,$$

(54)

$$v_3 = \beta n (a_2 b_1 - a_1 b_2)(\cos(\mathbf{a} \cdot \mathbf{x}) - \cos(\mathbf{b} \cdot \mathbf{x})) \sin nx_3,$$

where we now assume the normalizing factor

$$\beta = 2 \left((n^4 + (\mathbf{a} \cdot \mathbf{b})^2)(|\mathbf{a}|^2 + |\mathbf{b}|^2) + 2n^2((\mathbf{a} \cdot \mathbf{b})^2 + |\mathbf{a}|^2|\mathbf{b}|^2) \right)^{-1/2},$$

for which the r.m.s. flow velocity is again 1. Since this flow possesses all the symmetries of (49), the expression (50) for the minimum eddy diffusivity still applies.

Following algorithm II, for a sample flow (54) for

$$\mathbf{a} = (0,1,0), \qquad \mathbf{b} = (2,2,0), \qquad n = 1,$$

(55)

we have calculated with Mathematica 49 coefficients $\mathfrak{D}_{mk}^{(n)}$ of the series (40) up to order η^{-49}. The graph of the ratios $|(\mathfrak{D}_{mk}^l)^{(2n-1)}/(\mathfrak{D}_{mk}^l)^{(2n+1)}|^{1/2}$ for the five independent entries (see Figure 7a) of the eddy diffusivity tensor shows that the series (40) converge for $\eta \gtrsim 2.2$. The highest-order (for this set of coefficients) [25/24] approximants of the entries are free of Froissart doublets (see Figure 7b). They yield a satisfactory approximation of dependence on η of the minimum magnetic eddy diffusivity for $\eta \gtrsim 0.03$, which is roughly 70 times smaller than the bound obtained for convergence of the Taylor series (40) for $\mathfrak{D}_{mk}^l$ (see Figure 7c,d). However, their fidelity is insufficient to reproduce the singularity of the minimum eddy diffusivity observed in Figure 7c,d.

The symmetries of the flow (54), (55) imply that the neutral modes $\mathbf{s}_k$ reside in invariant subspaces of the magnetic induction operator, which can be categorized in terms of the Fourier harmonics $\mathbf{b}_\mathbf{n} e^{i\mathbf{n}\cdot\mathbf{x}}$, comprising the Fourier series for $\mathbf{s}_k$ (by virtue of the mode periodicity, the wave vectors $\mathbf{n}$ have integer components). Only the harmonics that have the following properties enter the Fourier series for $\mathbf{s}_k$:

- $\mathbf{b}_\mathbf{n} = \mathbf{b}_{-\mathbf{n}}$ are real;
- the numbers n_1 and $n_1/2 + n_2 + n_3$ are even;
- $\mathbf{s}_1$ and $\mathbf{s}_2$ are symmetric in x_3 (i.e., $b_\mathbf{n}^1 = b_{\mathbf{n}*}^1$, $b_\mathbf{n}^2 = b_{\mathbf{n}*}^2$, $b_\mathbf{n}^3 = -b_{\mathbf{n}*}^3$), and $\mathbf{s}_3$ is antisymmetric in x_3 (i.e., $b_\mathbf{n}^1 = -b_{\mathbf{n}*}^1$, $b_\mathbf{n}^2 = -b_{\mathbf{n}*}^2$, $b_\mathbf{n}^3 = b_{\mathbf{n}*}^3$), where $\mathbf{n}^* = (n_1, n_2, -n_3)$.

We have checked that for $\eta > \eta_{\text{cr}} = 0.00420516$, small-scale modes from the subspace, where $\mathbf{s}_1$ and $\mathbf{s}_2$ are located, are not generated, but at $\eta = \eta_{\text{cr}}$ the small-scale generation starts in the subspace, where $\mathbf{s}_3$ resides. It is known [20,21,23] that the point of the onset of the small-scale generation is typically associated with a singularity of the α-effect or eddy diffusivity tensors; this is the case for the flow under consideration. In Figure 7d, we observe that the least-squares fit by a hyperbola through 20 computed values of minimum eddy diffusivity at equispaced points in the interval $0.0045 \le \eta \le 0.014$ is very accurate (only 19 points out of the 20 are shown; for the smallest $\eta = 0.0045$, eddy diffusivity –20.217501, also well approximated by the hyperbola, is out of the vertical range of Figure 7d). The hyperbolic fit yields the location of the singularity at $\eta = 0.0041988$ (the vertical asymptote is shown by a dashed line in Figure 7d), which is very close to the point of the onset of the small-scale generation $\eta_{\mathrm{cr}} = 0.00420516$ computed by spectral methods; the hyperbola through the three smallest η from this interval yields a closer value 0.00420233.

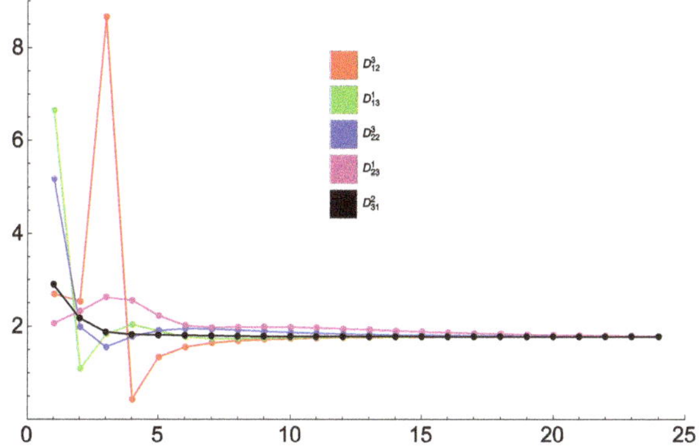

Figure 4. The ratios $|(\mathfrak{D}_{mk}^{l})^{(2n-1)} / (\mathfrak{D}_{mk}^{l})^{(2n+1)}|^{1/2}$ (vertical axis) versus $n > 0$ (horizontal axis) for five independent entries of the eddy diffusivity tensor for the cosine flow (49), (51).

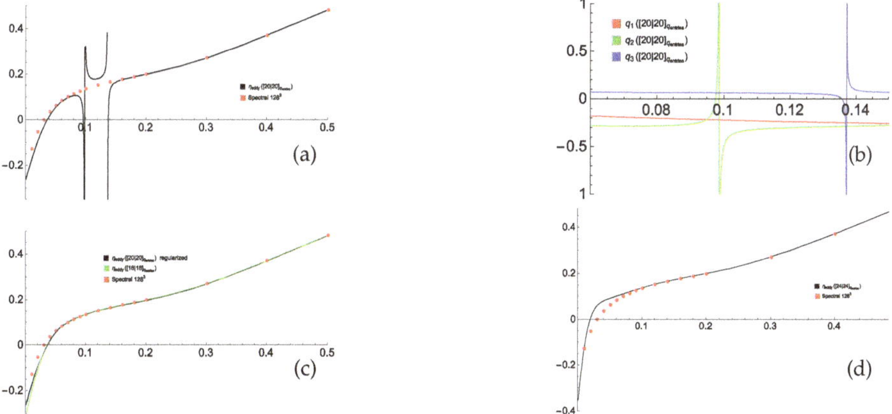

Figure 5. Minimum eddy diffusivity (53) (vertical axis) for the sample flow (49), (51) computed using Padé approximants of the quantities q_i (52) of orders [20/20] (**a**), [18/18] (green line) and regularized [20/20] (black line) (**c**), and [24/24] (**d**). Behavior of the [20/20] Padé approximants of q_i (vertical axis) near the points, where Froissart doublets are located (**b**). Horizontal axis: magnetic molecular diffusivity η. Red dots: minimum eddy diffusivity computed by spectral methods (resolution of 128^3 Fourier harmonics) individually for the respective η values.

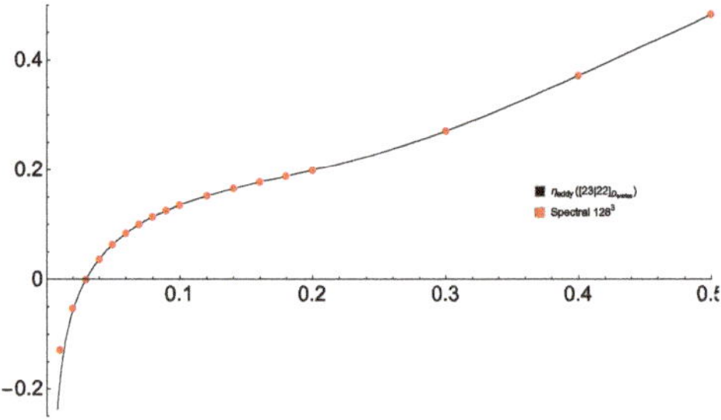

Figure 6. Minimum eddy diffusivity (53) (vertical axis) for the sample flow (49), (51) computed by Padé approximation of the individual entries of the eddy diffusivity tensor $\mathfrak{D}$ of orders [23/22] for varying magnetic molecular diffusivity η (horizontal axis).

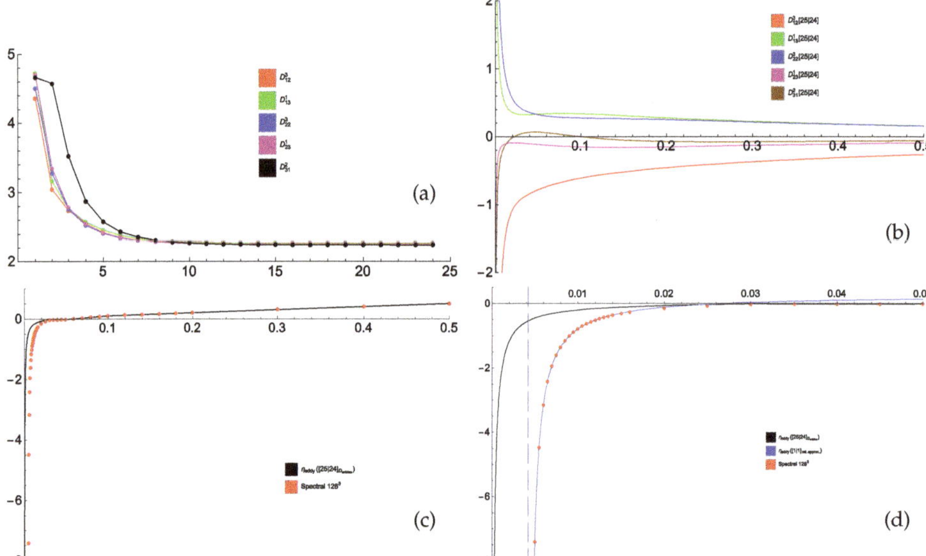

Figure 7. The ratios $\left|(\mathfrak{D}^l_{mk})^{(2n-1)}/(\mathfrak{D}^l_{mk})^{(2n+1)}\right|^{1/2}$ (vertical axis) versus $n > 0$ (horizontal axis) for five independent entries of the eddy diffusivity tensor for the flow (54), (55) (**a**). The [25/24] Padé approximants for the five entries $\mathfrak{D}^l_{mk}$ (vertical axis) versus η (horizontal axis) (**b**) and minimum eddy diffusivity (53) (vertical axis) computed using these approximants for varying η (horizontal axis) (**c**). Red dots: minimum eddy diffusivity computed by spectral methods (resolution of 128^3 Fourier harmonics) individually for the respective η values. Zoomed-in view of plot (**c**) for small η (black line) and a hyperbolic fit (blue line) through the 20 spectral eddy diffusivity values for 20 η points in the interval $0.0045 \leq \eta \leq 0.014$ step 0.0005 (**d**). The dashed line shows the vertical asymptote of the minimum eddy diffusivity at the onset of a small-scale dynamo in the symmetry subspace, where the neutral mode $\mathbf{s}_3$ resides.

5. Conclusions

We have tested Padé approximants of the α-effect and eddy diffusivity tensors, responsible for generation of large-scale fields, as functions of the respective molecular diffusivity: the viscosity ν when hydrodynamic perturbations are studied, and the magnetic diffusivity η when the kinematic dynamo problem is under scrutiny. We have tried different computational tools: Fortran codes relying on floating point arithmetics and Mathematica for symbolic and arbitrary precision calculations.

A relatively high (several dozens) order of Padé approximants is needed to obtain a reasonable accuracy of approximation of the tensor entries. For this, a high precision of computations (in particular, the quadruple precision in Fortran) has proved indispensable. For our sample flows, the Padé-approximated tensors yield large-scale magnetic field growth rates to satisfactory accuracy for η, several dozen times smaller than those for which power series in the inverse molecular diffusivity converge, for both large-scale generating mechanisms (the α-effect and negative magnetic eddy diffusivity).

The application of these techniques in computational fluid dynamics and magnetohydrodynamics seems natural for estimating transport coefficients quantifying the influence of small scales on the evolution of large-scale fields in the spirit of large eddy simulation methods. Our findings, while promising, suggest that to achieve this goal, additional algorithms are needed for the determination

- of Froissart doublets in approximants of tensor entries and their elimination (the approach of [36] may prove useful for monitoring the absence of the doublets);
- of the interval in molecular diffusivity, where the approximation is sufficiently accurate;
- of the realistic orders of a Padé approximant, for which the length of such intervals is close to the maximum.

It is relatively easy to perform these tasks manually by trial-and-error methods—the difficulty lies in performing them automatically.

Author Contributions: Conceptualization, S.M.A.G., R.C. and V.Z.; Investigation, S.M.A.G., R.C. and V.Z.; Software, S.M.A.G., R.C. and V.Z.; Writing—S.M.A.G., R.C. and V.Z.; Writing—review and editing, S.M.A.G., R.C. and V.Z.

Funding: This work was supported by CMUP (Centro de Matemática da Universidade do Porto, UID/MAT/00144/2019) (SG+VZ) and SYSTEC (Centro de Investigação em Sistemas e Tecnologias, POCI-01-0145-FEDER-006933/SYSTEC) (RC), which are funded by FCT with national (MCTES) and European structural funds through the FEDER (Fundo Europeu de Desenvolvimento Regional/European Regional Development Fund) program under the partnership agreement PT2020 and projects STRIDE [NORTE-01-0145-FEDER-000033] funded by FEDER–NORTE 2020 (SG+VZ+RC) and MAGIC [POCI-01-0145-FEDER-032485] funded by FEDER via COMPETE 2020–POCI (SG+VZ).

Conflicts of Interest: The authors declare no conflict of interest.

References

1. Forsythe, G.E.; Malcolm, M.A.; Moler, C.B. *Computer Methods for Mathematical Computations*; Prentice-Hall: Englewood Cliffs, UK, 1977.
2. Baker, G.A., Jr. *Quantitative Theory of Critical Phenomena*; Academic Press: Cambridge, MA, USA, 1990.
3. Baker, G.A., Jr.; Graves-Morris, P. *Padé Approximants*; Cambridge University Press: New York, NY, USA, 1996.
4. Gilewicz, J. *Approximants de Padé*; Lecture Notes in Mathematics; Springer: Berlin, Germany, 1978; Volume 667.
5. Press, W.H.; Teukolsky, S.A.; Vetterling, W.T.; Flannery, B.P. *Numerical Recipes in Fortran; The Art of Scientific Computing*, 2nd ed.; Cambridge University Press: New York, NY, USA, 1993.
6. Starr, V.P. *Physics of Negative Viscosity Phenomena*; McGraw-Hill: New York, NY, USA, 1968.
7. Moffatt, H.K. *Magnetic Field Generation in Electrically Conducting Fluids*; Cambridge University Press: New York, NY, USA, 1978.
8. Parker, E.N. Hydrodynamic dynamo models. *Astrophys. J.* **1955**, *122*, 293–314. [CrossRef]
9. Krause, F.; Rädler, K.-H. *Mean-Field Magnetohydrodynamics and Dynamo Theory*; Academic-Verlag: Berlin, Germany, 1980.
10. Steenbeck, M.; Krause, F.; Rädler, K.-H. A calculation of the mean electromotive force in an electrically conducting fluid in turbulent motion, under the influence of Coriolis forces. *Z. Naturforsch.* **1966**, *21*, 369–376.
11. Dubrulle, B.; Frisch, U. Eddy viscosity of parity-invariant flow. *Phys. Rev. A* **1991**, *43*, 5355–5364. [CrossRef] [PubMed]
12. Lanotte, A.; Noullez, A.; Vergassola, M.; Wirth, A. Large-scale dynamo by negative magnetic eddy diffusivities. *Geophys. Astrophys. Fluid Dyn.* **1999**, *91*, 131–146. [CrossRef]

13. Roberts, G.O. Spatially periodic dynamos. *Philos. Trans. R. Soc. Lond. A* **1970**, *266*, 535–558. [CrossRef]
14. Roberts, G.O. Dynamo action of fluid motions with two-dimensional periodicity. *Philos. Trans. R. Soc. Lond. A* **1972**, *271*, 411–454. [CrossRef]
15. Vishik, M.M. Periodic dynamo. In *Mathematical Methods in Seismology and Geodynamics*; (Computational Seismology, issue 19); Keilis-Borok, V.I., Levshin, A.L., Eds.; Nauka: Moscow, Russia, 1986; English Translation: *Computational Seismology*; Allerton Press: New York, NY, USA, 1987; Volume 19, pp. 176–209.
16. Vishik, M.M. Periodic dynamo. II. In *Numerical Modelling and Analysis of Geophysical Processes*; (Computational Seismology, issue 20); Keilis-Borok, V.I., Levshin, A.L., Eds.; Nauka: Moscow, Russia, 1987; pp. 12–22. English Translation: *Computational Seismology*; Allerton Press: New York, NY, USA, 1988; Volume 20, pp. 10–22.
17. Zheligovsky, V.A. *Large-Scale Perturbations of Magnetohydrodynamic Regimes: Linear and Weakly Nonlinear Stability Theory*; Lecture Notes Phys.; Springer: Heidelberg, Germany, 2011; Volume 829.
18. Chertovskih, R.; Zheligovsky, V. Large-scale weakly nonlinear perturbations of convective magnetic dynamos in a rotating layer. *Physica D* **2015**, *313*, 99–116. [CrossRef]
19. Andrievsky, A.; Brandenburg, A.; Noullez, A.; Zheligovsky, V. Negative magnetic eddy diffusivities from test-field method and multiscale stability theory. *Astrophys. J.* **2015**, *811*, 135. [CrossRef]
20. Rasskazov, A.; Chertovskih, R.; Zheligovsky, V. Magnetic field generation by pointwise zero-helicity three-dimensional steady flow of incompressible electrically conducting fluid. *Phys. Rev. E* **2018**, *97*, 043201. [CrossRef] [PubMed]
21. Andrievsky, A.; Chertovskih, R.; Zheligovsky, V. Pointwise vanishing velocity helicity of a flow does not preclude magnetic field generation. *Phys. Rev. E* **2019**, *99*, 033204. [CrossRef] [PubMed]
22. Andrievsky, A.; Chertovskih, R.; Zheligovsky, V. Negative magnetic eddy diffusivity due to oscillogenic α-effect. *Physica D* **2019**, in submitted.
23. Zheligovsky, V.A.; Podvigina, O.M.; Frisch, U. Dynamo effect in parity-invariant flow with large and moderate separation of scales. *Geophys. Astrophys. Fluid Dyn.* **2001**, *95*, 227–268. [CrossRef]
24. Frisch, U. *Turbulence: The Legacy of A.N. Kolmogorov*; Cambridge University Press: New York, NY, USA, 1995.
25. Lindborg, E. Can the atmospheric kinetic energy spectrum be explained by two-dimensional turbulence? *J. Fluid Mech.* **1999**, *388*, 259–288. [CrossRef]
26. Taylor, G.I. Eddy motion in the atmosphere. *Philos. Trans. R. Soc. A* **1915**, *215*, 1–26. [CrossRef]
27. Frisch, U.; She, Z.S.; Sulem, P.L. Large-scale flow driven by the anisotropic anisotropic kinetic alpha effect. *Physica D* **1987**, *28*, 382–392. [CrossRef]
28. Landau, L.D.; Lifshitz, E.M. *Theory of Elasticity*, 3rd ed.; Elsevier: New York, NY, USA, 2007.
29. Cummins, P.F.; Holloway, G. Reynolds stress and eddy viscosity in direct numerical simulations of sheared two-dimensional turbulence. *J. Fluid Mech.* **657**, 394–412, 2010.
30. Gama, S.; Vergassola, M.; Frisch, U. Negative eddy viscosity in isotropically forced two-dimensional flow: Linear and nonlinear dynamics. *J. Fluid Mech.* **1994**, *260*, 95–126. [CrossRef]
31. Vergassola, M.; Gama, S.; Frisch, U. Proving the existence of negative isotropic eddy viscosity. In *Theory of Solar and Planetary Dynamos*; Proctor, M.R.E., Matthews, P.C., Rucklidge, A.M., Eds.; Cambridge University Press: New York, NY, USA, 1993; pp. 321–327.
32. Hastings, C.; Mischo, K.; Morrison, M. *Hands-on Start to Wolfram Mathematica and Programming with the Wolfram Language*; Wolfram Media, Inc.: Champaign, IL, USA, 2015.
33. Podvigina, O.; Zheligovsky, V.; Frisch, U. The Cauchy–Lagrangian method for numerical analysis of Euler flow. *J. Comp. Phys.* **2016**, *306*, 320–342. [CrossRef]
34. Zheligovsky, V. Numerical solution of the kinematic dynamo problem for Beltrami flows in a sphere. *J. Sci. Comp.* **1993**, *8*, 41–68. [CrossRef]

35. Gonnet, P.; Güttel, S.; Trefethen, L.N. Robust Padé approximation via SVD. *SIAM Rev.* **2013**, *55*, 101–117. [CrossRef]
36. Beckermann, B.; Labahn, G.; Matos, A.C. On rational functions without Froissart doublets. *Numer. Math.* **2018**, *138*, 615–633. [CrossRef]

Article

Direct Numerical Simulation of a Warm Cloud Top Model Interface: Impact of the Transient Mixing on Different Droplet Population

Taraprasad Bhowmick [1],* and Michele Iovieno [2]

[1]   Department of Applied Science and Technology (DISAT), Politecnico di Torino, Corso Duca degli Abruzzi 24, 10129 Torino, Italy

[2]   Department of Mechanical and Aerospace Engineering (DIMEAS), Politecnico di Torino, Corso Duca degli Abruzzi 24, 10129 Torino, Italy

*   Correspondence: taraprasad.bhowmick@polito.it

Received: 27 May 2019; Accepted: 25 June 2019; Published: 1 August 2019

Abstract: Turbulent mixing through atmospheric cloud and clear air interface plays an important role in the life of a cloud. Entrainment and detrainment of clear air and cloudy volume result in mixing across the interface, which broadens the cloud droplet spectrum. In this study, we simulate the transient evolution of a turbulent cloud top interface with three initial mono-disperse cloud droplet population, using a pseudo-spectral Direct Numerical Simulation (DNS) along with Lagrangian droplet equations, including collision and coalescence. Transient evolution of in-cloud turbulent kinetic energy (TKE), density of water vapour and temperature is carried out as an initial value problem exhibiting transient decay. Mixing in between the clear air and cloudy volume produced turbulent fluctuations in the density of water vapour and temperature, resulting in supersaturation fluctuations. Small scale turbulence, local supersaturation conditions and gravitational forces have different weights on the droplet population depending on their sizes. Larger droplet populations, with initial 25 and 18 µm radii, show significant growth by droplet-droplet collision and a higher rate of gravitational sedimentation. However, the smaller droplets, with an initial 6 µm radius, did not show any collision but a large size distribution broadening due to differential condensation/evaporation induced by the mixing, without being influenced by gravity significantly.

Keywords: droplets; cloud micro-physics; transient evolution; warm cloud top mixing; direct numerical simulation

---

1. Introduction

Atmospheric clouds play an important role in the evolution of the weather, both locally and globally, by influencing incoming and outgoing radiations, amount of precipitations and transport of water vapour, heat and other advected scalars like pollutant concentrations. Low altitude stratocumulus and cumulus clouds are formed on top of the planetary boundary layer. Various physical processes inside these clouds, which span over a wide range of temporal and spatial scales, need to be carefully modelled to improve weather predictions and climate modelling (see Pruppacher and Klett (2010) [1], Stull (1988) [2] and Devenish et al. (2012) [3]). Therefore, many studies have been devoted to analyzing each single aspects of the clouds.

The presence of a wide range of scales in an atmospheric cloud adds a layer of complexity to the micro-physical processes, which are linked to the droplet evolution. Droplet condensation/evaporation, dispersion, preferential concentration and transport processes are the key phenomena of the the micro-physical processes [3,4]. Many of these processes occur at the smallest scales of the flow [5], therefore, such processes need to be parameterized when the whole cloud or large scale modelling is considered [6,7]. Particle resolving direct numerical simulations (DNS), which aim to solve all the dynamically significant scales of a turbulent environment, have recently become a tool to investigate some of the cloud micro-physical processes using idealized atmospheric flow configurations.

In these simulations, it is not only necessary to solve all the turbulent scales down to the diffusive scales but it is also necessary to track the motion and the growth of every single cloud droplet in the fluid flow. Therefore, due to computational limitations, only a small portion of a cloud could be considered in a DNS study. The basic model used in all such works was introduced by Vaillancourt et al. (2001) [8]. In this model the Navier-Stokes (NS) equations with the Boussinesq approximation are solved for the air phase on an Eulerian grid, where the suspended cloud droplets are considered as inertial variable-mass points. The droplet model takes into account the variation of the droplet size due to condensation or evaporation according to the local supersaturation condition [1]. Most of such works focused on the simulation of a small portion of the well mixed cloud-core and introduced an external forcing to reproduce the inflow of energy due to the larger-scale cloud motions [9–11]. Other works have extended this methodology to investigate the effects of non-uniformity in the underlying turbulent flow and its consequences on cloud droplet micro-physical evolution. Such non-uniformity was introduced in the form of a turbulent kinetic energy (TKE) gradient and in the supersaturation distribution [12]. Götzfried et al. (2017) [12] and Kumar et al. (2014, 2017) [13,14] considered the evolution of a tiny cloud layer to analyze the cloud edge affects on the cloud droplet population, whereas Andrejczuk et al. (2004, 2006, 2009) [15–17] considered decaying moist turbulence with worm-like structures of supersaturated and subsaturated regions. Gao et al. (2018) [18] investigated the differences between cloud micro-physical evolution for various initial configuration of cloud and clear air layers, both in the presence and absence of forcing.

Entrainment and mixing in between supersaturated and subsaturated air parcels have long been considered important processes which shape the size distribution of the cloud droplets and therefore contribute to enhancing the collision rate [3]. Therefore, the understanding of cloud droplet evolution inside the mixing zone could contribute to bridge the gap between the condensational growth (effective for smaller droplets, below 10 μm radius) and collisional growth (effective for larger droplets, above 40 μm radius) [5,19]. In the mixing zone, two mechanisms contribute to the droplet micro-physics: (a) the presence of large fluctuations in the supersaturation, which broadens the size distribution of the cloud droplets by condensation/evaporation and (b) large accelerations in some localized regions of the flow [4,20], which can cause occasional collisions. Different scenarios can occur, which range between the two limiting mixing situations: (a) homogeneous mixing, where entrained air and cloudy air quickly mix at the small scale, so that the entire droplet population evaporate/condensate at the same time and (b) inhomogeneous mixing, where only a portion of the volume remains subsaturated/supersaturated and therefore, part of the droplet population quickly evaporate/condensate, whereas, the other parts of the population remains almost unaffected [21].

In this work, we studied the transient evolution of cloud water droplet populations inside a simplified top interface model of a warm cloud. The objective of this work was to investigate the differences in the micro-physical transient evolution of various cloud water droplet populations with sizes that approach the lower bound of the *size gap* [15–40 μm radius] [5] and sizes below the *size gap* and their thermal feedback on the surrounding air. DNS using the Eulerian-Lagrangian model along with a collision detection algorithm (mostly neglected for this kind of study) was used for this work. Various thermodynamic processes, such as supersaturation, buoyancy, phase change of water vapour to liquid water and corresponding latent heat production, are interlinked by one-way coupling in between the cloud droplets and the surrounding fluid velocity and two-way coupling of the cloud droplets with the air temperature and the water vapour density. We investigated the transport inside a cloud interface characterized by a strong kinetic energy gradient (see also [12]). This configuration is known to generate a strong intermittent transient mixing layer [22–24] due to the entrainment/detrainment of fluid from/to the external regions [24,25]. This induces non-trivial complex behaviour in the droplet velocity distribution due to the interplay between larger scales and gravity, with little dependence on the Reynolds number [24]. Therefore, even if the Reynolds number which can be achieved in the direct numerical simulations is much smaller and not comparable with the real life cloud Reynolds number in the order of 10^4–10^5, we think such simulations can still provide information that can be useful to understand some of the real cloud phenomena. The present simulations undergo a transient decay in TKE [12,15,25]. The methodology and the simulation set-up is presented in Section 2. Results presented in Section 3 include statistics and visualizations on the transient evolution of the fluid flow and the droplet populations and a discussion on the role of condensation/evaporation, sedimentation and collision on the evolution of the different droplet populations in the mixing layer separating the lower cloudy region from the upper clear air region.

2. Numerical Methodology and Simulation Setup

In this study, the interactions in between atmospheric low altitude warm cloud top and the above-lying clear air regions and its impact on cloud water droplets were simulated using DNS. Two cubic boxes, each with $(0.256 \text{ m})^3$ volume, representing warm cloud and clear air properties were combined together to create the simulation domain (see Figure 1a). The portion of the domain representing the cloudy air was seeded with a mono-disperse cloud droplet population. The airflow was modelled as an incompressible fluid flow according to the Boussinesq approximation. The governing equations for the airflow (carrier fluid) and the cloud droplets (dispersed medium) are presented below.

2.1. Evolutive Equations for the Fluid Flow

The numerical model for the evolution of the fluid flow considers NS equations for humid air, which are coupled with the Lagrangian tracking of every cloud water droplet, introduced in the initial condition. The numerical model of this study followed the model used by Vaillancourt et al. (2001) [8], later used by Kumar et al. (2014) [13], Gotoh et al. (2016) [26], Götzfried et al. (2017) [12] and Gao et al. (2018) [18]. The air phase equations use Boussinesq-like approximation and include the continuity and momentum balance equations for the airflow velocity $\mathbf{u}$ and the temperature T and the water vapour density ρ_v equations. Temperature and water vapour density were considered as advected active scalars. The fluid flow equations are

$$\nabla \cdot \mathbf{u} = 0 \tag{1}$$

$$\frac{\partial \mathbf{u}}{\partial t} + \mathbf{u} \cdot \nabla \mathbf{u} = -\frac{1}{\rho_0} \nabla p + \nu \nabla^2 \mathbf{u} - B\mathbf{g} \tag{2}$$

$$\frac{\partial T}{\partial t} + \mathbf{u} \cdot \nabla T = \kappa \nabla^2 T + \frac{L}{\rho_0 c_p} C_d \tag{3}$$

$$\frac{\partial \rho_v}{\partial t} + \mathbf{u} \cdot \nabla \rho_v = \kappa_v \nabla^2 \rho_v - C_d \tag{4}$$

where $\partial/\partial t$ is the temporal derivative, ρ_0 is reference mass density of air at temperature T_0 and pressure p_0, ∇p is the pressure gradient, ν is the kinematic viscosity, $\mathbf{g}$ is the gravity acceleration, κ is the thermal diffusivity of air, L is the latent heat for condensation of water vapour, c_p is the specific heat at constant pressure and κ_v is the water vapour diffusivity. The "source" term in the momentum balance Equation (2) is B which represents the buoyancy force per unit volume due to small variations of temperature and water vapour density in the humid air. In the temperature (enthalpy) (3) and humidity (4) equations, the source term C_d represents the condensation rate, that is, the local condensating or evaporating water mass per unit time and unit volume.

These source terms are expressed as

$$B = \frac{T - T_0}{T_0} + \epsilon \frac{\rho_v - \rho_{v,e}}{\rho_0} \tag{5}$$

$$C_d = \frac{1}{V} \sum_i \frac{dm_i}{dt} = \frac{1}{V} \sum_i 4\pi \rho_L r_i^2 \frac{dr_i}{dt} \tag{6}$$

where T_0 is the reference temperature, $\rho_{v,e}$ is the reference water vapour density, $\epsilon = R_v/R_a - 1 = 0.608$ is a constant dependent on the gas constants R_v and R_a of the water vapour and the air respectively, ρ_L is the density of liquid water and r_i is the radius of the i-th cloud droplet. The sum in C_d is calculated on the droplets only within the (small) local volume V, where m_i is the mass of the i-th droplet (in simulation, this volume is the volume of a computational grid cell). Droplet volume fraction (volume of droplets/volume of carrier fluid) for the largest initial droplet (initial radius, $r_{in} = 25$ μm) population is 1.12×10^{-6}, which is considered as dilute suspension by Elghobashi (1991) [27], where the momentum feedback from the droplets (two-way momentum coupling) may be neglected. Therefore, in the computation of the buoyancy force, no momentum feedback from droplets to the fluid is considered, since in case of dilute suspension one-way coupling is prominent in between droplets and the surrounding fluid [27].

2.2. Lagrangian Equations for the Cloud Droplets

The Lagrangian descriptions of the cloud droplets consider them driven only by gravity and the Stokes drag law, since the droplet density is much higher than air density ($\rho_L/\rho_0 \sim 10^3$) and the Reynolds number of the droplet relative velocity with respect to the surrounding air is very small (after Pruppacher and Klett (1978) [1] pg 575 and Vaillancourt et al. (2001) [8], usually below 0.1). Moreover, their sizes change due to condensation/evaporation, driven by the local heat and water vapour diffusion around the droplets. Therefore, the equations for the i-th droplet are

$$\frac{d\mathbf{x}_i}{dt} = \mathbf{v}_i \tag{7}$$

$$\frac{d\mathbf{v}_i}{dt} = -\frac{\mathbf{v}_i - \mathbf{u}(\mathbf{x}_i, t)}{\tau_i} + \mathbf{g} \tag{8}$$

$$\frac{dr_i}{dt} = C\frac{\varphi(\mathbf{x}_i, t) - 1}{r_i}, \qquad \varphi(\mathbf{x}_i, t) = \frac{\rho_v(\mathbf{x}_i, t)}{\rho_{vs}(T)(\mathbf{x}_i, t)} \tag{9}$$

where $\mathbf{x}_i$ is the droplet position, $\mathbf{v}_i$ is the droplet velocity, φ is the local relative humidity, ρ_{vs} is the saturation vapour density and τ_i is the droplet response time, which is

$$\tau_i = \frac{2}{9} \frac{\rho_L}{\rho_0} \frac{r_i^2}{\nu}. \tag{10}$$

The droplet growth rate Equation (9) is dependent on the condition of surrounding relative humidity (Pruppacher and Klett (1978) [1], p. 511) and the proportionality coefficient C is defined as

$$C = \kappa_v \frac{\rho_{vs}(T_0)}{\rho_L} \left(1 + \frac{L^2 \rho_{vs}(T_0)}{R_v T_0^2} \frac{\kappa_v}{\lambda_T}\right)^{-1}$$

where λ_T is the thermal conductivity of the air. As in Kumar et al. (2013) [10], we approximated $L/(R_v T) \gg 1$ in the definition of C.

2.3. Numerical Method of the DNS

Model equations for the fluid flow phase are solved using the Fourier–Galerkin (FG) pseudospectral method as in Iovieno et al. (2001) [28], while temporal advancement is approximated using a low storage second order Runge-Kutta (RK2) method with exponential integration of the diffusive terms [29]. The numerical code uses one dimensional (1D) slab parallelization using Massage

Passing Interface (MPI) libraries. Dealiasing is carried out during data transposition. Discrete Fourier Transforms (DFT) are computed using FFTW subroutine library. Whereas, Lagrangian model equations for the droplets are solved in the physical space using the same RK2 method for temporal integration. After every time-step, droplets are exchanged among neighbouring processors according to their current respective positions. Fluid velocity, temperature and vapour density at particle positions are computed through a trilinear interpolation within each mesh cell, which is expected to be sufficient in the Lagrangian droplet model for $k_{max}\eta \gtrsim 2$. The feedback term C_d is computed through cell average (particle-in-cell method). Higher order interpolation methods, for example, third order B-spline method would be optimal according to [30], in the sense that the interpolation error would be less than the discretization error. However, as outlined by Sundaram and Collins (1996) [31], the interpolation and the reverse interpolation schemes must be symmetric in order to guarantee energy conservation in the domain, so that the linear particle-in-cell reverse interpolation should be replaced with an equivalent higher order method (see for example [32]). The pseudo-spectral code used triply-periodic boundary conditions for all the fluid flow variables. A non-periodic temperature profile is introduced in the vertical direction by the decomposition $T = \Gamma x_3 + T'$; where, the field T', which contains the temperature fluctuations, is triply periodic but the full temperature field T is not periodic in the vertical direction. The sign of the temperature gradient $\Gamma = \Delta T / L_{x_3}$ determines the stability of the flow. Such decomposition of the temperature field modifies the temperature Equation (3) into

$$\frac{\partial T'}{\partial t} + \mathbf{u} \cdot \nabla T' = \kappa \nabla^2 T' + \frac{L}{\rho_0 c_p} C_d - \Gamma u_3. \tag{11}$$

The code also includes a module for droplet-droplet binary collision detection, which assumes coalescence of the colliding droplet masses. After each time-step, the algorithm determines whether the distances between the pairs of droplet centers become smaller or equal to the sum of their redii by assuming a linear in–time variation of radius and position within the time-step. A collision between droplet i and j is considered to have occured if the first solution of $|\mathbf{x}_i(t) - \mathbf{x}_j(t)|^2 - (r_i(t) + r_j(t))^2 = 0$ lies between t and $t + \Delta t$ and their relative velocity is inward. Since the Weber numbers (ratio between the droplet kinetic energy and droplet surface energy) of the cloud water droplets are very small ($\ll 1$), each collision is modeled as a successful coalescence in agreement with the experimental investigation of water droplet collisions by Rabe et al. (2010) [33]. Conservation of mass and momentum are then used to determine the size, position and velocity of the new droplet emerged from the collision.

2.4. Simulation Setup

The simulation domain replicates a small portion of a warm cloud top and layer of clear air above it. All interactions in between the cloud and the clear air happens through the interface formed by the edges of both the cloud and the clear air regions. Simulation parameters, constants and domain specifications used in our simulations are tabulated in Table 1. The values of the basic referred states correspond to the reference height H_{ref} of 763 m, which is a typical height for the low altitude warm clouds [34].

Table 1. Simulation parameters, constants and domain specifications used for simulation setup.

Quantity	Symbol	Value	Unit
Reference temperature	T_0	283.16	K
Reference atmospheric pressure	p_0	92.4	kPa
Reference air density	ρ_0	1.13	$\mathrm{kg\,m^{-3}}$
Reference kinematic viscosity	ν	1.56×10^{-5}	$\mathrm{m^2\,s^{-1}}$
Gravitational acceleration	g	9.8	$\mathrm{m\,s^{-2}}$
Thermal conductivity of the air	λ_T	2.5×10^{-2}	$\mathrm{J\,K^{-1}\,m^{-1}\,s^{-1}}$
Thermal diffusivity of air	κ	2.2×10^{-5}	$\mathrm{m^2\,s^{-1}}$
Diffusivity of water vapour	κ_v	2.54×10^{-5}	$\mathrm{m^2\,s^{-1}}$
Specific heat of air at constant pressure	c_p	1005	$\mathrm{J\,kg^{-1}\,K^{-1}}$
Latent heat for condensation of water vapour	L	2.5×10^6	$\mathrm{J\,kg^{-1}}$
Gas constant for water vapour	R_v	461.5	$\mathrm{J\,kg^{-1}\,K^{-1}}$
Gas constant for air	R_a	286.84	$\mathrm{J\,kg^{-1}\,K^{-1}}$
Density of liquid water	ρ_L	1000	$\mathrm{kg\,m^{-3}}$
Reference saturated water vapour density at T_0	$\rho_{vs}(T_0)$	9.4×10^{-3}	$\mathrm{kg\,m^{-3}}$
Constant in Equation (9)	C	9.22×10^{-11}	$\mathrm{m^2\,s^{-1}}$
Simulation grid step	Δx	0.001	m
Simulation domain discretization	$N_1 \times N_2 \times N_3$	$256 \times 256 \times 512$	
Simulation domain size	$L_{x_1} \times L_{x_2} \times L_{x_3}$	$0.256 \times 0.256 \times 0.512$	$\mathrm{m^3}$

Table 2 summarizes the details of the simulation runs. In this simulation setup, an initial population of mono-disperse cloud droplets of three different size distribution are introduced in the cloudy volume of the simulation domain in three different simulation runs **R25**, **R18** and **R6**; whereas, the carrier airflow initial conditions remained the same for all these simulations. A sketch of the simulation domain is shown in Figure 1a. Gravitational force (as presented in Figure 1a) acts on both the fluid flow (in form of buoyancy forces B in Equation (2)) and on the momentum of the cloud droplets in Equation (8). In this cloud top simulation setup, gravity therefore acts in downwards direction, causing heavier droplets to settle down towards the bottom boundary of the cloudy volume.

Table 2. Details of simulation runs

	Simulation IDs		
Quantity	R25	R18	R6
Initial Droplet Radius r_{in} [μm]	25	18	6
Total number of initial droplets	286,240	286,240	286,240
Initial droplet number density $N_d(0)$ [$\mathrm{cm^{-3}}$]	17	17	17
Initial liquid water content lwc [$\mathrm{g\,m\,m^{-3}}$]	1.12	0.42	0.02
Initial Stokes number St	1.59	0.82	0.09
Initial rms of velocity fluctuations u' in cloud [$\mathrm{m\,s^{-1}}$]	0.268	0.268	0.268
Initial energy ratio E_{cloud}/E_{air}	20	20	20
Temperature difference between cloudy and clear air ΔT [K]	4	4	4
Initial integral scale L of cloud and air [m]	0.0235	0.0235	0.0235
Initial Taylor micro-scale Reynolds no. Re_λ of cloud	90	90	90
Initial Taylor micro-scale Reynolds no. Re_λ of air	20	20	20
Simulation time-step Δt [s]	1.224×10^{-4}	1.224×10^{-4}	1.377×10^{-5}
Total simulation duration [s]	2.1	2.1	2.1

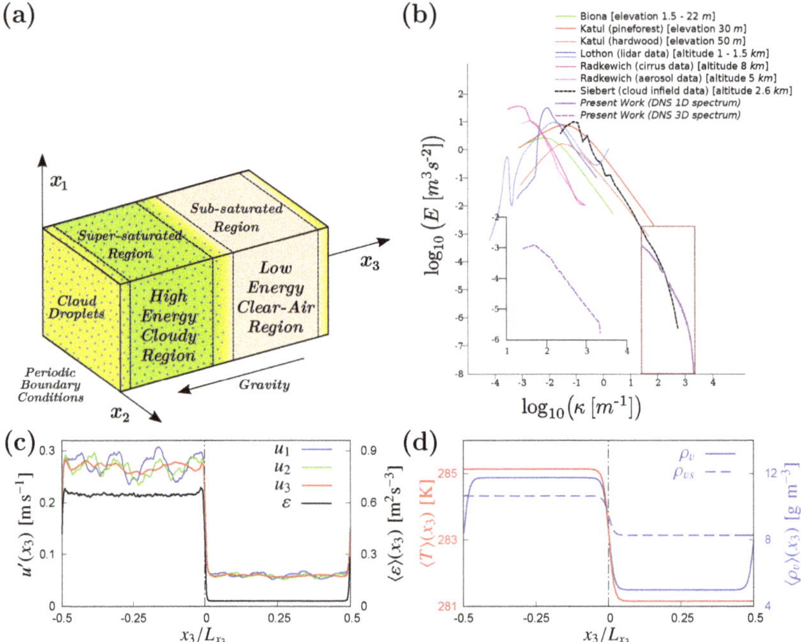

Figure 1. Simulation setup: (**a**) scheme of the three dimensional simulation domain and the boundary conditions; (**b**) comparison between the TKE spectrum $E(k)$ of present simulation with the infield measurements; (**c**) initial distribution of the rms of velocity fluctuations u' and TKE dissipation rate $\langle \varepsilon \rangle$; and (**d**) simulated initial profile of temperature $\langle T \rangle$ and water vapour density $\langle \rho_v \rangle$.

2.4.1. Initial Setup for the Fluid Flow

In field measurements of the cloud turbulent properties and associated clear air properties show different turbulent intensities, producing TKE gradients across the interfaces. Velocity fluctuations inside the cloudy regions mostly show higher TKE than that of the clear air region [34–36], which is mainly due to the instability of the moist air-mass, developed due to the buoyant updraft resulting in shear layer formation inside cloud [37]. This difference in kinetic energy is replicated by the average of velocity fluctuation root mean square (rms) distribution u' in the initial condition (Figure 1c) of present simulations. The clear air region of the simulation domain, shown in the right side region of Figure 1c is having lesser energy than the associated cloudy region (left side of the domain). The kinetic energy ratio between the cloudy domain and the clear air domain has been chosen to be around 20, which is in the range of the values which can be deduced from in-cloud measurements [34,35]. The thickness of the initial velocity fluctuations interface, measured as the distance between the horizontal planes where the difference of the TKE is 90% of the difference of kinetic energy between the cloudy and clear air regions, is about 0.006 m (see Figure 1c). Figure 1b presents few examples of spatial one-dimensional TKE spectra from infield measurements by Biona et al. (2001) [38], Katul et al. (1998) [39], Radkevich et al. (2008) [40], Lothon et al. (2009) [41] and Siebert et al. (2015) [42], to which the one dimensional initial TKE spectrum of present simulations has been superimposed (the three dimensional TKE spectrum is shown in the inset). Since in a DNS, we need to resolve Kolmogorov micro-scale η, which also plays important role for the micro-physical evolution of cloud droplets [3,5], only part of the inertial sub-range and the dissipation range (up to the last three decades of TKE spectrum in logarithmic wave-number space) can be reproduced. The initial velocity field is generated by the superposition of Fourier modes with random phases, whose amplitude is determined from the following model 3D-spectrum

$$E(k) = A\frac{(k/k_0)^\alpha}{1 + (k/k_0)^{\alpha+5/3}}f(k/k_{max}).$$

Coeffient α controls the low wavenumber slope of the spectrum ($\alpha = 2$ in present work). Coefficients A and k_0 control the variance and the initial correlation length and $f(k/k_{max})$ produces the exponential tail in the highest simulated wavenumbers approaching $k_{max} = \pi/\Delta x$. The transition of initial fluctuations for airflow velocity from higher values inside the cloudy region to the lower values inside the clear air region of the domain was carried out using a linear superposition of the two initial cloudy and clear air isotropic field (see Tordella and Iovieno, (2006) [43], supplemental material of Tordella and Iovieno, (2011) [22] and Iovieno et al. (2014) [23]).

For this study, we selected an unstably stratified temperature profile, which can also be locally observed from small scale local temperature profiles obtained from in-cloud measurements (see Figure 2 of Ref. [34]). Initial temperature and water vapour fields are uniform on horizontal planes. No fluctuations of temperature and water vapour density were introduced in the initial conditions, so that the supersaturation fluctuations were generated only due to the mixing or due to the condensation/evaporation of the droplets to a minor extent. Therefore, any particle size broadening can be attributed to the non-uniform supersaturation field generated by the mixing process. Constants T_0 and $\rho_{v,e}$ in Equation (5) were chosen as the mean values of T and ρ_v over the whole domain in the initial conditions. The thickness of the initial mean temperature and mean density of water vapour interface was slightly wider, around 0.007 m (Figure 1d). The initial mean density of water vapour and unstable mean temperature distribution inside the domain simulate supersaturated (relative humidity (RH) $\varphi = 1.1$, 10% supersaturation) cloudy volume and subsaturated ($\varphi = 0.6$, 40% subsaturation) clear air volume conditions. The supersaturated condition inside the cloudy region of the simulation domain will help the local cloud water droplets to grow by condensational deposition of water vapour on them; whereas, entrainment of the subsaturated clear air inside the cloudy volume can result in evaporative size reduction of the cloud droplets. These mean values of the supersaturation are around the upper estimation of the atmospheric measurements by Siebert and Shaw (2017) [44].

2.4.2. Initial Setup for the Cloud Droplets

A mono-disperse population of cloud water droplets of initial 25 μm, 18 μm and 6 μm radii was initially introduced for the three simulations at random positions inside the cloudy part of the domain, that is, in the supersaturated region of the domain. The initial velocity of the droplets was set equal to the interpolated flow velocity at the droplet position.

Boundary conditions for the cloud droplets also follow periodic boundary conditions in the two horizontal directions, that is, droplets which exit the domain from horizontal boundaries re-enter from the opposite side with the same velocity as the fluid flow. However, the droplets settling on the bottom boundary of the simulation domain are removed from the simulation considering that those droplets are no longer present inside the cloudy volume. In this way, they do not re-enter the simulation domain from the top in the clear air region above the cloud. This choice has some positive as well some negative impact on the statistics of the simulated cloud droplets. Due to gradual removal of the heavier droplets from the cloudy portion of the domain, the number of samples for averaged droplet quantities were reduced near the bottom zone of the cloudy volume. However, simultaneously the cloud droplets were prevented from reappearing inside the clear air part of the domain caused by the periodic boundary conditions. This setup prevented any spurious droplet feedback to be present on the fluid volume near the top of the clear air region. Evaporation of the liquid water from the droplets results in size reduction. In the case of a droplet reducing below 4% of their initial radius, it was assumed that the droplet was completely evaporated within one time-step. This avoids the numerical instability of the small droplets, which have time-scales much smaller than the Kolmogorov time-scale.

3. Results of Transient Evolution

3.1. Time Evolution of the Fluid Flow

Since no external forcing has been introduced, the only force that can amplify the velocity of the fluid flow is the buoyancy force generated by the variations of temperature and water vapour density in the humid air. Transient evolution of the volume averaged quantities, such as TKE E, its dissipation rate ε, Taylor micro-scale Reynold's number Re_λ and integral length scales L exhibit transient decay or growth with time as shown in Figure 2. Turbulent fluid statistics along anisotropic x_3 direction is carried out by plane averaging $\langle \cdot \rangle$ across homogeneous x_1, x_2 horizontal planes. For computation of volume averaged turbulent quantities $\langle \cdot \rangle_V$, the plane averaged quantities $\langle \cdot \rangle$ are averaged over the bulk of the cloudy region $[1/8 \leq x_3/L_{x_3} \leq 3/8]$ and the clear air region $[5/8 \leq x_3/L_{x_3} \leq 7/8]$. The definitions for $E, \varepsilon, \lambda, Re_\lambda$ and L are given as

$$\langle E \rangle = \frac{1}{2} \sum_{i=1}^{3} \langle u_i'^2 \rangle; \qquad \langle \varepsilon \rangle = \frac{1}{2} \nu \sum_{i,j=1}^{3} \langle (\frac{\partial u_i'}{\partial x_j} + \frac{\partial u_j'}{\partial x_i})^2 \rangle;$$

$$u_{avg}'^2 = \frac{1}{2}(\langle u_1'^2 \rangle + \langle u_2'^2 \rangle); \qquad \lambda = u_{avg}' \sqrt{\frac{15\nu}{\langle \varepsilon \rangle}}; \qquad Re_\lambda = \frac{u_{avg}'\lambda}{\nu};$$

$$B(r) = \langle u_i'(x_i)u_i'(x_i + r) \rangle; \qquad L = \frac{1}{B(0)} \int_0^\infty B(r)dr;$$

where u_{avg}' is the average of rms velocity fluctuations along two homogeneous directions (x_1, x_2) along which flow should remain homogeneous and isotropic, since the only sources of in-homogeneity and anisotropy are gravity and energy/temperature/density of water vapour gradient, which are acting along vertical x_3 direction; and $B(r)$ is the velocity correlation function [45]. Evolution of E and ε is plotted in Figure 2a using logarithmic scale in the both axes. More than the half of the initial E and ε inside the cloudy region is lost during the first 0.3 s, after which the evolution of the E and ε follows a power-law scaling with time (with scaling exponent of -1.25 for E and -2.25 for ε).

Figure 2b presents time evolution of Re_λ, which shows two phases in its evolution inside the cloudy region. During the first phase till 0.3 s of initial transient, turbulence develops from the initial random conditions. Initially a increase in Re_λ is observed due to the rapid increase in spatial scales (such as λ, L) compared to the decrease in E with time. This is followed by a sharp decrease in Re_λ, due to the rapid decrease in E with time which dominates over the the increase in λ. During the second phase, E decays at $t^{-1.25}$ and λ grows as $t^{0.5}$, which results in a gentle decay of Re_λ at $t^{-0.125}$. Since decrease in E inside the clear air region is much slower than the increase in its spatial scales and the detrainment of E happens from the cloudy to clear air region; the evolution of Re_λ in clear air shows increment for longer duration compared to the cloudy region, which is followed by a slow decrease.

For calculation of the integral length scales L, both longitudinal and transversal integral scales are computed along the two homogeneous directions and averaged. In Figure 2c, the transversal length scales are not exactly one half of the longitudinal length scales, which is an indication of anisotropic evolution of the flow across the domain with time. The effect of the numerical boundary of the domain also influenced this anisotropic evolution of the integral scale, since the scales can not grow beyond the domain size. The decay in kinetic energy also produces a growth of the Kolmogorov micro-scale $\eta = \sqrt[4]{\nu^3/\langle \varepsilon \rangle}$. The resolution of the simulation increases with time, since $k_{max}\eta$ varies from 1.0 at the beginning to 3.7 at the end of the simulation inside the cloudy region of the domain (inside clear air region, $k_{max}\eta \geq 2$ always). The growth of the η also produces a transient evolution in the average droplet Stokes number St and the settling parameter S_v of the different droplet populations in the domain, as shown in Figure 2d. St is a ratio between the droplet response time τ_i in Equation (10) and the Kolmogorov time scale $\tau_\eta = \sqrt{\nu/\langle \varepsilon \rangle}$ and S_v is a ratio between v_p and u_η, where $v_p = \tau_i g$ is the terminal velocity of a droplet and $u_\eta = (\langle \varepsilon \rangle \nu)^{1/4}$ is the Kolmogorov velocity. Due to decay in the kinetic energy (and the corresponding growth in η) in the domain, the droplets become gradually less and less sensitive to the turbulence (indicated by the transient growth of the S_v parameter), whereas, the droplet Stokes number St gradually reduces.

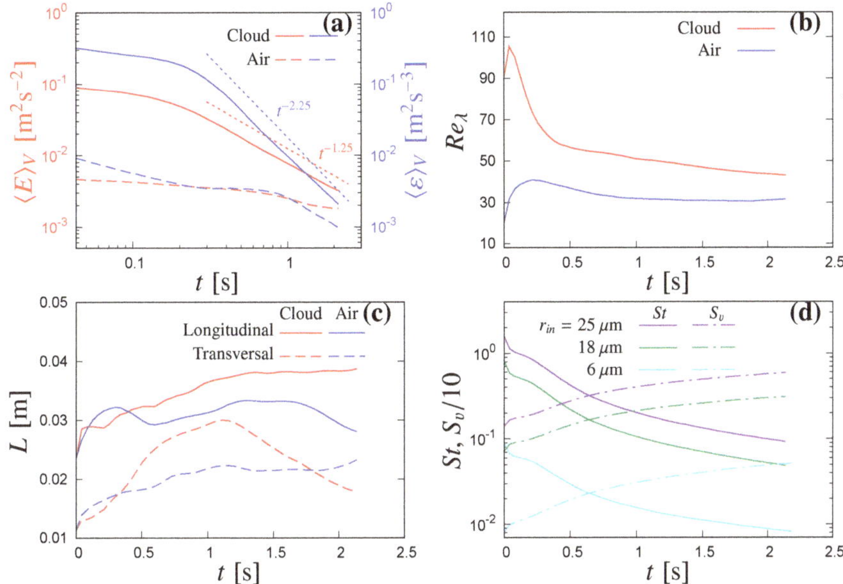

Figure 2. Transient evolution of the flow inside the clear air and cloudy region: (**a**) decay of TKE E and dissipation rate ε; (**b**) Taylor micro-scale Reynolds number Re_λ; (**c**) longitudinal and transversal integral length scales L; and (**d**) average cloud droplet Stokes number St and settling parameter S_v of the different droplet populations. Plots (**a–c**) are from simulation **R25**. Differences are small among the three simulations.

Figure 3 presents the time evolution of air statistics. Time has been re-scaled using the initial eddy turnover time $\tau_0 = L/u'$ taken at $t = 0.043$ s, which is 0.115 s, in order to reduce the influences of the initial evolution of the flow. Since all the three different simulations were initialized with same initial background fluid flow conditions, similar transient evolutions are observed for various fluid turbulent quantities. In transient evolution of $\langle E \rangle$ in Figure 3a, detrainment of TKE to clear air region of the domain is observed to happen through the interface (dotted line in the middle) and also through the top boundary on top of the clear air region of the domain (due to periodic boundary condition, which generates a secondary inhomogeneous layer). Detrainment of TKE and widening of the TKE interface towards the clear air region of the domain can be recognized by following the positions of the kurtosis peaks of vertical component of fluid velocity u_3 in the Figure 3b, which moves toward the core of the clear air region. Skewness $\mathcal{S}(\cdot) = \langle (\cdot)^3 \rangle / \langle (\cdot)^2 \rangle^{3/2}$ and kurtosis $\mathcal{K}(\cdot) = \langle (\cdot)^4 \rangle / \langle (\cdot)^2 \rangle^2$ of the fluid quantities are computed using horizontal plane averages $\langle \cdot \rangle$ as a function of vertical direction x_3. After 6 initial eddy turnover time ($t/\tau_0 = 6$), overall TKE and the amplitude of the kurtosis peaks are observed to become lower. During the final stage of the simulation ($t/\tau_0 = 18$), negligible TKE is left inside the fluid motion and the smaller peaks of kurtosis represent a well mixed stage in distribution of rms of velocity fluctuations throughout the domain. Decay in kinetic energy is reduced by the production of buoyancy inside the mixing layer. Unstably stratified temperature profile in the vertical direction amplifies vertical motion (see also Gallana et al. (2014) [46]) and the fluctuations of ρ_v and T produce buoyancy fluctuations (Equation (5)), which introduce energy into the vertical motion

(Equation (2)). This cumulative effect becomes visible after $t/\tau_0 = 6$ (see the growth of $\langle E \rangle$ in the mixing region in Figure 3a), when the flow lost most of its initial kinetic energy. This source of kinetic energy accelerates the growth of the mixing layer and reduces the extent of the undiluted regions as well. During the later stage of evolution, $t/\tau_0 > 12$, the initial configuration of two different regions is no more distinct and the flow begins to approach a homogenized state.

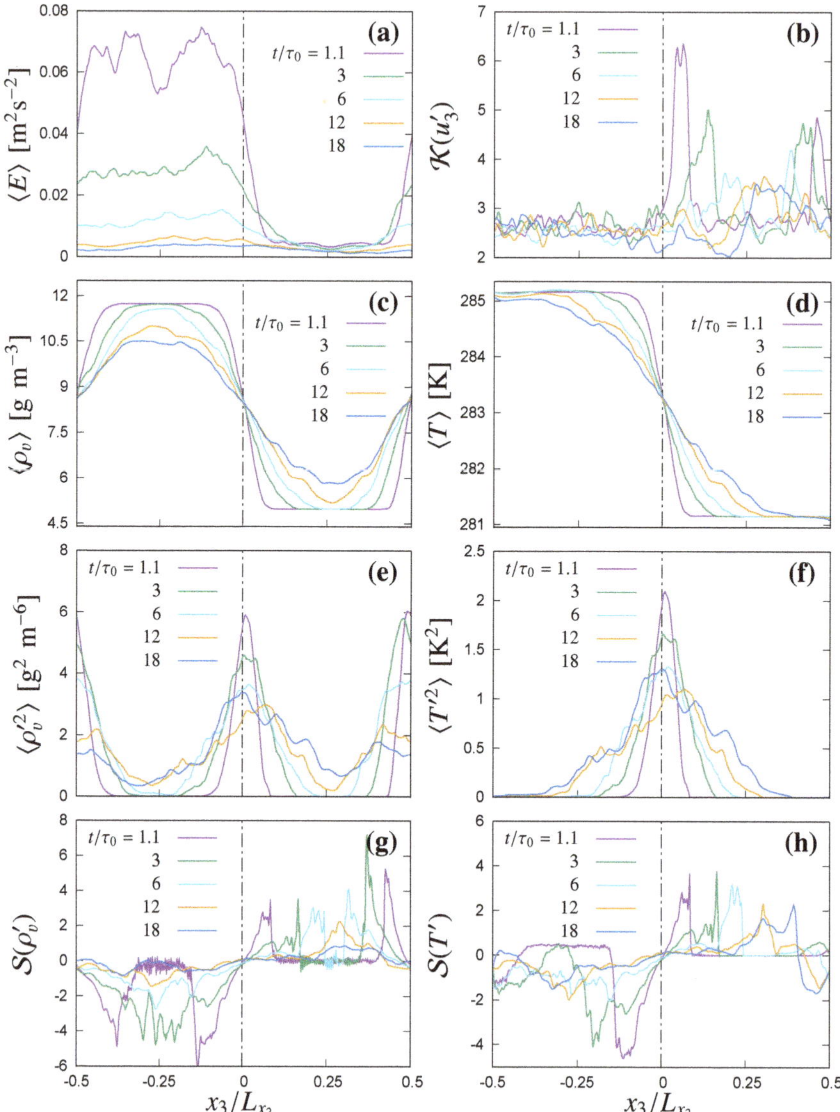

Figure 3. Transient evolution of flow statistics: (**a**) TKE $\langle E \rangle$; (**b**) kurtosis of vertical velocity fluctuations $\mathcal{K}(u'_3)$; (**c**) mean water vapour density $\langle \rho_v \rangle$; (**d**) mean temperature $\langle T \rangle$; (**e**) variance of water vapour fluctuations $\langle \rho'^2_v \rangle$; (**f**) variance of temperature fluctuations $\langle T'^2 \rangle$; (**g**) skewness of water vapour fluctuations $\mathcal{S}(\rho'_v)$; and (**h**) skewness of temperature fluctuations $\mathcal{S}(T')$ evolution. All plots present horizontal plane averaged quantities as in Figure 1c,d. The presented data are from simulation **R25**. Differences are small among the three simulations.

Figure 3c,d presents the time evolution of the mean of water vapour density $\langle \rho_v \rangle$ and temperature $\langle T \rangle$ profiles respectively. Mean of both the density of water vapour and temperature decrease inside the cloudy region of the domain, whereas, it increase inside the clear air region. The resulting profile of mean supersaturation $\langle S \rangle$ (which is initialized with a magnitude of 10% inside the cloudy region and -40% inside the clear air region) shows a decrease in its magnitude inside the supersaturated cloudy region and an increase inside subsaturated clear air region (Figure 4a). The mixing process tends to produce a uniform supersaturation profile, therefore, during the final stage ($t/\tau_0 = 18$), the mean supersaturation value remains positive ($\langle S \rangle > 0$) only in the central part of the cloudy region, whereas, most of the domain is subsaturated ($\langle S \rangle < 0$). Figure 3e,f present the time evolution of the variance of water vapour fluctuations $\langle \rho_v'^2 \rangle$ and temperature fluctuations $\langle T'^2 \rangle$. Resulting variance in plane averaged supersaturation fluctuations $\langle S'^2 \rangle$ is shown in Figure 4b. Though no fluctuations are introduced in the initial condition for the temperature and the density of water vapour, fluctuations are generated by the mixing in the interface region and are propagated inside the undiluted core of the cloudy or the clear air region gradually with the spreading of the mixing region. A minor source for the fluctuations in the fluid flow quantities is the droplet feedback term C_d in Equations (3) and (4). However, since the mean condensational time-scale is much larger than the eddy turnover time-scale, C_d gives an overall small contribution for generating fluctuations. In Figures 3e and 4b, two peaks (one in the interface in the middle of the domain and another one near the bottom and the top boundaries) are observed in the fluctuations of the density of water vapour and supersaturation, due to periodicity in the initial density of water vapour condition. However, since the initial temperature is non-periodic in the vertical direction and varies only at the mixing interface in the middle of the domain, Figure 3f shows only one peak in the variance of the temperature fluctuations. Temporal growth of the scalar (density of water vapour, temperature and supersaturation) mixing layers and widening of the scalar interfaces are evident from the shift in the peaks of the skewness and kurtosis of the scalars towards the undiluted central regions as shown in Figures 3g,h and 4c,d, which gradually decrease in magnitude with mixing spreading all across the domain.

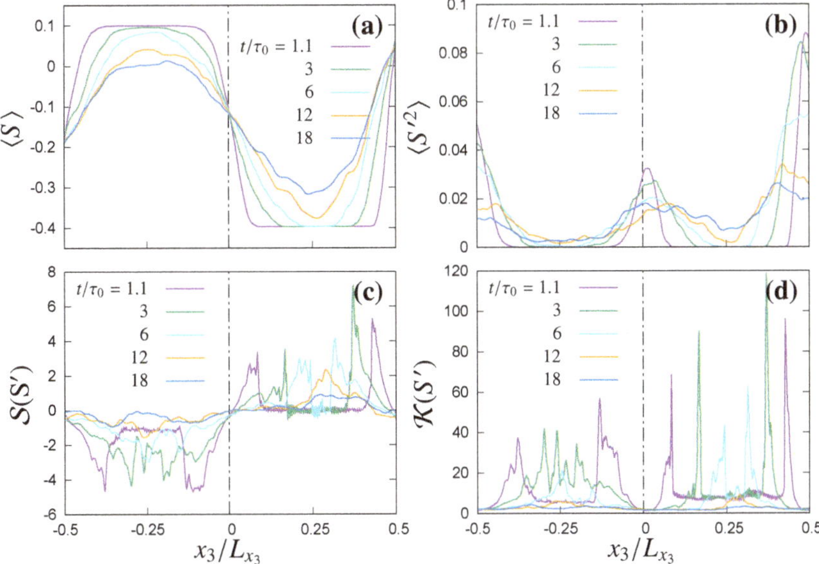

Figure 4. Transient evolution of supersaturation: (**a**) mean of supersaturation $\langle S \rangle$; (**b**) variance $\langle S'^2 \rangle$; (**c**) skewness $\mathcal{S}(S')$; and (**d**) kurtosis $\mathcal{K}(S')$ of supersaturation fluctuations. All plots present horizontal plane averaged quantities.

Figure 5a,b presents the time evolution of one dimensional (1D) horizontal spectrum of the vertical component of air velocity $E_{u'_3}(k)$ in the wavenumber space k, sampled at the middle horizontal plane of the initial configuration of the cloudy region and at the middle plane of the initial interface region (with 3 adjacent plane averaging in both cases). The 1D spectra in wavenumber space $E_f(k)$ distributed in the homogeneous x_1, x_2 directions is computed as

$$E_f(k) = \frac{1}{2}\left[\frac{1}{N_1}\sum_{j=0}^{N_1-1} \mid \hat{f}(k, x_2^j, x_3)\mid^2 + \frac{1}{N_2}\sum_{j=0}^{N_2-1} \mid \hat{f}(x_1^j, k, x_3)\mid^2\right]$$

where N_i is the number of grid points in x_i direction, $\hat{f}$ is the Fourier transform of quantity f. Since initially the cloudy and the clear air region were initialized with two homogeneous and isotropic turbulent cubic domains with a TKE ratio of 20 in between the two regions, the initial 1D transversal spectra for all the three components of velocity fluctuations looks almost similar (some differences can be observed in the lowest wavenumbers due to smaller number of samples). The interface region in the initial condition contains lower TKE than that of the cloudy region of the domain (due to linear interpolation in TKE magnitudes in between the cloudy and the clear air region), which can also be observed in the initial TKE spectra of the interface showing vertical shift downwards than that of the cloud core region. With time advancement, the dissipative wavenumber range from the initial condition shows transition towards smaller wavenumbers indicating growth in the Kolmogorov micro-scale η with time, gradually shrinking the inertial sub-range. Moreover, the spectra of different components of velocity fluctuations do not replicate each other during the later instances of the simulation, resembling an anisotropic time evolution of the fluid flow.

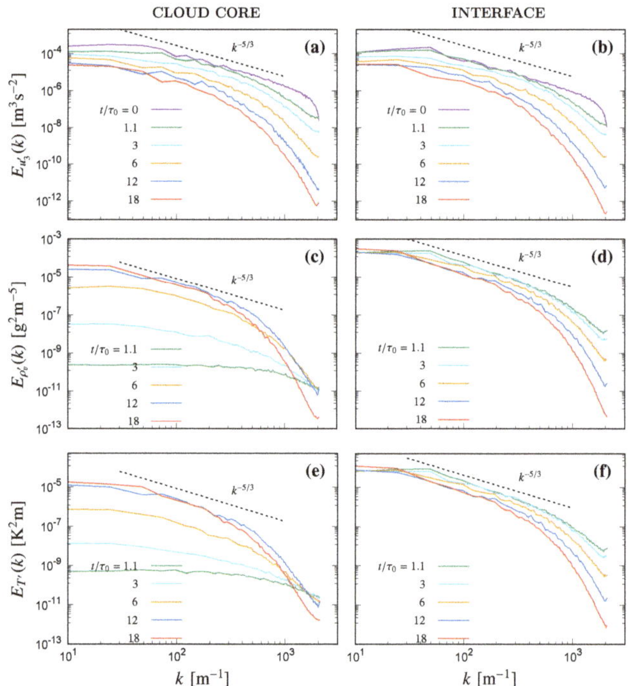

Figure 5. Transient evolution of one dimensional horizontal spectra: (**a,b**) transversal spectra of u'_3; (**c,d**) spectra of water vapour fluctuations ρ'_v; and (**e,f**) temperature fluctuations T' respectively at cloud core and at the interface.

In Figure 5c,d, the transient evolution of the water vapour density fluctuations spectra $E_{\rho_v'}(k)$ and in Figure 5e,f, temperature fluctuations spectra $E_{T'}(k)$ in wavenumber space k is presented, sampled at the cloud core and the interface. Presence of a mean gradient in the initial scalar profiles along the vertical direction creates a large variance of that scalar in the mixing layer (Figure 3), which is observed to spread towards the undiluted regions with time. Since the mean gradient in the mixing region is large enough to produce sufficient variance in the scalars to counter its dissipation and the turbulent transport, a well mixed region is observed to be created with a scalar spectrum quickly approaching $k^{-5/3}$ Kolmogorov inertial range [23]. Initially the only source of temperature and density of water vapour variance inside the undiluted cloudy region is the droplet condensation/evaporation. Therefore, initially the scalar spectra are not well developed but after $t/\tau_0 = 10$, the growth of the mixing layer gradually destroys the cloudy region,so that similar scalar spectra like the mixing region are replicated inside the cloud core as well.

3.2. Time Evolution of the Cloud Droplet Population

The three simulations of this study are initialized with three different mono-disperse cloud droplet populations as enlisted in Table 2. The droplet populations go trough distinct transient evolution according to their surrounding fluid flow conditions. In general, droplets in the cloud core experience an average condensational growth due to supersaturated ambient, whereas, the droplets exiting the cloud core region, tend to evaporate due to subsaturation. A visualization of the flow is shown in Figure 6, where enstrophy ($\mathcal{E} = |\nabla \times \mathbf{u}|^2$) of the fluid field across a vertical plane (plane (x_3, x_1)) is presented with superposition of the cloud droplets around that plane (thickness of droplets containing slice is 0.0025 m) after 6 initial eddy ($t/\tau_0 = 6$) turnover time and along with the supersaturation S field in contour lines. The line at $S = 0$ marks the extent of the cloudy region, where condensational growth occurs. In the region with $S \leq -0.2$, droplets would instead experience a quick evaporation. Although almost similar, small differences in the local enstrophy can be observed as a consequences of the cloud droplet feedback term C_d in Equations (3) and (4), which determines also the buoyancy term B in the momentum balance Equation (2). Since buoyancy is sensitive to the small local fluctuations in the density of the water vapour ρ_v and temperature T, differences in droplet feedback C_d due to different droplet sizes (Equation (6)) can result into differences in the local fluid velocity, although the initial fluid flow conditions are identical. Distribution of enstrophy in Figure 6 gradually decreases with time, as TKE distribution also decreased as shown in Figures 2a and 3a. Due to differences in the cloud droplet Stokes number St as shown in Figure 2d, droplets show different responses to the local enstrophy field. In Figure 6a, droplets of initial 25 µm mono-disperse distribution can be seen to preferentially concentrate away from the regions of higher enstrophy and often forming string like patterns and clustering in the areas of lower enstrophy. However, the gradual reduction of the average Stokes number with time reduces the tendency to cluster, while gravitational settling, droplet size broadening reduce correlation between the droplet concentration and the local strain. A similar but much milder tendency can also be observed in Figure 6b for the simulation with an initial 18 µm mono-disperse droplet population. On the contrary, a higher uniform concentration can be seen in Figure 6c for the droplet population with initial 6 µm radius.

At the same time, droplets undergo gravitational sedimentation which is only partially counterbalanced by turbulence. The relative importance of sedimentation is controlled by the dimensionless settling parameter S_v [19]. Since u_η decays as $t^{-(n+1)/4}, n = 1.25$ in these present simulations, the importance of gravitational sedimentation grows with time (see Figure 2d). Larger droplets (initial 25 µm radius population) begin to gather at the bottom of the domain from the beginning of the simulation and rarely enters the mixing layer (Figure 6a). Droplets with initial 18 µm radii have a comparatively slower rate of sedimentation and are observed to cross the cloudy region border through detrainment process (Figure 6b). On the contrary, smaller droplets (initial 6 µm radius population) do not show significant sedimentation. They are observed to easily diffuse in the clear air zone (Figure 6c), where, due to their shorter evaporation time-scale (proportional to $r^2/|S|$) and longer residence time (roughly proportional to L/u' but modified by the droplet settling velocity v_p), they can completely evaporate. Moreover, for the smaller droplet population, the presence of strong subsaturation near the bottom boundary and at the clear air region of the domain also removed droplets by complete evaporation. This evaporation contributes to cool down the subsaturated layer above the mixing region, increasing the negative buoyancy [7] and thus enhancing the mixing process. Whereas, for the larger droplets, after a few initial time-scales, this process of detrainment to subsaturated clear air zone and complete evaporation of the droplets is not present.

This is reflected in the transient evolution of normalized probability density functions (PDFs) of the droplet size, velocity and growth rate; which are presented in Figure 7. Figure 7a–c present the evolution of PDFs of the cloud droplet radius with time. Both the cloud droplet populations with initial 25 and 18 µm radius show limited broadening of their sizes due to condensation/evaporation and the presence of two secondary peaks which correspond to the collisions (Figure 7a,b). However for the droplet population with initial 6 µm radius, no collisional growth is observed to happen during the simulation duration. Whereas, the width of the DSD due to both the evaporation and condensation processes is observed to be wider in Figure 7c and certain number of droplets are observed to evaporate completely. The impact of condensational size growth or evaporative size reduction is more efficient for the smaller droplets (since droplet radius growth rate dr_i/dt is proportional to r_i^{-1}, see Equation (9)). PDFs of dr_i^2/dt, which indicates the growth rate in the droplet surface area, are presented in Figure 7d–f. As dr_i^2/dt from Equation (9) is proportional to the local supersaturation $S = (\varphi - 1)$, simulations of the three initial mono-disperse cloud population should exhibit similar transient evolution, since the background supersaturation spatial distribution is similar for the three simulations. However, droplets experience different supersaturation conditions due to their different paths, which also depends on their individual sizes and the local air conditions. Smaller droplets of initial 6 µm radius do not show the extreme negative tail of dr_i^2/dt as observed for larger droplets in Figure 7d,e, because this leads to a complete evaporation of the droplets (Figure 7f).

Moreover, since subsaturation can result in highly negative dr_i/dt for the sub-micron droplets from the initial 6 µm droplet population and since the numerical time-step for the sub-micron droplets needs to be very small and their micro-physics cannot be modelled using Equation (9); the droplets with sizes below 4% ($\leq$0.24 µm) are removed.

Due to gravity, vertical component of the cloud droplet velocity v_3 can exhibit different behaviour compared to the velocity components along the horizontal directions. Transient evolution of PDFs for v_3 is plotted in Figure 7g–i. During the early stage of evolution, the PDFs of v_3 shows wider distribution due to the presence of TKE inside the domain, influencing the droplet velocity as well. However the decay of TKE narrows the spectrum of u_3 and therefore v_3 with time. The gravitational settling is visible in the shift of the maxima of the PDFs of v_3 toward the negative values for the larger droplets. From Figure 7g, most of the cloud droplets of initial 25 µm population are observed to have higher negative v_3 during the later instances of the simulation. Free fall velocity for a 25 µm cloud water droplet is 0.077 m/s and the peak of the PDF of v_3 of this population is observed at 0.063 m/s after the first time-scale, which is implying dominance of gravitational settlement with velocities close to the free fall condition. Simulation with initial 18 µm population (Figures 6b and 7h) is observed to comparatively settle down slowly than the initial 25 µm population. However, the simulation with initial 6 µm population shows almost symmetric evolution of v_3 around the zero (Figure 7i), which indicates negligible effect of gravitational acceleration on this cloud droplet population.

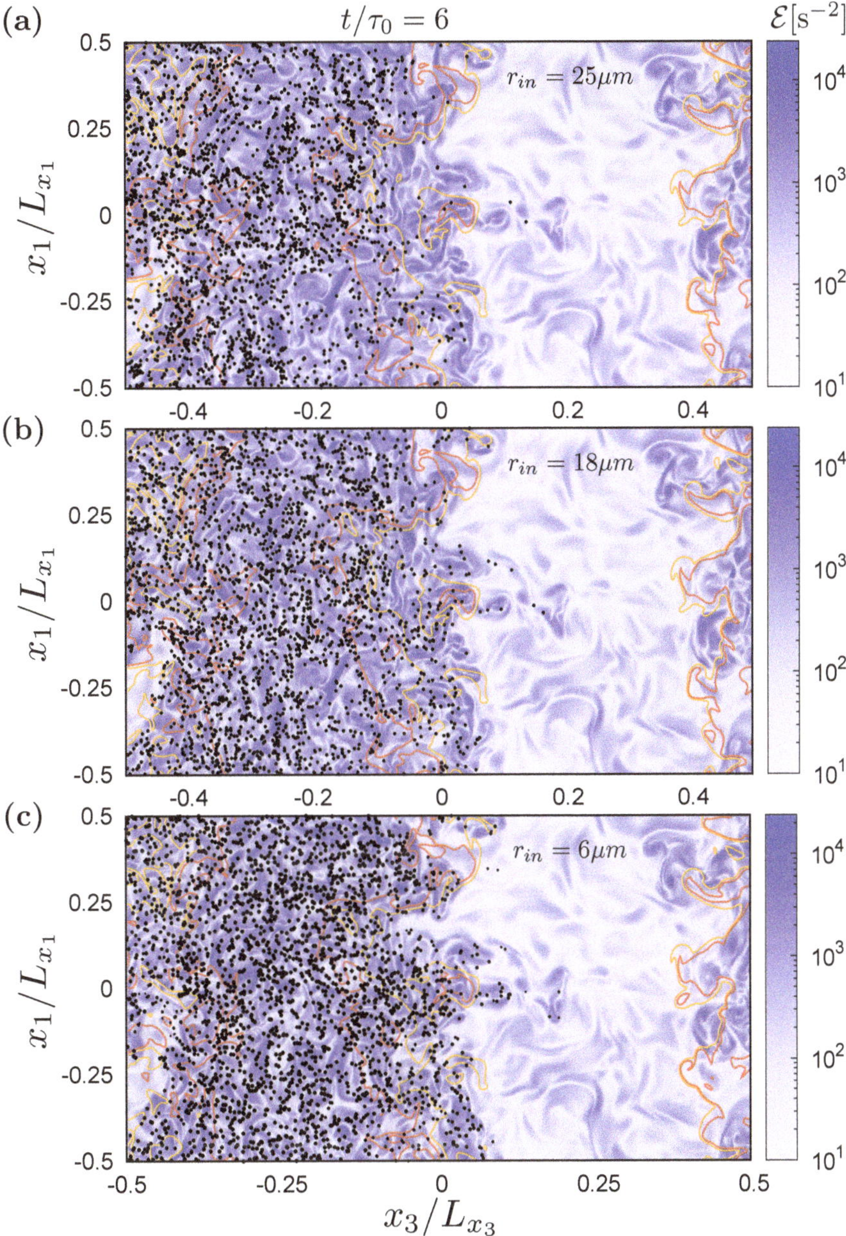

Figure 6. Visualization: Enstrophy $\mathcal{E}$ field across a vertical plane (plane (x_3, x_1)) is presented in superposition with the supersaturation S field in contour lines and cloud droplets around that plane (thickness of droplets containing slice is 0.0025 m) after 6 initial eddy ($t/\tau_0 = 6$) turnover time. Colourbar represents magnitude of the enstrophy in the flow field, red and yellow contour lines represent saturated ($S = 0$) and subsaturated ($S = -0.2$) conditions respectively. The sizes of the droplets are proportional to r/r_{in} for each population. The panels show the simulations with the droplet populations of (**a**) 25 μm; (**b**) 18 μm; and (**c**) 6 μm initial radius.

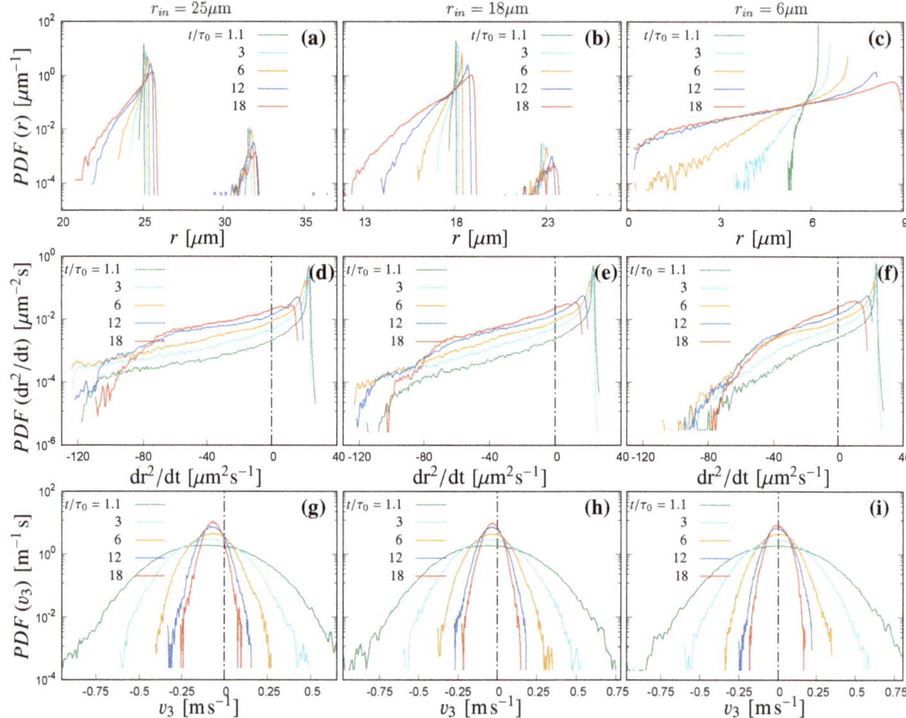

Figure 7. Probability density functions (PDFs): PDFs of (**a–c**) cloud droplet radius r; (**d–f**) droplet surface area growth rate dr^2/dt; and (**g–i**) vertical component of droplet velocity v_3 for 25 µm, 18 µm, and 6 µm initial droplet size populations respectively.

The number density of the droplets inside the simulation domain can change with time due to collisions or complete evaporation of the liquid water or due to gravitational sedimentation of the droplet out of the domain bottom boundary. To quantify the relative importance of condensation, evaporation, collision and gravitational sedimentation; the transient evolution in the number density of droplets is presented in Figure 8. Evolution in number density of all droplets $N_d(t)$ (normalized by initial droplet number density $N_d(0)$) is presented in Figure 8a, where most significant reduction in total number of droplets can be observed for the initial 25 µm droplet population and comparatively less for the initial 18 µm droplet population and much lesser for the initial 6 µm droplet population. The most active physical process to result in reduction of the total number of cloud droplets for the initial 25 and 18 µm droplet populations is the gravitational settling and subsequent removal of the droplets falling below the bottom boundary of the domain. However, the most active physical process for the initial 6 µm droplet population, reducing the total number of droplets is the complete evaporation. In Figure 8b, time evolution in the normalized number density of the droplet population remaining to its initial radius or exhibiting size growth due to condensational water vapour deposition $N_{cond}(t)/N_d(0)$ is presented ($r_{in} \leq r_i < \sqrt[3]{2}r_{in}$, where $\sqrt[3]{2}r_{in}$ corrospond to radius of a droplet after first collision with a similar sized droplet). Though, the initial 6 µm droplet population did not exhibit any collisional growth but some fraction of the population grew more than $\sqrt[3]{2}r_{in}$ size due to higher degree of condensational growth. Transient evolution in the normalized number density of the cloud droplets experiencing evaporative size reduction $N_{evap}(t)/N_d(0)$ is presented in Figure 8c. For the initial 25 µm droplet population, droplets with a radius smaller or equal to 24.5 µm are considered for counting the number density of evaporating droplets ($r_i \leq 17.5$ µm for initial 18 µm population and $r_i \leq 5.5\ \mu m$ for initial 6 μm population). The effect of complete evaporation for the initial 6 μm droplet population is evident from Figure 8b where the number density of the droplets that is equal to, or larger than, the initial 6 μm size is observed to decrease with time but in Figure 8c the number density of evaporating droplets during the later stage of the simulation is almost steady, implying some physical process is resulting in the removal of droplets in the evaporating range. In Figure 7c, the absence of sub-micron droplets is seen to happen from 12 to 18 initial eddy turnover time, which also confirms the presence of the complete evaporation of sub-micron droplets for initial 6 μm droplet population. The initial 25 and 18 μm droplet population shows growth in the DSD due to occurrence of collision in different size ranges. Figure 8d presents normalized number density of the droplets in size ranges corresponding to collision between two similar sized droplets $N_{coll}(t)/N_d(0)$ (secondary peaks in Figure 7a,b). For this transient evolution of the number density of colliding droplets, the source is the occurrence of collisions but the sink is the gravitational sedimentation of the droplets out of the domain. For both the initial 25 and 18 μm droplet population in Figure 8d, during the initial 4 initial eddy turnover time, occurrence of collision dominates over the gravitational sedimentation but later the droplets with 1 collision for 25 μm initial droplet population are removed from the domain very rapidly, whereas for the 18 μm initial droplet population, the number of droplets with 1 collision remains almost steady. The occurrence of collision in between larger sized droplets (already with one collision) to the smaller droplets resulting in droplets with two or more collisions was very rare and happened mostly in the case of 25 μm initial droplet population.

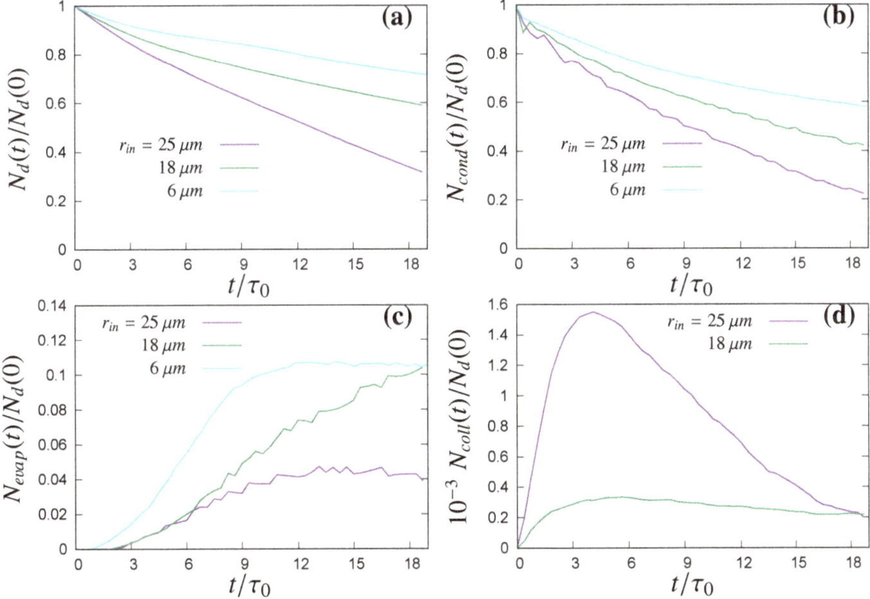

Figure 8. Transient evolution in normalized number density of droplets: (**a**) number density of total number of cloud droplets $N_d(t)$; (**b**) number density of droplets went through condensational growth $N_{cond}(t)$; (**c**) number density of droplets undergone evaporative size reduction $N_{evap}(t)$; and (**d**) number density of droplets with one collision $N_{coll}(t)$ as observed for 25 μm, 18 μm and 6 μm initial droplet size populations. No collisional growth is observed for 6 μm initial droplet populations.

Time evolution of the one-point correlation between fluid flow and cloud droplet $B(a,b) = \langle a'b' \rangle / (\langle a'^2 \rangle \langle b'^2 \rangle)^{1/2}$ (where a and b are fluid and droplet quantities respectively) is presented in Figure 9. Since the droplet distribution is not uniform, for the calculation of these correlation parameters, both the fluctuations in the cloud droplets and fluid flow quantities are plane averaged in horizontal directions (x_1, x_2), considering only grid cells containing cloud droplets withing its Δx^3 volume and the droplet quantities are averaged to the corresponding grid points. In the correlation between the fluctuations in the vertical component of fluid velocity u_3' and droplet velocity v_3' in Figure 9a–c, u_3' and v_3' are very well correlated for initial 6 μm droplet population but less correlated for 18 μm and lesser for 25 μm initial droplet population during the initial instances. Spurious fluctuations in correlation parameters are observed in the interface region, where number of droplet samples are much smaller. Since TKE inside the domain during later instances was much smaller and the Stokes numbers decreases (Figure 2d), as well for all the populations, the velocity fluctuations for both the fluid u_3' and the droplet v_3' tend to correlate more with time advancement. In Figure 9d–f, correlation between the supersaturation fluctuations S' and fluctuations in liquid water content lwc' is presented. Due to particle clustering and high fluctuations in the size of the statistical samples, bezier smoothing has been applied to the correlation in between S' and lwc'. This smoothing significantly modifies the data only in the clear air region of the domain, where number of droplets are very small. Improved statistics could be obtained by considering ensamble averaging between different simulations with independent initial conditions. Since initially inside the undiluted cloudy part of the domain, S' was 0 and gradually the fluctuations picked up, the widening of the interface mixing region can be witnessed in these correlation plots. With positive S', positive lwc' is observed, which shows highest positive correlation for initial 6 μm droplet population and the correlation is less for 18 μm and lesser for the 25 μm initial droplet population. In general, almost no correlation is observed in between the fluid enstrophy $\mathcal{E}'$ and v_3'. However, for the two larger cloud droplet population (initial 25 and 18 μm radii), increase in negative correlation was observed to happen with time.

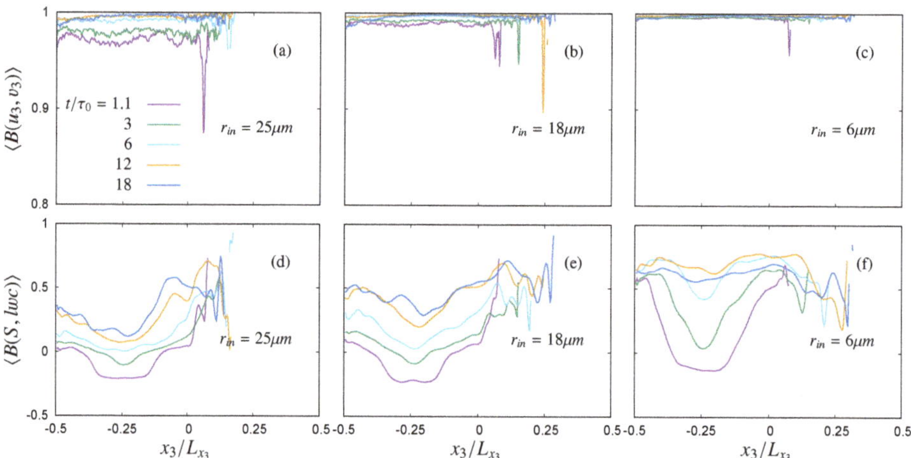

Figure 9. Correlation between fluid and droplet: (**a–c**) correlation between vertical component of fluid velocity fluctuations u_3' and droplet velocity fluctuations v_3' ; and (**d–f**) between supersaturation fluctuations S' and fluctuations in liquid water content lwc'. Vertical panels from the left to right present correlation diagrams for 25 μm, 18 μm and 6 μm initial droplet size populations respectively.

In Figure 10, time evolution of three sample droplets from the three different simulations reaching a specific region in the initial clear air portion of the domain (see the box in the clear air region of panel (**a**), (**c**), (**e**) of Figure 10) after 3 initial eddy ($t/\tau_0 = 3$) turnover time is presented. These droplets were transported to the subsaturated clear air region due to detrainment from the near interface region of the cloudy part of the domain. Due to subsaturation, only the droplets from simulations with initial 25 and 18 μm droplet populations are observed to survive the entire simulation. The impact of gravitational settlement is observed to be very pronounced for the larger droplet population, leading to a short residence time in the subsaturated area (two out of the three droplets were back to the cloudy supersaturated region of the domain almost immediately, see Figure 10a). Whereas, the other remaining droplet was trapped in some eddy to follow lateral movement inside the clear air region. In Figure 10b, these droplets are observed not to follow the fluid velocity exactly but rather show negative v_3 indicating stronger influences of gravitational forces on these droplets. The sample droplets from the simulation with initial 18 μm droplet population shows comparatively less influence under gravitational forces and remains entrapped in the eddies inside the clear air region of the domain (Figure 10c), which produces a continuous size reduction due to local subsaturation (Figure 10d). Whereas, local subsaturation played most important impact on the samples of the droplets from the simulation with initial 6 μm droplet population. After being detrained to the subsaturated clear air region, these droplets cloud not return back to the saturated cloudy part of the domain due to decay in TKE inside the domain (Figure 10e) and eventually evaporated completely in the middle of the simulation duration (Figure 10f).

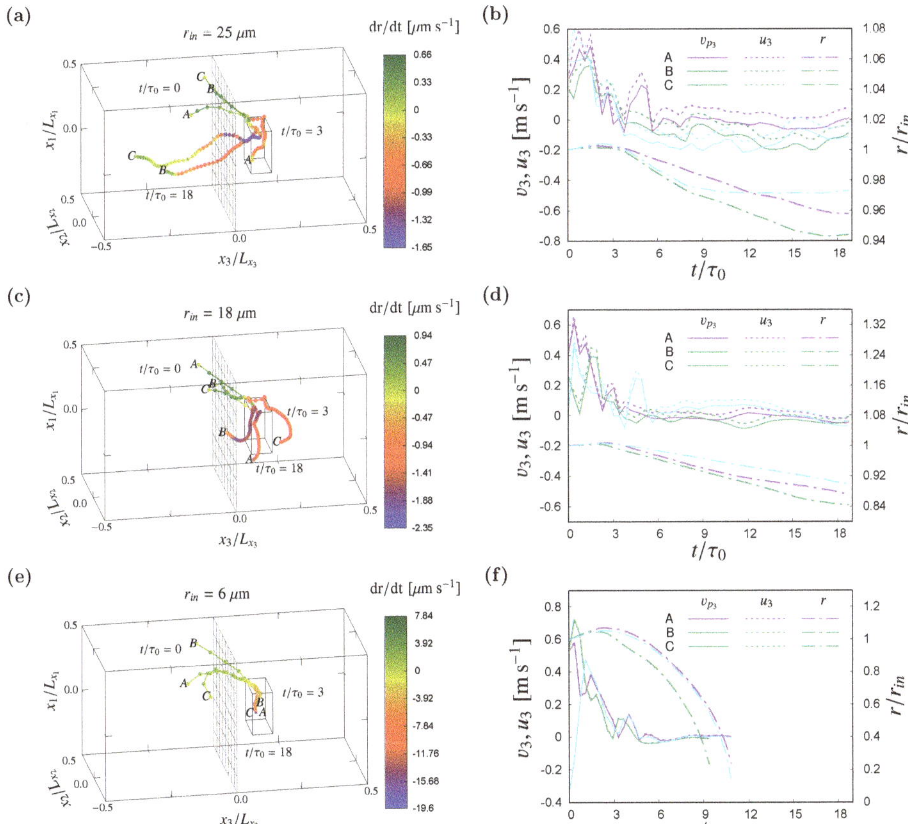

Figure 10. Lagrangian trajectories: A sample of few individual cloud droplet trajectories. (**a,c,e**) visualization of the time evolution of droplet positions; and (**b,d,f**) Lagrangian history of the vertical component of droplet velocity v_3, fluid velocity at that droplet position u_3 and their normalized droplet radius r/r_{in} up to the end of simulation duration ($t/\tau_0 = 18$) or to the end of the droplet life till being completely evaporated (panels **e** and **f**) are presented. Colourbar for the left panels represents droplet radius growth rate dr/dt, thus indicating the droplet positions where condensation or evaporation occur. Size of the droplets are proportional to normalized droplet radius r/r_{in}.

4. Discussions and Concluding Remarks

Understanding of the growth of inertial cloud droplets in transient mixing of horizontal interface is extended in this research by inclusion of the impact of gravitational sedimentation and the impact of collision on the cloud water droplets, along with the condensational/evaporative growth/shrink in size. Three mono-disperse cloud water droplet populations of radii 25 µm, 18 µm and 6 µm were simulated with the same initial background airflow. This flow represents a transient mixing in between a warm cloud top and above-lying clear air. We simulated transient initial value problem, where we initialized the TKE inside the domain following the infield measurements of TKE spectra in the ranges of inertial sub-range and dissipation range. The initial condition of the temperature and the water vapour density replicate the in-cloud measurements of the same quantities, however no fluctuations of temperature or fluctuations in the density of water vapour are introduced in the initial condition. The mixing in between the cloudy and the clear air region of the domain produces fluctuations in the scalar quantities, such as, temperature, density of water vapor and therefore fluctuations on the saturation ratio. Entrainment of subsaturated clear air inside the cloudy region and detrainment of supersaturated cloudy air is observed to happen during this transient mixing phenomenon, which widened the initial thickness of the kinetic energy as well as the scalar interfaces. Initial isotropic homogeneous turbulence inside the cloudy and the clear air region of the simulation domain gradually becomes anisotropic due to mixing, which is evident from the transient growth of the correlation scales.

Depending on the initial size of the droplet population, they are observed to undergo different transients when initialized with the same background flow condition. This study attempts to investigate the differences in between the cloud droplet growth in the *size gap* from 15 µm to 40 µm of radius and for the droplets smaller than 15 µm of radius. We observed that the small 6 µm radius droplets do not grow by collision but droplets inside the *size gap* grow significantly by droplet droplet collision and coalescence. The mixing produces a size broadening of the initial mono-disperse population due to supersaturation fluctuations, which is more evident for the smaller population. In the larger droplet populations of both the 25 µm and 18 µm radii, collisional growth becomes important and multiple collisions occurred in between different sizes of droplets. Since the flow is decaying with time, gravitational settling becomes more and more important for the larger population as the simulation evolves, leading to a gradual removal of falling droplets from the simulation domain and show higher decorrelation in their vertical velocity from that of the fluid velocity. In contrary, the reduction in total droplet count for the 6 µm initial size population happened mostly due to complete evaporation of the sub-micron sized droplets of this population, which were very sensitive to local subsaturation due their very small size. This smaller droplet population has a small Stokes number, which makes them follow the fluid velocity almost perfectly and a very small settling parameter due to negligible terminal velocity, which prevents them from sedimentation. With the decay in TKE, these droplets are observed to remain dispersed in the domain with very negligible vertical velocity.

To extend this research, we foresee studies using DNS with initial poly-disperse droplet population representative of the in-cloud droplet size measurements, which can help in understanding the process of rapid broadening of the droplets inside the *size gap* due to collisional growth. A further analysis could introduce a constant rate of TKE inflow inside both the cloudy region as well as the clear air region of the domain, so that the total TKE inside the simulation domain remains constant, which could be an attempt to simulate precipitating clouds.

Author Contributions: Conceptualization, T.B. and M.I.; methodology, T.B. and M.I.; software, T.B. and M.I.; simulation, T.B.; investigation, T.B. and M.I.; visualization, T.B. and M.I.; writing-original draft, T.B. and M.I.; writing-review and editing, T.B. and M.I.; supervision, M.I.

Funding: This research was funded by the Marie - Skłodowska Curie Actions (MSCA) under the European Union's Horizon 2020 research and innovation programme (grant agreement No. 675675).

Acknowledgments: Computational resources were provided by HPC@POLITO, a project of Academic Computing within the Department of Control and Computer Engineering at the Polytechnic of Turin (http://www.hpc.polito.it). First author would like to acknowledge Daniela Tordella, coordinator of project COMPLETE network and David Codoni.

Conflicts of Interest: The authors declare no conflict of interest.

References

1. Pruppacher, H.R.; Klett, J.D. *Microphysics of Clouds and Precipitation*, 2nd ed.; Springer: Dordrecht, The Netherlands, 2010. [CrossRef]
2. Stull, R.B. *An Introduction to Boundary Layer Meteorology*; Kluwer Academic: Dordrecht, The Netherlands, 1988.
3. Devenish, B.J.; Bartello, P.; Brenguier, J.L.; Collins, L.R.; Grabowski, W.W.; IJzermans, R.H.A.; Malinowski, S.P.; Reeks, M.W.; Vassilicos, J.C.; Wang, L.P.; et al. Droplet growth in warm turbulent clouds. *Q. J. R. Meteorol. Soc.* **2012**, *138*, 1401–1429. [CrossRef]
4. Shaw, R.A. Particle-turbulence interactions in atmospheric clouds. *Ann. Rev. Fluid Mech.* **2003**, *35*, 183–227. [CrossRef]
5. Grabowski, W.W.; Wang, L.P. Growth of Cloud Droplets in a Turbulent Environment. *Ann. Rev. Fluid Mech.* **2013**, *45*, 293–324. [CrossRef]
6. Bodenschatz, E.; Malinowski, S.P.; Shaw, R.A.; Stratmann, F. Can We Understand Clouds Without Turbulence? *Science* **2010**, *327*, 970–971. [CrossRef] [PubMed]
7. de Lozar, A.; Mellado, J.P. Cloud droplets in a bulk formulation and its application to buoyancy reversal instability. *Q. J. R. Meteorol. Soc.* **2014**, *140*, 1493–1504. [CrossRef]
8. Vaillancourt, P.A.; Yau, M.K.; Grabowski, W.W. Microscopic Approach to Cloud Droplet Growth by Condensation. Part I: Model Description and Results without Turbulence. *J. Atmos. Sci.* **2001**, *58*, 1945–1964. [CrossRef]
9. Vaillancourt, P.A.; Yau, M.K.; Bartello, P.; Grabowski, W.W. Microscopic Approach to Cloud Droplet Growth by Condensation. Part II: Turbulence, Clustering, and Condensational Growth. *J. Atmos. Sci.* **2002**, *59*, 3421–3435. [CrossRef]
10. Kumar, B.; Schumacher, J.; Shaw, R.A. Cloud microphysical effects of turbulent mixing and entrainment. *Theor. Comput. Fluid Dyn.* **2013**, *27*, 361–376. [CrossRef]
11. Lanotte, A.S.; Seminara, A.; Toschi, F. Cloud Droplet Growth by Condensation in Homogeneous Isotropic Turbulence. *J. Atmos. Sci.* **2009**, *66*, 1685–1697. [CrossRef]
12. Götzfried, P.; Kumar, B.; Shaw, R.A.; Schumacher, J. Droplet dynamics and fine-scale structure in a shearless turbulent mixing layer with phase changes. *J. Fluid Mech.* **2017**, *814*, 452–483. [CrossRef]
13. Kumar, B.; Schumacher, J.; Shaw, R.A. Lagrangian Mixing Dynamics at the Cloudy–Clear Air Interface. *J. Atmos. Sci.* **2014**, *71*, 2564–2580. [CrossRef]
14. Kumar, B.; Bera, S.; Prabha, T.V.; Grabowski, W.W. Cloud-edge mixing: Direct numerical simulation and observations in Indian Monsoon clouds. *J. Adv. Model. Earth Syst.* **2017**, *9*, 332–353. [CrossRef]
15. Andrejczuk, M.; Grabowski, W.W.; Malinowski, S.P.; Smolarkiewicz, P.K. Numerical Simulation of Cloud–Clear Air Interfacial Mixing. *J. Atmos. Sci.* **2004**, *61*, 1726–1739. [CrossRef]
16. Andrejczuk, M.; Grabowski, W.W.; Malinowski, S.P.; Smolarkiewicz, P.K. Numerical Simulation of Cloud–Clear Air Interfacial Mixing: Effects on Cloud Microphysics. *J. Atmos. Sci.* **2006**, *63*, 3204–3225. [CrossRef]
17. Andrejczuk, M.; Grabowski, W.W.; Malinowski, S.P.; Smolarkiewicz, P.K. Numerical Simulation of Cloud–Clear Air Interfacial Mixing: Homogeneous versus Inhomogeneous Mixing. *J. Atmos. Sci.* **2009**, *66*, 2493–2500. [CrossRef]
18. Gao, Z.; Liu, Y.; Li, X.; Lu, C. Investigation of Turbulent Entrainment-Mixing Processes with a New Particle-Resolved Direct Numerical Simulation Model. *J. Geophys. Res. Atmos.* **2018**, *123*, 2194–2214. [CrossRef]
19. Vaillancourt, P.A.; Yau, M.K. Review of Particle-Turbulence Interactions and Consequences for Cloud Physics. *Bull. Am. Meteorol. Soc.* **2000**, *81*, 285–298. [CrossRef]

20. Siebert, H.; Gerashchenko, S.; Gylfason, A.; Lehmann, K.; Collins, L.; Shaw, R.; Warhaft, Z. Towards understanding the role of turbulence on droplets in clouds: In situ and laboratory measurements. *Atmos. Res.* **2010**, *97*, 426–437. [CrossRef]

21. Lehmann, K.; Siebert, H.; Shaw, R.A. Homogeneous and Inhomogeneous Mixing in Cumulus Clouds: Dependence on Local Turbulence Structure. *J. Atmos. Sci.* **2009**, *66*, 3641–3659. [CrossRef]

22. Tordella, D.; Iovieno, M. Small-Scale Anisotropy in Turbulent Shearless Mixing. *Phys. Rev. Lett.* **2011**, *107*, 194501. [CrossRef]

23. Iovieno, M.; Savino, S.D.; Gallana, L.; Tordella, D. Mixing of a passive scalar across a thin shearless layer: Concentration of intermittency on the sides of the turbulent interface. *J. Turbul.* **2014**, *15*, 311–334. [CrossRef]

24. Ireland, P.J.; Collins, L.R. Direct numerical simulation of inertial particle entrainment in a shearless mixing layer. *J. Fluid Mech.* **2012**, *704*, 301–332. [CrossRef]

25. Cimarelli, A.; Cocconi, G.; Frohnapfel, B.; De Angelis, E. Spectral enstrophy budget in a shear-less flow with turbulent/non-turbulent interface. *Phys. Fluids* **2015**, *27*, 125106. [CrossRef]

26. Gotoh, T.; Suehiro, T.; Saito, I. Continuous growth of cloud droplets in cumulus cloud. *New J. Phys.* **2016**, *18*, 043042. [CrossRef]

27. Elghobashi, S. Particle-laden turbulent flows: direct simulation and closure models. *Appl. Sci. Res.* **1991**, *48*, 301–314. [CrossRef]

28. Iovieno, M.; Cavazzoni, C.; Tordella, D. A new technique for a parallel dealiased pseudospectral Navier–Stokes code. *Comput. Phys. Commun.* **2001**, *141*, 365–374. [CrossRef]

29. Hochbruck, M.; Ostermann, A. Exponential integrators. *Acta Numer.* **2010**, *19*, 209–286. [CrossRef]

30. van Hinsberg, M.A.T.; Boonkkamp, J.H.M.t.T.; Toschi, F.; Clercx, H.J.H. Optimal interpolation schemes for particle tracking in turbulence. *Phys. Rev. E* **2013**, *87*, 043307. [CrossRef]

31. Sundaram, S.; Collins, L.R. Numerical Considerations in Simulating a Turbulent Suspension of Finite-Volume Particles. *J. Comput. Phys.* **1996**, *124*, 337–350. [CrossRef]

32. Carbone, M.; Iovieno, M. Application of the nonuniform fast fourier transform to the direct numerical simulation of two-way coupled particle laden flows. *WIT Trans. Eng. Sci.* **2018**, *120*, 237–253. [CrossRef]

33. Rabe, C.; Malet, J.; Feuillebois, F. Experimental investigation of water droplet binary collisions and description of outcomes with a symmetric Weber number. *Phys. Fluids* **2010**, *22*, 047101. [CrossRef]

34. Jen-La Plante, I.; Ma, Y.; Nurowska, K.; Gerber, H.; Khelif, D.; Karpinska, K.; Kopec, M.K.; Kumala, W.; Malinowski, S.P. Physics of Stratocumulus Top (POST): Turbulence characteristics. *Atmos. Chem. Phys.* **2016**, *16*, 9711–9725. [CrossRef]

35. Malinowski, S.P.; Gerber, H.; Jen-La Plante, I.; Kopec, M.K.; Kumala, W.; Nurowska, K.; Chuang, P.Y.; Khelif, D.; Haman, K.E. Physics of Stratocumulus Top (POST): Turbulent mixing across capping inversion. *Atmos. Chem. Phys.* **2013**, *13*, 12171–12186. [CrossRef]

36. Siebert, H.; Franke, H.; Lehmann, K.; Maser, R.; Saw, E.W.; Schell, D.; Shaw, R.A.; Wendisch, M. Probing Finescale Dynamics and Microphysics of Clouds with Helicopter-Borne Measurements. *Bull. Am. Meteorol. Soc.* **2006**, *87*, 1727–1738. [CrossRef]

37. Grabowski, W.W. Dynamics of cumulus entrainment. In *Geophysical and Astrophysical Convection*; Fox, P., Kerr, R.M., Eds.; Gordon and Breach Science Publishers: Sydney, Australia, 2000; pp. 107–127.

38. Biona, C.B.; Druilhet, A.; Benech, B.; Lyra, R. Diurnal cycle of temperature and wind fluctuations within an African equatorial rain forest. *Agric. For. Meteorol.* **2001**, *109*, 135–141. [CrossRef]

39. Katul, G.G.; Geron, C.D.; Hsieh, C.I.; Vidakovic, B.; Guenther, A.B. Active Turbulence and Scalar Transport near the Forest—Atmosphere Interface. *J. Appl. Meteorol.* **1998**, *37*, 1533–1546. [CrossRef]

40. Radkevich, A.; Lovejoy, S.; Strawbridge, K.B.; Schertzer, D.; Lilley, M. Scaling turbulent atmospheric stratification. III: Space-time stratification of passive scalars from lidar data. *Q. J. R. Meteorol. Soc.* **2008**, *134*, 317–335. [CrossRef]

41. Lothon, M.; Lenschow, D.H.; Mayor, S.D. Doppler Lidar Measurements of Vertical Velocity Spectra in the Convective Planetary Boundary Layer. *Bound.-Layer Meteorol.* **2009**, *132*, 205–226. [CrossRef]

42. Siebert, H.; Shaw, R.A.; Ditas, J.; Schmeissner, T.; Malinowski, S.P.; Bodenschatz, E.; Xu, H. High-resolution measurement of cloud microphysics and turbulence at a mountaintop station. *Atmos. Meas. Tech.* **2015**, *8*, 3219–3228. [CrossRef]

43. Tordella, D.; Iovieno, M. Numerical experiments on the intermediate asymptotics of shear-free turbulent transport and diffusion. *J. Fluid Mech.* **2006**, *549*, 429–441. [CrossRef]

44. Siebert, H.; Shaw, R.A. Supersaturation Fluctuations during the Early Stage of Cumulus Formation. *J. Atmos. Sci.* **2017**, *74*, 975–988. [CrossRef]
45. Monin, A.S.; Yaglom, A.M. *Statistical Fluid Mechanics: Mechanics of Turbulence*; The MIT Press: Cambridge, MA, USA, 1981; Volume 2.
46. Gallana, L.; Savino, S.; Santi, F.; Iovieno, M.; Tordella, D. Energy and water vapor transport across a simplified cloud-clear air interface. *J. Phys. Conf. Ser.* **2014**, *547*. [CrossRef]

Article

Assessment of Solution Algorithms for LES of Turbulent Flows Using OpenFOAM

Santiago López Castaño [1,*,†], Andrea Petronio [2] and Giovanni Petris [1] and Vincenzo Armenio [1,*]

[1]   Dipartimento di Ingegneria e Architettura, Università di Trieste, Via Valerio 6, 34127 Trieste, Italy; petrisgiovanni@gmail.com

[2]   IE-FLUIDS S.r.l, 34127 Trieste, Italy; andrea.petronio@gmail.com

*   Correspondence: santilopez01@gmail.com (S.L.C.); vincenzo.armenio@dia.units.it (V.A.)

†   Current address: Flanders Hydraulics Research, Berchemlei 115, 2140 Antwerpen-Borgerhout, Belgium.

Received: 26 June 2019; Accepted: 8 September 2019; Published: 12 September 2019

Abstract: We validate and test two algorithms for the time integration of the Boussinesq form of the Navier—Stokes equations within the Large Eddy Simulation (LES) methodology for turbulent flows. The algorithms are implemented in the OpenFOAM framework. From one side, we have implemented an energy-conserving incremental-pressure Runge–Kutta (RK4) projection method for the solution of the Navier–Stokes equations together with a dynamic Lagrangian mixed model for momentum and scalar subgrid-scale (SGS) fluxes; from the other side we revisit the PISO algorithm present in OpenFOAM (pisoFoam) in conjunction with the dynamic eddy-viscosity model for SGS momentum fluxes and a Reynolds Analogy for the scalar SGS fluxes, and used for the study of turbulent channel flows and buoyancy-driven flows. In both cases the validity of the anisotropic filter function, suited for non-homogeneous hexahedral meshes, has been studied and proven to be useful for *industrial* LES. Preliminary tests on energy-conservation properties of the algorithms studied (without the inclusion of the subgrid-scale models) show the superiority of RK4 over pisoFoam, which exhibits dissipative features. We carried out additional tests for wall-bounded channel flow and for Rayleigh–Bènard convection in the turbulent regime, by running LES using both algorithms. Results show the RK4 algorithm together with the dynamic Lagrangian mixed model gives better results in the cases analyzed for both first- and second-order statistics. On the other hand, the dissipative features of pisoFoam detected in the previous tests reflect in a less accurate evaluation of the statistics of the turbulent field, although the presence of the subgrid-scale model improves the quality of the results compared to a correspondent coarse direct numerical simulation. In case of Rayleigh–Bènard convection, the results of pisoFoam improve with increasing values of Rayleigh number, and this may be attributed to the Reynolds Analogy used for the subgrid-scale temperature fluxes. Finally, we point out that the present analysis holds for hexahedral meshes. More research is need for extension of the methods proposed to general unstructured grids.

Keywords: OpenFOAM; Runge–Kutta (RK4); PISO; wall-resolved Large Eddy Simulation (LES); Rayleigh–Bènard convection; channel flow

---

1. Introduction

Since the seminal works of Chorin [1] and Temam [2], different variants of the fractional-step method have been proposed and used for the integration of the incompressible form of the unsteady Navier–Stokes Equations (NSE). Over the years, different algorithms have been developed using multistep time advancement techniques, conceived to be used with generalized grid topologies and different locations of the flow variables (co-located or staggered). Algorithms developed for the analysis of unsteady laminar flows have been successfully employed for the Direct Numerical Simulations (DNS) and for the Large Eddy Simulation (LES) of turbulent flows. The latter, by requiring some forms of parametrization for the smallest scales of turbulence, is computationally less demanding. In principle, even non-conservative or dissipative numerical algorithms may give reasonable results when used for DNS, provided that the time step of the simulation and the cell size are smaller than the Kolmogorov scales, so that the truncation error remains confined in the insignificant part of the power spectrum. This is not true for LES though, where a low-pass filter is applied to the NSE to separate the large energy-carrying scales of motion from the small and more isotropic ones; here the truncation error necessarily affects both the resolved and the modeled part of the spectrum, hence the importance of using energy-conserving and non-dissipative algorithms for LES of turbulent flow.

In a sequence of papers [3,4] numerical algorithms for the solution of the NSE in conservative form have been proposed, using Cartesian staggered grids. There, 2nd-order accurate central difference schemes were used for the discrete spatial operators, second-order time accurate schemes for the diffusive terms, and, at least second-order accurate schemes for time integration of the non-linear term (i.e., Adams–Bashforth or Runge–Kutta). The need to move toward complex grid topologies has pushed the development of algorithms using curvilinear coordinate transformations or unstructured grids. A very popular curvilinear-grid fractional-step algorithm is that of Zang et al. [5]: they used a semi-implicit time advancement scheme in conjunction with a second-order accurate explicit Adams–Bashforth scheme for the time integration of the non-linear term. Since the algorithm co-locates the flow variables on the cell centroids it requires a non-conservative formulation of the NSE [6] thus, high-order upwind flux schemes were used for the spatial discretization of the advective term of momentum and scalar transport. The algorithm was successfully employed for the study of LES of turbulent flows over curvilinear geometries. Later, the algorithm was modified using central differences for the calculation of non-linear momentum and scalar fluxes in several papers (see [7] for a general discussion). On the other side, energy-conserving, second-order fractional-step algorithms for finite-differences/volumes have been proposed and used for the LES of turbulent flows in unstructured grids [8].

Given the *Courant-Friedrich-Lewis* (CFL) restrictions imposed by the semi-explicit methods mentioned earlier, implicit multistep methods for the solution of the NSE were sought. In its incompressible version, the NSE present some difficulties in enforcing mass conservation and in writing the discrete non-linear operator of velocities when numerically integrated using implicit multistep methods. The most widely used algorithm for the implicit integration of the incompressible NSE is the PISO algorithm, proposed by Issa [9], which is first-order accurate in time. As it is a non-iterative predictor-corrector method, the momentum equation (predictor) is solved once and, afterward, the non-solenoidal velocities obtained are 'corrected' (at least twice) in order to enforce conservation of mass, by solving the pressure equation and an explicit algebraic corrector equation. It is important to note that for the algorithm to be implicit, the advective term must be linearized. Details on such linearization will be revisited in later sections.

`OpenFOAM` [10–12] is a numerical library based on the idea of offering an unified mathematical framework for the description of systems of hyperbolic Partial Differential Equations (PDE) in a natural way, where the finite-volume (FV) approach is used for the 'discretization' of spatial derivative operators and where multistep time integration techniques are present. In particular, the PISO algorithm present in `OpenFOAM`, referred also as `pisoFoam`, has become a tool of widespread use in industry as well as in academia for the solution of the incompressible NSE. In academia, the code has been largely used for unsteady simulations using Detached Large Eddy, wall-resolving Large Eddy (A simulation where the near-wall turbulent structures are fully resolved and a no-slip boundary condition is applied at the wall), implicit Large Eddy (ILES) (a simulation where the SGS model is replaced by the truncation error of the numerical algorithm) models and more recently for DNS. There, it has proven to be the *go-to* code in many groups for the simulation of a wide class of flows of interest for industrial and environmental applications.

Special attention has been given to ILES in the early years of the code for use in compressible and incompressible flows, given that low-order FV methods are amenable to use within the Monotone

Implicit LES, or MILES, framework (for a review, see [13]). In general, the use of `pisoFoam` along with MILES for the numerical simulation of high-Reynolds number flows has proven to be accurate in a wide range of scenarios [13,14]. However, it is difficult to determine *a-priori* whether for a given grid resolution (and type of grid elements) MILES would resolve accurately the backscattering effects of turbulence or, more importantly, whether the numerical diffusion may be considered representative of the unresolved scales of turbulence: a finite-scale analysis proves the existence of a 'scale-similar term' for *structured* grids, but not so for more general, unstructured, grids composed of non-hexahedral elements. In other words, if special care is not taken, solutions obtained using MILES may tend to be *over-dissipative* particularly in cases where non-linear effects are predominant or needed to be accurately modeled.

More recently wall-resolving and Wall-Modeled LES models (WMLES) (a LES where the near-wall dynamics is not resolved and the wall shear stress is parametrized using a model), although present since the early stages of `OpenFOAM`, have begun to be used for the simulation of turbulent flows. However, most studies have focused on high-Reynolds external, or wall-bounded, flows with massive separation, i.e., airfoils, wind turbines, flows over complex topography, where over-dissipation may not be noticeable. A more recent account on low-Reynolds LES on channel flows using `pisoFoam` was made by Vuorinen et al. [15]. There, it was shown that DNS results for a wall-bounded channel flow obtained using `pisoFoam` are over-dissipative with respect to the results obtained for the same case using a Runge–Kutta Navier–Stokes Equation solver, implemented within the `OpenFOAM` framework. In an attempt to remedy the over-dissipative properties of `pisoFoam`, various authors [15,16] have implemented non-incremental projection methods for the solution of the incompressible NSE in `OpenFOAM`. All such works consider the Rhie–Chow interpolation for the momentum fluxes projected onto the faces, in order to guarantee velocity-pressure coupling on the discrete PDE system. However, note that by not considering the pressure gradient in the predictor step (non-incremental), the pressure obtained is only first-order accurate in time, depending on the Reynolds number. The latter term is often referred to as *computational pressure* since it differs from the physical one by a term proportional to the computational parameters. Furthermore, boundary conditions for the pressure equation are not trivial: an accurate definition of the pressure gradient in the no-slip boundaries require the projection of the momentum equations onto the boundary which, at the corrector step, are unknown. A *deferred-correction* approach can be used for determining the projection of the NSE onto the no-slip boundaries. However, the cited works fail to mention how the boundary conditions for the pressure equation are treated.

A more accurate account on the over-dissipative properties of `pisoFoam` and other commercial codes was made by Komen et al. [17]. An analysis of the turbulent kinetic energy budget shows that for the resolved and *residual* dissipation rates there is only an 8% difference, for quasi-DNS turbulent channel flows. Furthermore, explicit LES calculations on channel flows show that subgrid scale contributions are *at least* 3.5 times lower compared to residual contributions due to numerical dissipation, and such ratio varies little with Re_τ. The authors argue that no clear improvement is gained when using an explicit LES model compared to just running an under-resolved DNS when using `pisoFoam`. This issue will be exploited in a successive Section.

Incidentally, the work of Tuković et al. [18] has revisited the PISO implementation present in `pisoFoam` and shown that the time integration order of pressure and velocity is time-dependent, passing from first-order accuracy for small time-steps to second-order accuracy for bigger time-steps, and errors on pressure tend to accumulate in time for very low CFL numbers. There, the authors propose to project in time the face fluxes of momentum, in an attempt to mitigate the error caused by the splitting of the non-linear term in PISO. Please note that the former remark is not new: the original work of Issa [9] shows that the operator-splitting in time is only first-order accurate.

The present work focuses on the analysis of the overall performance of the standard implementations of `pisoFoam` compared to those of an incremental-pressure correction version of the Runge–Kutta algorithm proposed by Le and Moin [4] (hereinafter RK4). Specifically, first, two benchmark cases are run to test energy-conservation properties of the two algorithms: (1) a 2-D Taylor vortex, in which boundary condition consistency can be verified; (2) the calculation of the energy decay rate of a 2D Tollmien–Schlichting wave in a channel, to determine the time integration order of the algorithm.

Successively, the algorithms are validated for LES of neutral and buoyant flows. In this context, filter consistency and associated boundary conditions are also tested and compared with the standard solvers available in OpenFOAM. In particular, pisoFoam is used in conjunction with the *Dynamic Smagorinsky* model for turbulent neutral and buoyancy flows available in the framework. To be noted that in OpenFOAM the subgrid-scale (SGS) scalar fluxes are parametrized using the Reynolds Analogy, namely assuming a constant value for the SGS Prandtl number. Our implementation of RK4 is used with a *Dynamic Mixed Lagrangian Smagorinsky* model [19,20] implemented in the present work. Also, a new model for the calculation of the turbulent SGS scalar fluxes, in the spirit of the one proposed by Armenio and Sarkar [21], is implemented and compared with results obtained using the Reynolds Analogy for the SGS scalar fluxes present in OpenFOAM for simulations where buoyancy is active.

The verification of the LES models along with the fluid flow solver for turbulence flows uses two cases: (1) a turbulent neutral Poiseuille flow, where the consistency of the filters and of the SGS model can be evaluated; and (2) Rayleigh–Benard convection between infinite plates, which may serve as a test of the overall performance in the presence of active scalars.

The paper is organized as follows: Section 2 reports the mathematical formulation of the different algorithms just discussed, including the description of the SGS model for the momentum and scalar transport equations; Section 3 reports a description of the two algorithms tested in this paper, namely PISO and RK4; Section 4 reports results of tests aimed at verifying the energy-conservation properties of the two algorithms; Section 5 reports verification tests of the two algorithms for LES of turbulent flows and, finally, concluding remarks and a discussion are given in Section 6.

2. Mathematical Formulation

In this section, a description of the Boussinesq form of the unsteady NSE for the filtered flow variables is presented. We recall that the Boussinesq form allows the study of systems with variable density, only when the density variations are much smaller than the bulk density of the flow. Spatial filtering over the NSE produces unresolved terms (SGS fluxes) which need to be modeled.

The filtered Boussinesq form of the Navier–Stokes equations read as:

$$\frac{\partial \bar{u}_i}{\partial x_i} = 0, \tag{1}$$

$$\frac{\partial \bar{u}_i}{\partial t} + \frac{\partial \bar{u}_j \bar{u}_i}{\partial x_j} = -\frac{1}{\rho_0}\frac{\partial \bar{p}}{\partial x_i} + \nu\frac{\partial^2 \bar{u}_i}{\partial x_j \partial x_j} - \frac{\partial \tau_{ij}^r}{\partial x_j} - \bar{\rho} g \delta_{i,3}, \tag{2}$$

$$\frac{\partial \bar{\rho}}{\partial t} + \frac{\partial \bar{u}_j \bar{\rho}}{\partial x_j} = \kappa\frac{\partial^2 \bar{\rho}}{\partial x_j \partial x_j} - \frac{\partial \eta_j^r}{\partial x_j}, \tag{3}$$

where u_i is the velocity component in the i-direction, ρ is the density perturbation with respect to the bulk density ρ_0, p is the hydrodynamic pressure, g is gravity, ν is the kinematic viscosity, and κ the scalar diffusivity. The quantities

$$\tau_{ij}^r = \overline{u_i u_j} - \bar{u}_i \bar{u}_j, \tag{4}$$

$$\eta_j^r = \overline{\rho u_j} - \bar{\rho}\, \bar{u}_j, \tag{5}$$

are the SGS, or residual, fluxes of momentum and density.

The frame of reference has x_1 and x_2 over the streamwise and spanwise horizontal directions, and x_3 vertical upward. The streamwise, spanwise, and vertical directions may be also referred to as (x, y, z); similarly, for the velocity components we also use (u, v, w), depending on the context.

2.1. Turbulence Modeling: LES

This apart will provide a description of the LES models used in this work. The *Dynamic Lagrangian Mixed* Smagorinsky Model (DLMM) for scalar and momentum turbulence will be described first, along with the filter used. Afterwards, a brief description of the *Dynamic 'Mean'* Smagorinsky Model (DMM) for momentum turbulence and the *Reynolds Analogy* (RA) for modeling the scalar SGS fluxes will be made. Notice that the latter methodologies (DMM and RA) are also available in `foam-extend`, a fork of `OpenFOAM`.

2.1.1. Dynamic Lagrangian Mixed Smagorinsky for Scalar and Momentum Turbulence

By applying a second filter, $\widehat{(\cdot)}$ of size $\widehat{\Delta} = 2\overline{\Delta}$ on Equation (4) and after some algebraic manipulations, a relation for the residual stresses, referred to as the Germano Identity, is obtained:

$$\widehat{\tau}_{ij}^{r} = T_{ij} - L_{ij}, \tag{6}$$

$$L_{ij} = \widehat{\overline{u}_i \overline{u}_j} - \widehat{\overline{u}}_i \widehat{\overline{u}}_i, \tag{7}$$

$$T_{ij} = \widehat{\overline{u_i u_j}} - \widehat{\overline{u}}_i \widehat{\overline{u}}_i. \tag{8}$$

The same holds for the residual scalar fluxes, just by replacing u_i with ρ.

The SGS fluxes of momentum can be expressed as the sum of a scale-similar part and a Smagorinsky eddy-viscosity term:

$$\tau_{ij}^{r} \approx m_{ij} = \underbrace{\left(\overline{\overline{u}_i \overline{u}_j} - \overline{\overline{u}}_i \overline{\overline{u}}_j \right)}_{\tau_{ij}^{SS}} - 2c_s \overline{\Delta}^2 |\overline{S}| \overline{S}_{ij}, \tag{9}$$

where $|\overline{S}|$ is the contraction of the strain rate tensor of the filtered field, and $c_s \overline{\Delta}^2 \sim l^2$ is a measure of the mixing length for the unresolved eddies. By using identity (6), one can obtain an expression for c_s. By writing the mixed Smagorinsky model for the NSE using the filter $\widehat{(\cdot)}$, one obtains

$$T_{ij} = \left(\widehat{\overline{\overline{u}_i \overline{u}_j}} - \widehat{\overline{\overline{u}}}_i \widehat{\overline{\overline{u}}}_i \right) - 2c_s \widehat{\overline{\Delta}}^2 |\widehat{\overline{S}}| \widehat{\overline{S}}_{ij},$$

which, once substituted in Germano's identity Equation (6), gives:

$$H_{ij} + c_s M_{ij} = L_{ij}, \tag{10}$$

$$M_{ij} = -2\widehat{\overline{\Delta}}^2 |\widehat{\overline{S}}| \widehat{\overline{S}}_{ij} + 2\overline{\Delta}^2 \widehat{|\overline{S}| \overline{S}_{ij}}, \tag{11}$$

where c_s is assumed constant at the filter level.

The system (10) is overdetermined. A least-square approach for finding c_s leads to the following final expression:

$$c_s = \frac{\langle M_{ij}(L_{ij} - H_{ij}) \rangle}{\langle M_{ij} M_{ij} \rangle}. \tag{12}$$

Please note that the brackets $\langle (\cdot) \rangle$ indicate some form of averaging. By time-averaging along Lagrangian trajectories (i.e., pathlines), one obtains a *mixed version* of the *Dynamic Lagrangian* Model [19] where

$$c_s = \frac{\mathcal{F}_{lm}}{\mathcal{F}_{mm}}, \tag{13}$$

$$\frac{D\mathcal{F}_{lm}}{Dt} = \frac{1}{T}\left[M_{ij}(L_{ij} - H_{ij}) - \mathcal{F}_{lm}\right], \tag{14}$$

$$\frac{D\mathcal{F}_{mm}}{Dt} = \frac{1}{T}(M_{ij}M_{ij} - \mathcal{F}_{mm}), \tag{15}$$

$$T = \frac{3}{2}\overline{\Delta}(\mathcal{F}_{mm}\mathcal{F}_{lm})^{-1/8}. \tag{16}$$

For the unresolved SGS fluxes of density, one can define the *subgrid buoyancy flux* in a similar way to the subgrid stress tensor, as follows:

$$\eta_j^r \approx n_j = (\overline{\overline{\rho}\,\overline{u}_j} - \overline{\overline{\rho}}\,\overline{\overline{u}}_j) - c_c\overline{\Delta}^2|\overline{S}|\frac{\partial\overline{\rho}}{\partial x_j} \tag{17}$$

By analogy, the same procedure for the derivation of the LES model for the momentum equation follows, thus one can determine the constant c_c in a least-square sense:

$$c_c = \frac{\langle M_j^c(L_j^c - H_{ij}^c)\rangle}{\langle M_j^c M_j^c\rangle} \tag{18}$$

where

$$H_j^c = (\overline{\overline{\rho}\,\overline{u}_j} - \overline{\overline{\rho}}\,\overline{\overline{u}}_j) - (\widehat{\overline{\overline{\rho}}\,\overline{\overline{u}}_j} - \widehat{\overline{\overline{\rho}}}\,\widehat{\overline{\overline{u}}}_j), \tag{19}$$

$$L_j^c = \widehat{\overline{\rho}\,\overline{u}_j} - \widehat{\overline{\rho}}\,\widehat{\overline{u}}_j, \tag{20}$$

$$M_j^c = -\widehat{\overline{\Delta}}^2|\widehat{\overline{S}}|\frac{\partial\widehat{\overline{\rho}}}{\partial x_j} + \overline{\Delta}^2|\overline{S}|\frac{\widehat{\partial\overline{\rho}}}{\partial x_j}. \tag{21}$$

Notice that the averaging procedure proposed in Equation (18) can be made also along Lagrangian trajectories, therefore:

$$c_c = \frac{\mathcal{F}_l^c}{\mathcal{F}_m^c}, \tag{22}$$

$$\frac{D\mathcal{F}_l^c}{Dt} = \frac{1}{T}\left[M_j^c(L_j^c - H_j^c) - \mathcal{F}_l^c\right], \tag{23}$$

$$\frac{D\mathcal{F}_m^c}{Dt} = \frac{1}{T}(M_j^c M_j^c - \mathcal{F}_m^c) \tag{24}$$

$$T = \frac{3}{2}\overline{\Delta}\,\sigma_c(\mathcal{F}_m^c\mathcal{F}_l^c)^{-1/4}, \tag{25}$$

where σ_c is the standard deviation of the scalar concentration. Notice that the particular choice of scalar turbulence timescale goes in correspondence with the timescale proposed for the residual momentum fluxes; that is, T is reduced in regions where the scalar gradients or velocity-scalar correlations are large, and vice-versa. The dynamic evaluation of the SGS fluxes for the scalar equation, instead of using the Reynolds Analogy for the SGS Prandtl number ($Pr_r = \nu_r/\kappa_r$), is particularly advantageous when studying stable stratified turbulence (see [21] for a discussion).

Here we use the non-uniform Laplacian filter [22], i.e., a Laplace filter where the filter width is non-constant. It is a generalization of the Laplacian filter proposed by Germano [23]. Such filter is a Reynolds operator in space, i.e., it commutes with derivatives, it preserves constants, and is self-adjoint. For structured grid one has the following:

$$\bar{f} = f + \frac{1}{24}\frac{\partial}{\partial x_k}\left(\Delta_k \frac{\partial f}{\partial x_k}\right), \tag{26}$$

where f is the field being filtered, and Δ_i is the filter width along the i-direction. In structured grids, the filter width is just a multiple of the width, height, or length of the volume cell where the field is being filtered. The filtered field does not necessarily 'inherit' the boundary conditions of the unfiltered field: here we will assume that they do, under the assumption that $\Delta_k \rightarrow 0$ as one approaches the boundaries.

Finally, please note that for the family of Gaussian filters the following identity

$$\widehat{\overline{\Delta}}^2 = \overline{\Delta}^2 + \widehat{\Delta}^2$$

is valid. Previous works [5,6] using finite-difference codes have considered the filter width at the test level as

$$\widehat{\overline{\Delta}} = 2\overline{\Delta}, \tag{27}$$

which is only valid for spectral cutoff filters. Such inconsistency in finite-volume and finite-difference algorithms is noted by Vreman et al. [20]. The standard implementation of the Smagorinsky models in OpenFOAM has said inconsistency.

2.1.2. Dynamic 'Mean' Smagorinsky for Momentum and Reynolds Analogy for Scalar Turbulence

A difference in the implementation of the Dynamic Smagorinsky Model is present in OpenFOAM, where the parameter c_s is volume-averaged across the domain instead of being averaged along homogeneous directions as proposed by Germano et al. [24], i.e., starting from Equation (9), the calculation of the dynamic coefficient for the *sgs* mixing length is as follows:

$$c_s = \left\langle \frac{M_{ij}L_{ij}}{M_{ij}M_{ij}} \right\rangle_v,$$

where $\langle \cdot \rangle_v$ means volume-averaging (hence the term 'Mean'). Notice that the calculation of c_s does not consider the H_{ij} term, since is not a mixed formulation ($\tau_{ij}^{SS} = 0$). The terms M_{ij} and L_{ij} are calculated directly from Equations (11) and (7), respectively. As noted previously, the test filter width in this formulation is taken as $\widehat{\overline{\Delta}} = 2\overline{\Delta}$.

The Reynolds Analogy for the scalar SGS fluxes simply considers that the turbulent mixing length for scalars l_c scales linearly with that of momentum l_s, where such scaling factor $Pr_{SGS} = l_c/l_s$ (the ratio between turbulent mixing lengths) remains constant. This assumption leads to the expression:

$$\eta_j^r \approx n_j = \frac{\nu_r}{Pr_{SGS}}\frac{\partial \bar{\rho}}{\partial x_j}. \tag{28}$$

3. Numerical Formulations

In this section, we briefly describe two algorithms for the numerical solution of the incompressible NSE. First, we describe the one present in the OpenFOAM framework, PISO, in its implicit form. Such form (pisoFoam) is a generalization of the time integration in PISO. Some particularities on the PISO implementation for implicit Euler time integration are discussed as well. These peculiarities appertain only the implicit Euler time integration scheme implemented in OpenFOAM for PISO. Afterward, we give a description of our semi-implicit three-stage Runge–Kutta incremental projection method.

3.1. PISO Algorithm

A finite-volume implementation of the incompressible NSE is carried out implicitly, using the PISO algorithm, which will be described next. Hereafter, c denotes the index of an arbitrary velocity node, λ is the index that indicates neighboring points, n is a superscript denoting the current time iteration, m denotes the inner iterations within each n iteration, Q any source term that may be function of velocity (i.e., the Coriolis force) or not (i.e., the buoyancy force), and A is the coefficient matrix resulting from the linearization of spatial operators. The momentum equation is advanced in time by means the following implicit-in-time linear system:

$$A_c(u_i)u_{i,c}^{n+1} + \sum_{\lambda} A_{\lambda}(u_i)u_{i,\lambda}^{n+1} = Q^{n+1}(u_i) - \left(\frac{\delta P^{n+1}}{\delta x_i}\right)_c,$$

where P is the hydrodynamic pressure divided by the bulk density ρ_0. Please note that this system is non-linear for $\sum_{\lambda} A_{\lambda}(u_i)u_{i,\lambda}^{n+1}$ thus one has to resort to inner iterations for which the coefficient matrix A is no longer constant. Also, the mass conservation constraint cannot be directly enforced in the system. By projecting the momentum equation in a solenoidal space one can enforce mass conservation, i.e., at a certain time step n one solves a *predictor equation* for an intermediate velocity u_i^*:

$$A_c(u_i)u_{i,c}^{*,m} + \sum_{\lambda} A_{\lambda}(u_i)u_{i,\lambda}^{*,m} = Q^{m-1}(u_i) - \left(\frac{\delta P^{m-1}}{\delta x_i}\right)_c,$$

which is not solenoidal. In order to enforce the mass conservation, a *corrector step* is applied to the intermediate velocity to obtain a solenoidal intermediate velocity $\widetilde{u}_i^*$:

$$\widetilde{u}_i^{*,m} = u_i^{*,m} - \frac{1}{A_c(u_i)}\left(\frac{\delta P^{m-1}}{\delta x_i}\right)_c.$$

Please note that the final solenoidal velocity, after m inner iterations, is obtained using

$$u_i^{m} = \widetilde{u}_i^{*,m} - \frac{1}{A_c(u_i)}\left(\frac{\delta P^{m}}{\delta x_i}\right)_c,$$

where $\delta P^m / \delta x_i$ is obtained by solving the pressure Poisson equation

$$\frac{\delta}{\delta x_i}\left[\frac{1}{A_c(u_i)}\left(\frac{\delta P^{m}}{\delta x_i}\right)\right]_c = \left[\frac{\delta \widetilde{u}_i^{*,m}}{\delta x_i}\right]_c.$$

The OpenFOAM framework implements the PISO algorithm for arbitrary grid topologies, and various time-integration schemes where the flow variables are co-located at the element's centroid. For this purpose, the Rhie–Chow [25] interpolation is made to guarantee velocity-pressure coupling. If one defines the off-diagonal coefficient flux matrix:

$$H_{\lambda,i}(u_i^{n+1}) = Q_{\lambda}^{m}(u_i) - \sum_{\lambda} A_{\lambda}(u_i)u_{i,\lambda}^{n+1},$$

According to the Rhie–Chow method, one can calculate the momentum fluxes, U_{λ}, onto the faces with surface area normal S_{λ} of each volume cell, in the following way

$$U_{\lambda} = \frac{H_{\lambda,i}(u_i^{n+1})}{A_{\lambda}(u_i)}S_{\lambda,i} - \left(\frac{\frac{\delta p^{n+1}}{\delta x_i}}{A_c(u_i)}\right)_{\lambda} S_{\lambda,i}.$$

In an attempt to increase robustness of the Euler-Implicit time scheme in OpenFOAM, a relaxation on the face momentum fluxes is made, in the following manner

$$U_\lambda^{n+1} = U_\lambda + U_{\lambda,\text{relax}}, \tag{29}$$

$$U_{\lambda,\text{relax}} = \frac{K_c}{\Delta t}\left(U_\lambda^n - (u_i^n)_\lambda S_\lambda\right)\left(A_c(u_i)\right)_\lambda,$$

$$K_c = 1 - \min\left(\frac{|U_\lambda^n - (u_i^n)_\lambda S_\lambda|}{|U_\lambda| + \epsilon^2}, 1\right),$$

where Δt is the time step. Please note that as the simulation arrives to a stationary state (in a RANS sense), $K_c \to 0$. In the case of unsteady simulations, it might be interpreted as a correction for interpolation errors due to non-orthogonality of the grid. However, note that this correction introduces a $O(1/\Delta t)$ error in the time integration of the advective term in the case of having structured grids.

3.2. Runge–Kutta Algorithm

Here we present a finite-volume implementation of an incremental-pressure projection method, where the advective term is projected in time using a classical RK4 scheme whereas the diffusive term is projected using Euler-Implicit scheme. The linearized system is as follows

$$\frac{u_i^{*,m} - u_i^{m-1}}{\Delta t} = (\alpha_m + \beta_m)\mathcal{L}(u_i^{m-1})$$
$$- \beta_m \mathcal{L}(u_i^{*,m} - u_i^{m-1})$$
$$- \gamma_m C(u_i^{m-1}) - \xi_m C(u_i^{m-2})$$
$$- (\alpha_m + \beta_m)\frac{\delta P^{*,m}}{\delta x_i} \tag{30}$$

$$\frac{u_i^m - u_i^{*,m}}{\Delta t} = -\frac{\delta(\delta P)}{\delta x_i}, \tag{31}$$

where the discrete diffusion operator is indicated by $\mathcal{L}$, and the advective operator by C. Please note that the explicit treatment of the advective term avoids the linearization of $u_i u_j$. Mass conservation is enforced via Poisson equation for pressure:

$$\frac{\delta}{\delta x_i}\left[\frac{\delta(\delta P)}{\delta x_i}\right] = \frac{1}{\Delta t}\frac{\delta u_i^{*,m}}{\delta x_i} \tag{32}$$

$$P^{*,m} = P^{*,m-1} + \delta P \tag{33}$$

Please note that each time iteration n is split in three substeps m, where $m = 1$ gives $m - 1 = n$ and $m = 3 = n + 1$. The Runge–Kutta coefficients are the following:

$$\alpha_1 = \beta_1 = 4/15, \ \alpha_2 = \beta_2 = 1/15, \ \alpha_3 = \beta_3 = 1/6$$
$$\gamma_1 = 8/15, \ \gamma_2 = 5/12, \ \gamma_3 = 3/4$$
$$\xi_1 = 0, \ \xi_2 = -17/60, \ \xi_3 = -5/12$$

To preserve the velocity-pressure coupling in a co-located finite-volume mesh, a Rhie–Chow interpolation is used.

As mentioned earlier, in order to guarantee consistency in the calculation of the pressure, appropriate boundary conditions for the Poisson equation are needed [1,3,5,26]. Classical non-incremental projection methods obtain the following Neumann boundary condition for pressure on non-periodic boundaries, by projecting the momentum equations onto said boundary with surface-normal n_i:

$$\frac{\partial P}{\partial x_i}\, n_i = -\frac{u_i^{*,m} - u_i^{m-1}}{\Delta t}\, n_i \tag{34}$$

Please note that the right-hand-side of Equation (34) is unknown at the corrector stage of the algorithm, so deferred extrapolation methods for projecting the velocity components normal to the boundaries are needed [3]. In the present case, note that the Poisson equation is written in terms of *pressure increments*, δP. Thus, it is valid to use a homogeneous Neumann boundary condition [27] for δP. This strategy is also used in PISO.

4. Verification of Conservation Properties

Here a study on the energy-conservation properties of the algorithms discussed above is presented. Specifically, we make a formal comparison with known analytic solutions and benchmark cases. First, we consider the case of the unsteady and inviscid 2D Taylor vortex but where the boundary conditions for velocity are imposed from the analytic solution. Then we consider the energy growth of one unstable mode for the case of Poiseuille flows, for which an analytic expression exists for the first phase of the growth of a linear disturbance; in this case the flow is confined between two solid walls and, differently from the previous case, the diffusion term becomes part of the solution algorithm. In the latter case only the RK4 algorithm is tested for reasons that will be explained later in this section. For both algorithms, the discretization of the divergence and gradient operators are set to be Gauss linear, which corresponds to the central differences schemes; also, fourth-order schemes are used for the Laplacian operator. Finally, unless explicitly specified, hereinafter the time integration used for PISO will be the implicit Crank-Nicolson method to avoid the relaxation function present in the Euler algorithm in OpenFOAM.

4.1. The Two-Dimensional Inviscid Taylor Vortex

Previous studies have reviewed the low-dissipative properties of the fully explicit Accelerated Runge–Kutta and classic Runge–Kutta projection methods for incompressible flows in OpenFOAM [15,16]. To show that, the two-dimensional Taylor vortex has been used as benchmark due to its simplicity and the existence of a viscous analytic solution. In the case of finite viscosity, the velocity and pressure fields in a domain $x \in [0, 2\pi]$, $y \in [0, 2\pi]$ are the following:

$$u(x,y,t) = \sin x \cos y\, e^{-vt}$$
$$v(x,y,t) = -\sin y \cos x\, e^{-vt}$$
$$p(x,y,t) = \frac{1}{4}(\cos 2x + \cos 2y)\, e^{-2vt}$$

The present array of vortices, in the absence of viscosity, must yield zero dissipation and remain stationary, i.e., $dE/dt = 0$. From the definition of kinetic energy, an analytic solution for volume-averaged kinetic energy, E, is the following:

$$E(t) = \frac{1}{4} e^{-8vt} \tag{35}$$

In the inviscid case, it is clear that $E(t) = E_0 = 1/4$. The kinetic energy resulting from the numerical integration of the aforementioned solution using PISO and RK4 is shown in Figure 1. Here, a grid of size 32×32 with uniform spacing is used; additionally, a second-order flux interpolation is used for the calculation of the advective term. The RK4 conserves the energy of the system, while *pisoFoam* exhibits dissipative properties with an error proportional to the time step. From the results shown in Figure 1 is easy to show that the error increases linearly with Δt. According to what was previously stated, such error in the PISO algorithm is not due to the relaxation term of Equation (29) neither from the spatial schemes for the different operators used nor to the Rhie–Chow interpolation, since these are also used for RK4. This error originates from the treatment of the advective term, as will be shown next.

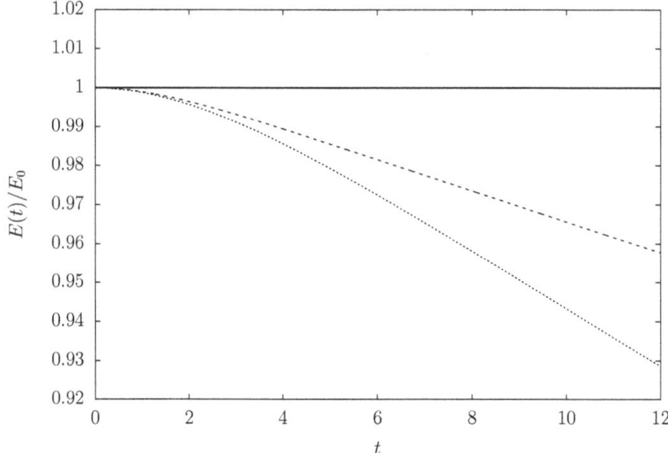

Figure 1. Normalized volume-averaged kinetic energy for the inviscid two-dimensional Taylor vortex: Solid line, RK4 with Cou = 0.8; dashed line, *pisoFoam* with Cou = 0.2; dotted line, *pisoFoam* with Cou = 0.8.

The class abstraction for the multiphysics finite-volume framework present in OpenFOAM is thought around the premise of writing hyperbolic PDEs in a natural way, more specifically transport equations. To this end, the following assumption must be made for the time integration of the transport equation of some quantity ϕ:

$$\left(\frac{\partial u_j \phi}{\partial x_j}\right)^{n+1} \approx \frac{\partial u_j^n \phi^{n+1}}{\partial x_j} + O(\Delta t^\alpha), \ \alpha > 1$$

Please note that the error induced by this time-splitting depends on the degree of coupling that exists between the velocity and the transported variable ϕ. In the case of momentum transport, such splitting introduces an additional approximation, since a Taylor expansion around the time step yields the following

$$u_j^{n+1} u_i^{n+1} = u_j^n u_i^{n+1} + u_j^{n+1} u_i^n - u_j^n u_i^n + O(\Delta t^2) \tag{36}$$

Please note that the formulation in OpenFOAM only retains one or two of the three terms shown in the right-hand side of the previous equation, depending on the time scheme selected. Such approximation is remedied by the addition of inner iterations in the corrector cycle of PISO, and overall, the algorithm behaves as being first-order accurate in time. On the other hand, in RK4 this operation is not required, and it makes the algorithm more accurate than *pisoFoam*.

4.2. Hydrodynamic Instabilities: 2D Tollmien–Schlichting Waves

Two-dimensional disturbances occurring in laminar Poiseuille flow, causing transition to turbulence, may be determined using the Orr-Sommerfeld/Squire equations. Such transition, being the least stable mode of the aforementioned equations, is often called K-type transition. This case is of particular interest in the verification of energy-conserving Navier–Stokes solvers since analytic expressions exist for the energy growth rate

$$\log\left(\frac{E(t)}{E_0}\right) = 2|\alpha c_i|t.$$

Please note that the wave velocity c is a complex number, defined as

$$c = c_r + c_i\, i = -\frac{\lambda}{i\alpha},$$

where λ is the eigenvalue of the least stable mode for the laminar Poiseuille flows. The least stable mode is then chosen by obtaining the eigenspectrum of the Orr-Sommerfeld equation for a certain *critical Reynolds number*, Re_{cr}, based on the centerline velocity, and a streamwise wavelength, α. We use the same parameters used by [28], i.e., we use $Re_{cr} = 5000$ and $\alpha = 1.12$ which according to the neutral stability curve for Poiseuille flows, is stable. The resulting eigenspectrum (wave celerities) and eigenvectors for the least stable mode are shown in Figure 2. As a reminder, the velocities (u, v) relate to the eigenvectors $(\hat{u}, \hat{v})$ in the following way:

$$u(x, y, t) = \hat{u}(y)\exp[i(\alpha x + \omega t)],$$
$$v(x, y, t) = \hat{v}(y)\exp[i(\alpha x + \omega t)].$$

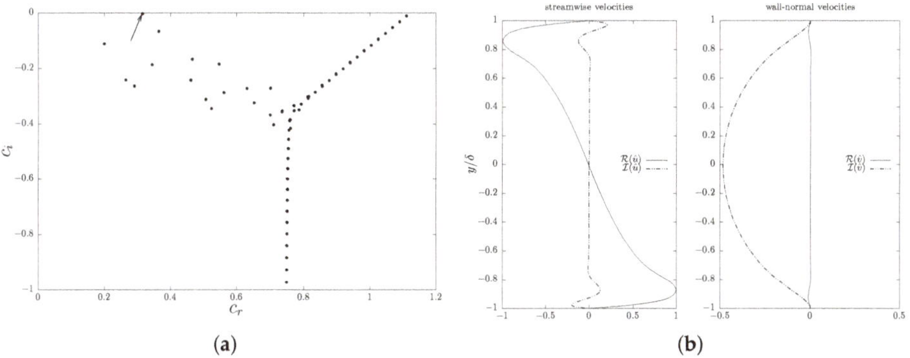

(a)   (b)

Figure 2. (a) Eigenspectrum of the Orr-Sommerfeld/Squire equations for a 2D TS wave of $Re_{cr} = 5000$ and $\alpha = 1.12$. The arrow indicates its least stable wave celerity $c = 0.2817524273 - 0.0024847328i$; (b) streamwise (**left**) and vertical (**right**) velocity perturbation coefficients for the least stable mode.

The computational domain is of dimensions $(2\pi/\alpha) \times 2$, and the resolution of the coarsest mesh is 64×100. Notice that due to the linearity and 2-dimensionality of the present initial field, the spanwise direction does not need to be refined nor solved. The perturbation velocities are used as initial fields for the viscous simulation using an amplitude equal to 3% the centerline velocity, in order to guarantee linearity of the solution for the first few iterations.

As a reference the energy growth rate for the first few seconds of the simulation is shown in Figure 3, where time is normalized by $T = 2\pi/(\alpha\lambda_r)$. Notice that the present method over-predicts the energy growth, situation common when using 4th-order spatial stencils [28], whereas PISO predicts a decay, meaning that the algorithm behaves as simulating a smaller value of Re which does not lie on a stability curve.

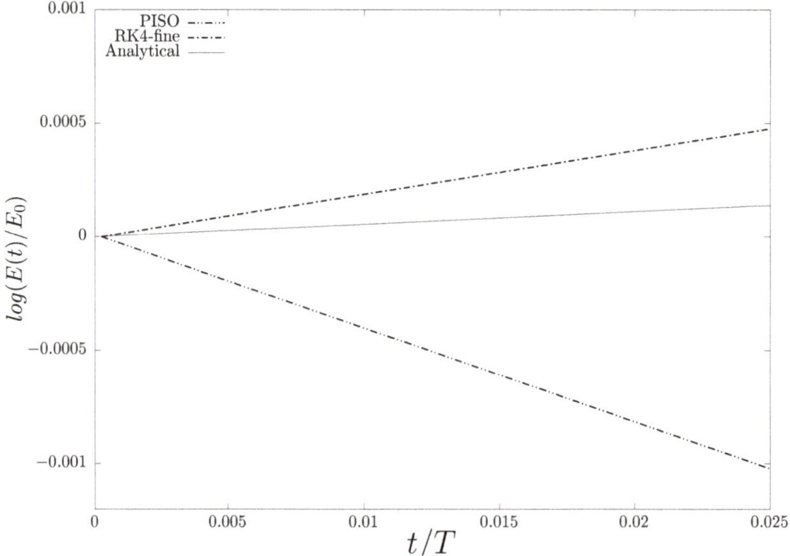

Figure 3. Energy growth rate obtained for the finest grid using RK4, and for PISO.

An order-of-accuracy analysis is made for which two additional grids are constructed by just doubling the vertical resolution of the previous one. The rate of error reduction can be calculated as

$$p \log (2) = \log \left(\frac{\phi_{medium} - \phi_{coarse}}{\phi_{fine} - \phi_{medium}} \right),$$

where ϕ represents a value over which the error is to be calculated, and p the order of error reduction. For the RK4 solver, using the data shown in Table 1, a value of $p \approx 2$ is obtained. The reduction order in this case corresponds to the second-order interpolation used for the discretization of the advective operator. Please note that the error-reduction analysis is made only for the RK4 algorithm, since for PISO it would not make sense in this case.

Table 1. Computed energy growth at $t/T = 0.025$.

Mesh Refinement	$E(t/T = 0.025) \times 10^{-5}$
coarse	1.4945492
medium	1.4947583
fine	1.4948110

5. LES of Turbulent Flows

In this Section we report results for LES of two classical turbulent problems. First we consider the canonical plane turbulent channel flow, where the two algorithms are validated for the case of neutral stratification. Successively, we study the Rayleigh–Bènard convection between two horizontal plates

RK4 is used with the Dynamic Lagrangian Mixed Model (DLMM) already described, while the simulations with PISO use the Dynamic 'Mean' Model (DMM) available in OpenFOAM. Note that in both cases we employ the non-uniform Laplacian filter described in Equation (26). Furthermore, for the Rayleigh–Bènard problem, the dynamic evaluation of the sgs scalar fluxes is used except otherwise stated.

Here, the following convention will be used to distinguish between *resolved* and *sgs* intensities (or fluctuation) of a certain quantity, ϕ,

$$\phi(x,y,z,t) = \langle\phi\rangle(x,y,z) + \phi''(x,y,z,t) + \phi^{sgs}(x,y,z,t),$$

where $\langle\cdot\rangle$ denotes a mean quantity, ϕ'' are the resolved fluctuations, and ϕ^{sgs} are the *sgs* contributions. Thus, the *total* fluctuations of ϕ are

$$\phi'(x,y,z,t) = \phi''(x,y,z,t) + \phi^{sgs}(x,y,z,t). \tag{37}$$

5.1. LES of Turbulent Poiseuille Flow: Physical Analysis

We consider a turbulent plane channel flow driven by a constant pressure gradient. The frame of reference has x in the streamwise direction, y along the cross-stream direction and z as the wall-normal direction. Here we use the DNS database reported in [29] for a channel flow at $Re_\tau = (u_\tau\delta)/\nu = 800$, where the friction velocity is $u_\tau = \sqrt{\tau_{wall}/\rho_0}$, the half height of the channel is δ, and the kinematic viscosity is ν. The grid spacing is uniform in the $x - y$ plane, whereas the grid is stretched along the z-direction to guarantee adequate resolution near the solid walls. A geometric growth function is used in which the first off-the-wall cell centroid is at $\Delta z^+ \approx 0.34$. In order to guarantee a minimum of eight grid points within $z^+ = 10$ a stretching ratio of 7% is used, following Komen et al. [17].

A total of seven simulations were run using different grid distributions in the horizontal direction, and different solvers/LES models. Table 2 summarizes the cases with their respective characteristics. A first set of cases, hereinafter baseline cases (RK4M, PISOM, PISONM, PISOuDNS), use the horizontal grid spacing constraints proposed by Choi and Moin [30], which are commonly used in the literature of wall-bounded turbulence using LES [24,31–33]. Notice that an additional control group composed of three fine grid cases (RK4Mf, PISOMf, PISONMf), using the grid spacing constraints of Komen et al. [17], is studied in order to follow the recommendations made by the aforementioned work regarding resolved LES using *commercial* codes.

Table 2. Test and control cases for the verification of the proposed methods in turbulent channel flows.

Case	Grid Dimensions	Domain Size	$(\Delta x^+, \Delta y^+)$	Solver	LES Model
RK4Mf				RK4	DLMM
PISOMf	$100 \times 100 \times 60$	$\frac{4}{3}\pi \times \frac{2}{3}\pi \times 2$	$(33.5, 16.76)$	PISO	DLMM
PISONMf					DMM
RK4M				RK4	DLMM
PISOM					DLMM
PISONM	$80 \times 80 \times 60$	$\frac{4}{3}\pi \times \frac{2}{3}\pi \times 2$	$(41.9, 20.9)$	PISO	DMM
PISOuDNS					None

Figure 4 shows the non-dimensional velocity profiles obtained with the two algorithms and different *sgs* models, together with the uDNS and the DNS reference profiles. The left panel contains data for the baseline grid, the right panel refers to the fine grid. The results obtained using the RK4 algorithm exhibit a very good agreement with reference data; on the other hand, solutions using PISO overestimate the velocity profile, although the results substantially improve with respect to the uDNS. In this kind of simulations, given a constant longitudinal pressure gradient, one obtains the nominal wall shear stress as given by the integral balance of momentum, and a dissipative algorithm overestimates the velocity profile due to an incorrect curvature in the buffer layer. For PISO algorithm, the results show that in the presence of coarser grid, the velocity profile is rather insensitive to the SGS model; on the other hand, when using a finer grid the velocity profile obtained with the eddy-viscosity dynamic model appears somewhat worse than that obtained with the dynamic mixed model.

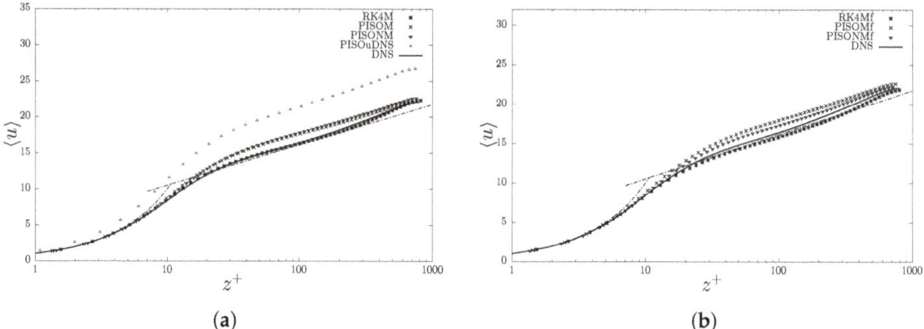

Figure 4. Mean velocity profile scaled using the friction velocity as a function of the distance from the wall in wall units. The dash-dot line represents the log-law profile $u^+ = (1/\kappa)\log y^+ + 4.9$ and the viscous boundary layer velocity profile $u^+ = y^+$. (**a**) Baseline cases; (**b**) control, or fine grid, cases.

The *rms* of the diagonal terms of the total (resolved plus SGS) Reynolds stress tensor are shown in Figure 5 for the two grids. The RK4 together with the DLMM gives satisfactory results for the three statistics. PISO exhibits a higher level of fluctuations of the streamwise velocity and smaller level of fluctuations for the other velocity components. Results similar to the ones obtained here are reported also in Komen et al. [17]; in general the over-prediction of the *rms* of the streamwise velocity component (and under-prediction of the *rms* of the other components) is due to an abnormal energy redistribution through the pressure-strain correlation of turbulent fluctuations among the three directions. The results are even worse in case of uDNS. Finally, note that the peak of the streamwise component of the Reynolds stresses for the solution obtained with PISO is slightly displaced towards the center of the channel: this may offer a reason for the incorrect resolution of the buffer layer, as previously commented. Overall, the presence of the SGS model in PISO improves the results with respect to the uDNS case, showing that the numerical dissipation does not overwhelm the effect of the model. Also, the dynamic mixed model gives second-order statistics in better agreement when compared with the performance of the dynamic eddy diffusivity model. This is particularly true in the presence of coarse grids.

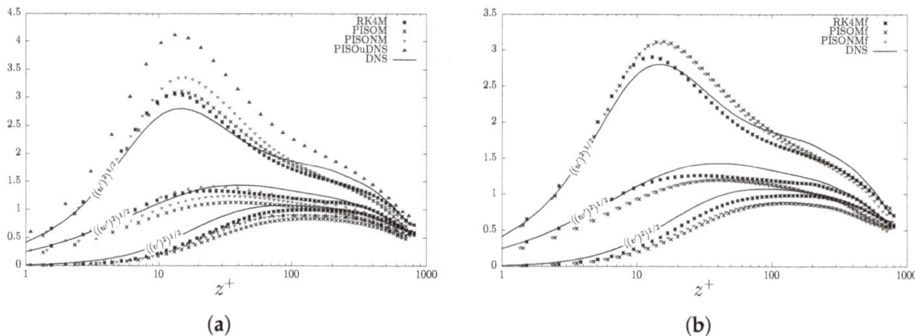

Figure 5. Total root-mean-square profiles of the velocity components (u, v, w) made non-dimensional with the friction velocity. (**a**) Baseline cases; (**b**) control, or fine grid, cases.

The mean Reynolds shear stress profiles shown in Figure 6 exhibit behavior within the ranges expected for LESs of channel flow. No particularity can be drawn between the results obtained with PISO using the turbulence model, except for a slight underestimation of τ_{xz} in the outer region of the flow for PISO with LES model. This justifies the mismatch in the mean velocity profile due to the fact that the molecular part of the shear stress appears slightly overestimated by PISO. Furthermore, the uDNS Reynolds stress may give the impression of attaining results similar to DNS but, from the velocity profiles it is clear that such results corresponds to an artificially lower Re_τ. Once again, the SGS model in the PISO algorithm improves the quality of the results compared to an under-resolved (or equivalently ILES) solution. As for the second-order statistics, the baseline cases using the DLMM seem more accurate than the results obtained with DMM; on the other hand, results obtained for the fine grid cases using DLMM are more accurate, specially for RK4. To be mentioned that the authors have run also simulations using RK4 and the DMM (not shown), obtaining similar results to those obtained with PISO and the DMM.

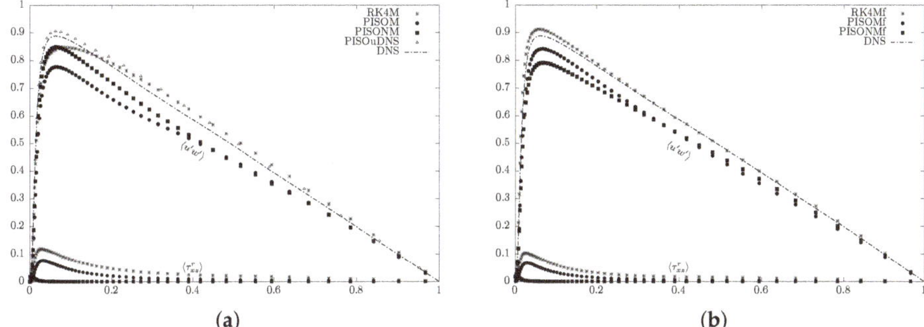

(a)   (b)

Figure 6. Total and residual (SGS) Reynolds shear stress profiles. (**a**) Baseline cases; (**b**) control, or fine grid, cases.

With the aim to analyze the behavior of the filters/LES models herein employed, we carried out an additional analysis which gives an estimation of the growth of the residual eddy viscosity and of the scale-similar shear stress close to the wall. Considering that

$$u(x,y,z,t) = a(x,y,t)z, \tag{38}$$

$$v(x,y,z,t) = b(x,y,t)z, \tag{39}$$

$$w(x,y,z,t) = c(x,y,t)z^2, \tag{40}$$

If one filters the $u-w$ components of the velocity fields using Equation (26), the following relations are obtained:

$$\overline{u} = \overline{a}z + \frac{a}{12}\left(\frac{\partial \Delta_z}{\partial z}\right),$$

$$\overline{w} = \overline{c}z^2 + \frac{c}{12}\left(\frac{\partial \Delta_z}{\partial z}z + \Delta_z\right).$$

Now, if one takes the definition of the residual stress component τ_{13}^r as

$$\tau_{13}^r = \overline{uw} - \overline{u}\,\overline{w}, \tag{41}$$

The terms on the RHS are written as:

$$\overline{uw} = \overline{ac}z^3 + \frac{3ac}{24}\left(\frac{\partial \Delta_z}{\partial z}z^2 + 2\Delta_z z\right),$$

$$\overline{u}\,\overline{w} = \frac{1}{12^2}\left(a\frac{\partial \Delta_z}{\partial z} + 12\overline{a}z\right)\left(c\Delta_z + c\frac{\partial \Delta_z}{\partial z}z + 12\overline{c}z^2\right).$$

From the same empirical relations expressed in Equations (38)–(40) it can be shown that in the vicinity of the wall [34]:

$$\tau_{13}^r = 2\nu_r \overline{S}_{13} \sim 2\nu_r \overline{a},$$

(42)

Meaning that

$$\frac{\nu_r}{\nu} \sim O(z^3 + z^2 + z).$$

(43)

The same scaling applies for the scale-similar SGS shear stress. Notice that the previous relation holds for the filter herein used; the relation may change if different filters are used, whether filtering is made only along certain directions, or whether the filtering is made in physical space (for a more complete discussion see [35]). For instance, the top-hat filter when used following the approach proposed by Jordan [36] for finite-difference/volume methods leads to the relation

$$\frac{\nu_r}{\nu} \sim O(z^3).$$

The near-wall scaling profiles of the residual viscosity as well of the scale-similar shear stress are shown in Figure 7. Please note that both the eddy diffusivity and the scale-similar contribution are in good agreement with the scaling estimate presented in Equation (43), for the DLMM algorithm proposed in the present work. The normalized turbulent viscosity profiles obtained for the DMM algorithm do not match the scaling estimates presented earlier, and seems to remain roughly constant across the channel. Notice that the sgs contribution of the DMM is about two orders of magnitude less compared to the contributions obtained with DLMM in the logarithmic region.

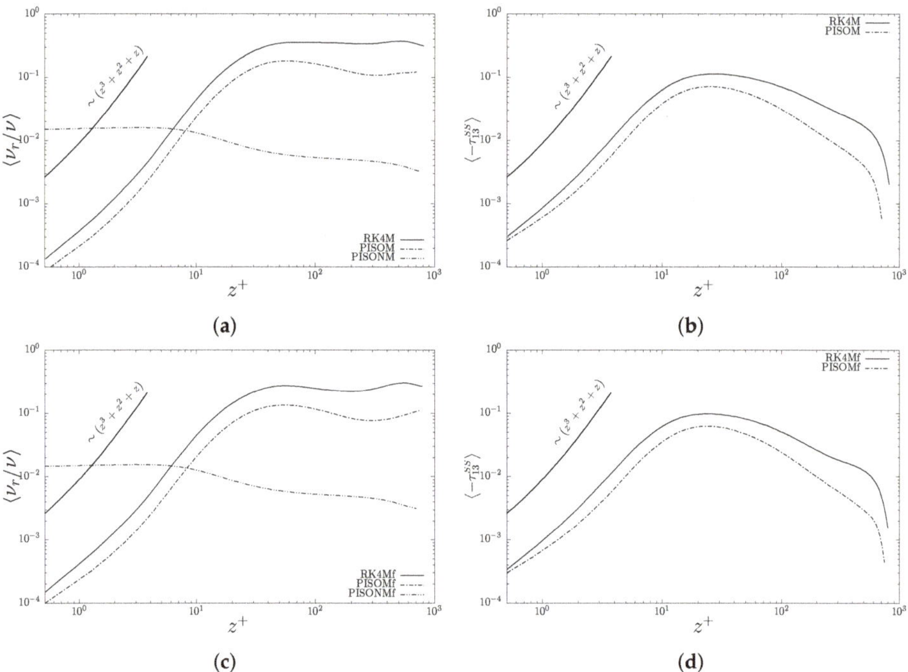

Figure 7. (**a,c**) Mean SGS viscosity profile normalized by the molecular viscosity; (**b,d**) Mean scale-similar contribution to the residual Reynolds stress.

5.2. LES of Rayleigh–Benard Convection

In this section, we analyze the performance of the two algorithms in a problem of wide interest in the scientific community, namely Rayleigh–Benard Convection (RBC). In canonical RBC the fluid is confined between two horizontal plates, the lower of which is hotter than the upper one. The Rayleigh number rules the intensity of buoyancy with respect to momentum and heat diffusion

$$Ra = \frac{U_f^2 H^2}{\nu \kappa} = \frac{g \alpha \Delta T H^3}{\nu \kappa}$$

where α is the thermal expansion coefficient, ΔT is the imposed temperature difference between lower and upper plate and H is the gap between the plates; κ is the thermal diffusivity of the fluid and $U_f = \sqrt{g \alpha \Delta T H}$ the free-fall velocity of a fluid parcel. Other parameters ruling convective processes are the Prandtl number $Pr = \nu / \kappa$ and the aspect ratio of the cell $\Gamma = L/H$ where L is the horizontal length scale of the domain. In the atmosphere, the convective motions [37] can be very energetic, with values of the Rayleigh number of the order of 10^{18}. Physical experiments with Rayleigh numbers close to that of the atmosphere have been carried out by [38]. Numerical experiments of RBC are typically carried out using Direct Numerical Simulation (DNS) or by LES. The former can afford low-to-moderate values of Rayleigh number, the latter can push the limit of Ra to larger values. LES of RBC in cubic cavities have been performed by [39] and by [40] in the presence of grooves.

Here, particular attention is paid the study of the relation between Ra, a measure of the forcing acting on the flow, and Nu which is a measure of the heat exchange properties of the flow. Such relation may be expressed in the form

$$Nu = \gamma Ra^\beta. \tag{44}$$

Such power law is not universal [41]: the γ, β coefficients may be function of other non-dimensional parameters. Universality may be assumed in cases where the aspect ratio is unimportant (namely infinite plates), the Rayleigh Number is not very large ($< 10^{11}$), and for Prandtl numbers larger than > 0.4. Comparisons with power laws calibrated via DNS [42] and physical experiments [43] will be made, in order to test the performance of the algorithms discussed in the present paper.

Here we study turbulent convection between two infinite horizontal plates, so that the aspect ratio of the domain is not a free parameter of the problem. The computational domain is $L_x = L_y = 6H, L_z = H$ and the grid is composed of $307,200$ hexahedral cells. The spacing along the horizontal directions is set constant, while the mesh is stretched along the vertical direction, in order to adequately reproduce the thermal boundary layer, δ_θ. Following the meshing strategy proposed by Verzicco and Camussi [41], at least 6 grid points have been placed within the thermal boundary layer defined as $\delta_\theta \simeq H/2 Nu$. Notice that for $Pr \leq 1$, the Kolmogorov scale, η, is lower or equal to the Batchelor length scale, η_T, but for the $Pr = 1$ herein considered these can be regarded as equal. This implies that for wall-resolving LES, the meshing criteria may be based on an accurate resolution of the thermal boundary layer, for which estimates are readily available. The distribution of the grid points is made using the hyperbolic tangent function, clustering them near the plates. The first cell node off the wall is located at $\Delta z/H = 0.001$, which is much smaller than $\delta_\theta/H \sim 1/2Nu = 0.025$ for the largest Rayleigh number herein examined ($Ra = 2 \times 10^7$). Periodic boundary conditions are imposed on the horizontal direction (to mimic infinite plates) and no-slip condition on the plates. A summary of the different settings used for the simulations ran in this section are presented in Table 3. The results of three different Rayleigh numbers are analyzed $Ra = 6.3 \times 10^5, 2 \times 10^6, 2 \times 10^7$. For the sake of clarity, normalization of the wall-normal coordinate, temperature and velocity are made in the following way:

$$\Theta = \frac{T - T_{cold}}{\Delta T},$$

$$(u, v, w) = (u_1, u_2, u_3)/U_f,$$

$$z = x_3/H,$$

where T_{cold} refers to the temperature in the cold plate, and U_f the free-fall velocity. A first level of analysis of the results obtained using the proposed algorithms can be made by calculating Nu along the vertical. The integral relation

$$Nu = \sqrt{\frac{Ra}{Pr}} \langle w\Theta + \eta_z^c \rangle - \frac{\partial \langle \Theta \rangle}{\partial z} \tag{45}$$

Dictates that the Nusselt number must remain constant along the vertical. Please note that the above relation considers the SGS scalar fluxes product of the LES formulation used.

Table 3. Test cases for the verification of the proposed methods in Rayleigh Bernard Convection.

Case	Grid Dimensions	Domain size	$(\Delta x, \Delta y, \Delta z_{\min})$	Solver	LES Model
RK4DLMM	$80 \times 80 \times 48$	$6 \times 6 \times 1$	$(0.075, 0.075, 0.001)$	RK4	DLMM
PISODLMM	$80 \times 80 \times 48$	$6 \times 6 \times 1$	$(0.075, 0.075, 0.001)$	PISO	DLMM
RK4DLMMRA	$80 \times 80 \times 48$	$6 \times 6 \times 1$	$(0.075, 0.075, 0.001)$	RK4	DLMM + RA

Vertical profiles for each of the terms just discussed, for the Ra considered in this work, are plotted in Figure 8a–c. These results for Nu lie between the ranges calculated using different empirical relations, obtained either by physical or numerical experiments, proposed in the literature [42,43]. Additionally, Figure 8d show the results obtained running the DLMM for momentum and using the RA for the scalar. Near-wall predictions of the Nusselt number are similar to the other simulations, although predictions in the core of the channel show higher SGS fluctuations than expected leading to a non-constant Nusselt profile. The latter results will not be discussed further.

A second level of analysis involves the calculation of the temperature *rms* profiles. As shown in Figure 8e, the resolved rms profiles are underestimated compared to the DNS results of [44] for the lowest of Ra. This is not surprising since the SGS temperature fluctuations are not present in the statistical quantity. The trend of temperature fluctuations seems to physically represent the thinning of the temperature boundary layer as Ra increases. In general, the solutions obtained with PISO exhibit a higher state of temperature fluctuations, as expected for over-dissipative algorithms.

For an accurate estimation of the γ, β parameters of the function $Nu = f(Ra)$, several other ways of calculating Nu are proposed in the literature:

$$Nu = 1 + \sqrt{Ra\, Pr} \langle \epsilon \rangle_{\textit{eff}}, \tag{46}$$

$$Nu = \sqrt{Ra\, Pr} \langle N \rangle, \tag{47}$$

$$Nu = \frac{\partial \langle \Theta \rangle}{\partial z}\bigg|_{wall} \tag{48}$$

where $\langle \epsilon \rangle_{\textit{eff}}$ is the *effective* averaged energy dissipation rate (*sgs* contributions included), and $\langle N \rangle$ the *resolved* averaged temperature variance dissipation rate. By volume-averaging each of the quantities just presented at every iteration, one may obtain for each Ra a distribution of Nu in time. The sampling, and time-averaging, is performed over a period of $300H/U_f$, starting from a fully developed flow field.

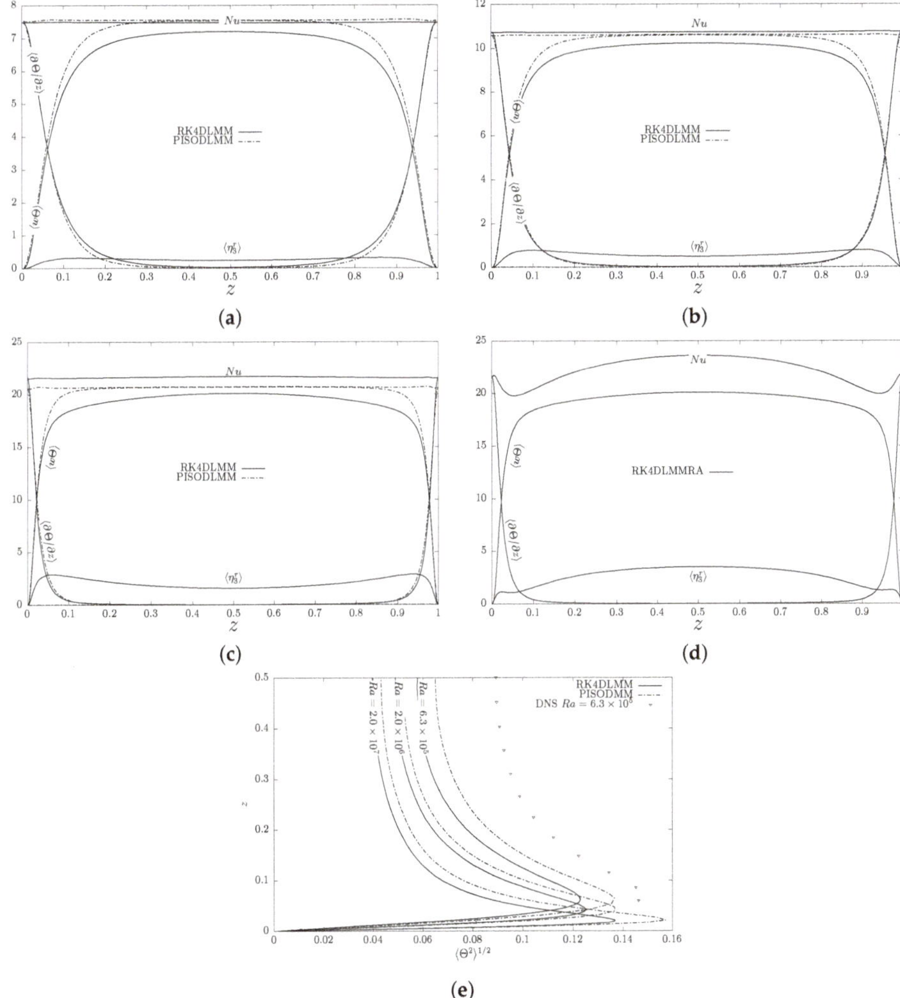

Figure 8. Vertical profiles of the *Nusselt* number, Nu, the viscous heat flux contribution $\partial\Theta/\partial z$, the turbulent contribution $\langle w\Theta \rangle$, and the *sgs* contributions for: (**a**) $Ra = 6.3 \times 10^5$; (**b**) $Ra = 2 \times 10^6$; (**c**) $Ra = 2 \times 10^7$; and (**d**) $Ra = 2 \times 10^7$ using the Reynolds Analogy; using RK4 and `pisoFoam`; Additionally, the (**e**) temperature RMS for the Ra considered are presented.

Results obtained both with `pisoFoam` and RK4, each using its corresponding turbulence models, is shown in Figure 9. The results obtained using RK4 with the Lagrangian mixed model show increasing variability in the calculation of Nu as Ra increases, whereas for PISO using the Reynolds Analogy such variability is not as strong as it can also be seen in Table 4. A least-square fit, considering the variance of Ra, for the RK4 data gives a slope β which is very similar to experiments, and the theoretical value $\beta_t = 2/7$. On the other hand the slope β for PISO is slightly higher, indicating the presence of a small amount of 'artificial' heat exchange caused by the overall model. However, the rather strong variability shown for the highest Ra using the different approaches for Nu is expected in this case, given the relative under-resolution of the mesh in the horizontal directions.

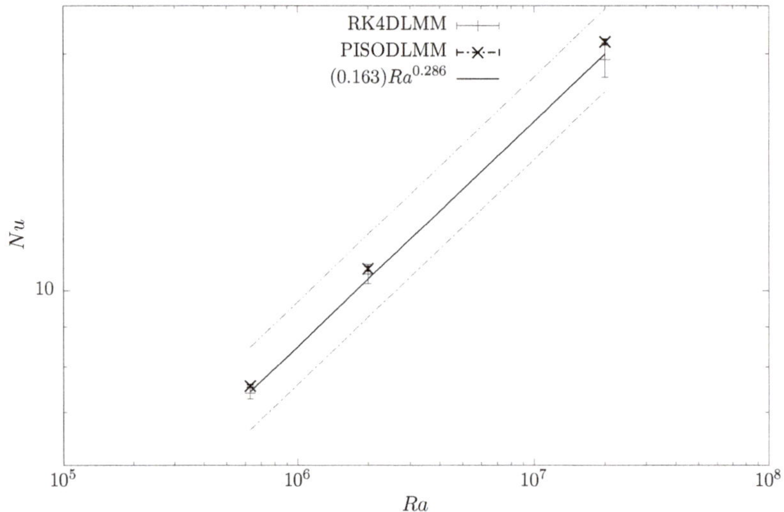

Figure 9. Nusselt number as a function of the Rayleigh number for the results obtained using PISO and RK4. Please note that the results obtained are presented with their corresponding error-bars, taken as one standard deviation of the Nu distributions obtained using different definitions. Line-dots: From [43]; Line-double dots: From [42].

Table 4. Nusselt number as a function of the Rayleigh number for the results obtained using PISO and RK4, in which the error is taken as the time standard deviation.

Ra	$Nu \pm \langle \sigma^2 \rangle^{1/2}$	
	PISO	**RK4**
6.3×10^5	7.561356 ± 0.029477	$7.4051439417 \pm 0.1218679281$
2.0×10^6	10.660694 ± 0.071748	$10.4987945729 \pm 0.2913541913$
2.0×10^7	20.700736 ± 0.168435	$19.6569584688 \pm 0.9787957580$

Typically, the Reynolds Analogy considers that mixing of temperature occurs at the same rate as mixing of momentum, namely that the turbulent Prantdl number $Pr_r = \nu_r / \kappa_r \sim O(1)$. This assumption is made explicit by the relation

$$Pr_{SGS} = \nu_{SGS} / \kappa_{SGS} = 1,$$

In the context of LES modeling. Nevertheless, the work of [21], and literature therein cited, shows that this quantity is not constant along the vertical.

The SGS Prandtl number for the three RBC flows obtained using the present algorithms is shown in Figure 10. Please note that as the Ra increases, the Pr_{SGS} increases in the core region. Nevertheless, such ratio does not remain constant across the channel height, and its variation is not monotonic. In the core of the channel, as Ra increases, SGS turbulent momentum mixing tends to be higher than its scalar counterpart. Such scenario may explain the trend of the PISO predictions shown in Figure 10: the assumption $Pr_{SGS} = 1$ under-predicts scalar turbulent mixing for the core portion of the channel, as Ra increases.

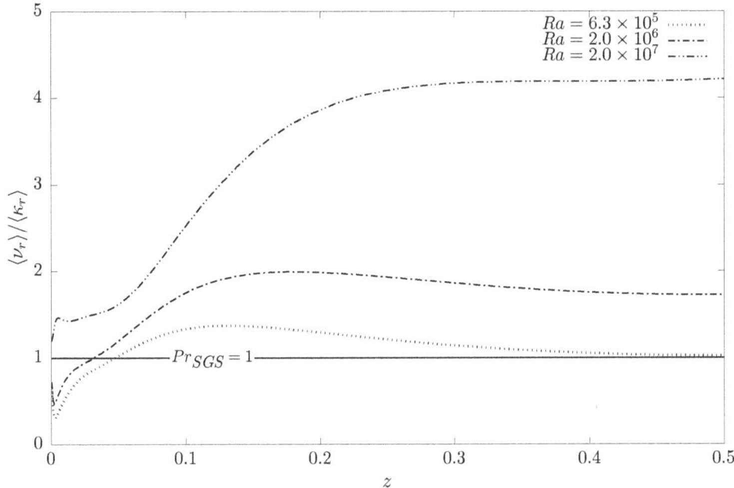

Figure 10. Profiles of Pr_{SGS} calculated using the present RK4 LES model for the scalar.

6. Conclusions

In the present paper, we compare two different solution methods for the time integration of the Boussinesq form of the NSE, namely the RK4 and PISO. The PISO along with the DMM algorithms are present within the OpenFOAM library, whereas the RK4 and the DLMM models shown in this work are not.

We first consider the standard implementation of the algorithms in the framework and study the conservation properties using literature test cases, namely the two-dimensional unsteady Taylor vortex and the growth of the linear disturbance in a plane Pouseuille flow under critical conditions. Results of these tests reveal the superiority of the RK4 algorithm over pisoFoam, since the former predicts better the time evolution of the non-linear term in both cases, whereas the latter exhibits dissipative features which may have negative consequences in the study of transitional flows.

Successively, we test the two algorithms within the Smagorinsky LES philosophy: the Dynamic Lagrangian Mixed Model (DLMM), on the other hand the Dynamic 'Mean' Model present in OpenFOAM/foam-extend. Also, in case of buoyant flows, the SGS Prandtl number is either evaluated dynamically by the DLMM, or just by setting Pr_{SGS} constant. The studies involving the DLMM and DMM use the anisotropic Laplace filter.

The use of the RK4 algorithm in conjunction with the DLMM for momentum turbulence show better results compared to those obtained using pisoFoam along with DMM. RK4 together with the DLMM provides good first- and second-order statistics for the neutral wall-bounded flow whereas the results obtained with pisoFoam, although reasonable, are less accurate. This has been attributed both the dissipative character of pisoFoam and to the particularities in the SGS models present in the distributions of OpenFOAM. Interestingly, the use of the SGS model substantially improves the performance of pisoFoam, compared to corresponding coarse DNS (or equivalently ILES) solutions, regardless the LES model. This means that although PISO being more dissipative than RK4, numerical dissipation is not able to overwhelm the effect of the SGS model, and this is probably the feature which has made the algorithm successful in the scientific and engineering communities.

In case of buoyant flows, the Nusselt number predicted using RK4 behaves somewhat better than pisoFoam, although it exhibits some deviations when calculated using different methods. Such deviation is the result of taking into account quantities for which no SGS model is implemented (i.e., the temperature variance.) The analysis of the behavior of the SGS Prandtl number, calculated dynamically within the RK4 LES methodology shows that it tends to 1 in the core region.

This clearly shows that the assumption of constant SGS Pr number may produce higher inaccuracies as the *Ra* number increases, as the cases studied here have shown.

Overall, although `pisoFoam` exhibits some dissipative features, our tests show that a consistent SGS model is able to improve the results when compared to under-resolved DNS and, as such, it can be still considered a reasonable good algorithm for LES studies. It is robust and provides good results even in the presence of unstructured non-hexahedral meshes, needed when studying very complex geometrical configuration. The RK4 algorithm appears more accurate because of its energy-conserving properties, but its own efficiency in the presence of very stable stratification or in the presence of non-hexahedral meshes has still to be verified. Also, an accurate and robust filter function to be used for non-hexahedral meshes is needed. These issues will be analyzed in upcoming research.

Author Contributions: S.L.C.: Conceptualization, Software, Validation, Writing–Original Draft, Formal Analysis. V.A. & A.P.: Supervision, Writing–Review & Editing. V.A.: Investigation, Formal analysis, Funding acquisition. G.P.: Validation.

Funding: This project received no external funding.

Acknowledgments: The present research has been supported by Region FVG-POR FESR 2014-2020, Fondo Europeo di Sviluppo Regionale, Project PRELICA "Metodologie avanzate per la progettazione idroacustica dell'elica navale".

Conflicts of Interest: The authors declare no conflict of interest.

References

1. Chorin, A.J. Numerical Solution of the Navier-Stokes Equations. *Math. Comp.* **1968**, *23*, 341–362. [CrossRef]
2. Temam, R. Une methode d'approximation de la solution des equations de Navier-Stokes. *Bull. Soc. Math. Fr.* **1968**, *96*, 115–152. [CrossRef]
3. Kim, J.; Moin, P. Application of a fractional-step method to incompressible Navier-Stokes equations. *J. Comput. Phys.* **1985**, *59*, 308–323. [CrossRef]
4. Le, H.; Moin, P. An improvement of fractional step methods for the incompressible Navier-Stokes equations. *J. Comput. Phys.* **1991**, *92*, 369–379. [CrossRef]
5. Zang, Y.; Street, R.L.; Koseff, J.R. A non-staggered Grid, fractional step method for time-dependent incompressible Navier-Stokes equations in Curvilinear Coordinates. *J. Comp. Phys.* **1994**, *114*, 18–33. [CrossRef]
6. Zang, Y.; Street, R.L.; Koseff, J.R. A dynamic mixed subgrid-scale model and its application to turbulent recirculating flows. *Phys. Fluids A Fluid Dyn.* **1993**, *5*, 3186–3196. [CrossRef]
7. Armenio, V. Large Eddy Simulation in Hydraulic Engineering: Examples of Laboratory-Scale Numerical Experiments. *J. Hyd. Eng.* **2017**, *143*. [CrossRef]
8. Mahesh, K.; Constantinescu, G.; Moin, P. A numerical method for large-eddy simulation in complex geometries. *J. Comput. Phys.* **2004**, *197*, 215–240. [CrossRef]
9. Issa, R.I. Solution of the implicitly discretised fluid flow equations by operator-splitting. *J. Comput. Phys.* **1986**, *62*, 40–65. [CrossRef]
10. Jasak, H. Error Analysis and Estimation for the Finite Volume Method with Applications to Fluid Flows. Ph.D. Thesis, University of London, London, UK, 1996.
11. Chen, G.; Xiong, Q.; Morris, P.J.; Paterson, E.G.; Sergeev, A.; Wang, Y. OpenFOAM for computational fluid dynamics. *Not. AMS* **2014**, *61*, 354–363. [CrossRef]
12. Weller, H.G.; Tabor, G.; Jasak, H.; Fureby, C. A tensorial approach to computational continuum mechanics using object-oriented techniques. *Comput. Phys.* **1998**, *12*, 620–631. [CrossRef]
13. Grinstein, F.F.; Margolin, L.G.; Rider, W.J. *Implicit Large Eddy Simulation: Computing Turbulent Fluid Dynamics*; Cambridge University Press: Cambridge, UK, 2007.
14. Lysenko, D.A.; Ertesvåg, I.S.; Rian, K.E. Large-eddy simulation of the flow over a circular cylinder at Reynolds number 3900 using the OpenFOAM toolbox. *Flow Turbul. Combust.* **2012**, *89*, 491–518. [CrossRef]
15. Vuorinen, V.; Keskinen, J.P.; Duwig, C.; Boersma, B. On the implementation of low-dissipative Runge–Kutta projection methods for time dependent flows using OpenFOAM®. *Comput. Fluids* **2014**, *93*, 153–163. [CrossRef]

16. D'Alessandro, V.; Binci, L.; Montelpare, S.; Ricci, R. On the development of OpenFOAM solvers based on explicit and implicit high-order Runge–Kutta schemes for incompressible flows with heat transfer. *Comput. Phys. Commun.* **2018**, *222*, 14–30. [CrossRef]
17. Komen, E.; Camilo, L.; Shams, A.; Geurts, B.J.; Koren, B. A quantification method for numerical dissipation in quasi-DNS and under-resolved DNS, and effects of numerical dissipation in quasi-DNS and under-resolved DNS of turbulent channel flows. *J. Comput. Phys.* **2017**, *345*, 565–595. [CrossRef]
18. Tuković, Ž.; Perić, M.; Jasak, H. Consistent second-order time-accurate non-iterative PISO-algorithm. *Comput. Fluids* **2018**, *166*, 78–85. [CrossRef]
19. Meneveau, C.; Lund, T.S.; Cabot, W.H. A Lagrangian dynamic subgrid-scale model of turbulence. *J. Fluid Mech.* **1996**, *319*, 353–385. [CrossRef]
20. Vreman, B.; Geurts, B.; Kuerten, H. On the formulation of the dynamic mixed subgrid-scale model. *Phys. Fluids* **1994**, *6*, 4057–4059. [CrossRef]
21. Armenio, V.; Sarkar, S. An Investigation of stably stratified flows using large-eddy simulation. *J. Fluid Mech.* **2002**, *459*, 1–42. [CrossRef]
22. Vasilyev, O.V.; Lund, T.S.; Moin, P. A general class of commutative filters for LES in complex geometries. *J. Comput. Phys.* **1998**, *146*, 82–104. [CrossRef]
23. Germano, M. Turbulence: The filtering approach. *J. Fluid Mech.* **1992**, *238*, 325–336. [CrossRef]
24. Germano, M.; Piomelli, U.; Moin, P.; Cabot, W.H. A dynamic subgrid-scale eddy viscosity model. *Phys. Fluids A Fluid Dyn.* **1991**, *3*, 1760–1765. [CrossRef]
25. Rhie, C.; Chow, W.L. Numerical study of the turbulent flow past an airfoil with trailing edge separation. *AIAA J.* **1983**, *21*, 1525–1532. [CrossRef]
26. Guermond, J.L.; Minev, P.; Shen, J. An overview of projection methods for incompressible flows. *Comput. Methods Appl. Mech. Eng.* **2006**, *195*, 6011–6045. [CrossRef]
27. Gresho, P.M.; Sani, R.L. On Pressure boundary conditions for the incompressible Navier-Stokes Equations. *Intl. J. Num. Meth. Fluid* **1987**, *7*, 1111–1145. [CrossRef]
28. Kaltenbach, H.J.; Driller, D. Phase-error reduction in large-eddy simulation using a compact scheme. In *Advances in LES of Complex Flows*; Springer: Berlin/Heidelberg, Germany, 2002; pp. 83–98.
29. Tanahashi, M.; Kang, S.J.; Miyamoto, T.; Shiokawa, S.; Miyauchi, T. Scaling law of fine scale eddies in turbulent channel flows up to $Re\tau$ = 800. *Int. J. Heat Fluid Flow* **2004**, *25*, 331–340. [CrossRef]
30. Choi, H.; Moin, P. Grid-point requirements for large eddy simulation: Chapman's estimates revisited. *Phys. Fluids* **2012**, *24*, 011702. [CrossRef]
31. Rasam, A.; Brethouwer, G.; Schlatter, P.; Li, Q.; Johansson, A.V. Effects of modelling, resolution and anisotropy of subgrid-scales on large eddy simulations of channel flow. *J. Turbul.* **2011**, *12*, N10. [CrossRef]
32. Armenio, V.; Piomelli, U. A Lagrangian Mixed Subgrid-Scale Model in Generalized Coordinates. *Flow Turbul. Combust.* **2000**, *65*, 51–81. [CrossRef]
33. Chong, M.S.; Monty, J.P.; Marusic, I. The topology of surface of surface skin friction and vorticity fields in wall-bounded flows. *J. Turbul.* **2011**, *13*, N6. [CrossRef]
34. Pope, S.B. *Turbulent Flows*; Cambridge University Press: Cambridge, UK, 2000.
35. Piomelli, U. High Reynolds number calculations using the dynamic subgrid-scale stress model. *Phys. Fluids A Fluid Dyn.* **1993**, *5*, 1484–1490. [CrossRef]
36. Jordan, S.A. A large-eddy simulation methodology in generalized curvilinear coordinates. *J. Comput. Phys.* **1999**, *148*, 322–340. [CrossRef]
37. Chillà, F.; Schumacher, J. New perspectives in turbulent Rayleigh-Bénard convection. *Eur. Phys. J. E* **2012**, *35*, 58. [CrossRef] [PubMed]
38. Niemela, J.; Skrbek, L.; Sreenivasan, K.; Donnelly, R. Turbulent convection at very high Rayleigh numbers. *Nature* **2000**, *404*, 837. [CrossRef] [PubMed]
39. Foroozani, N.; Niemela, J.J.; Armenio, V.; Sreenivasan, K.R. Influence of container shape on scaling of turbulent fluctuations in convection. *Phys. Rev. E* **2014**, *90*, 063003. [CrossRef] [PubMed]
40. Foroozani, N.; Niemela, J.; Armenio, V.; Sreenivasan, K. Turbulent convection and large scale circulation in a cube with rough horizontal surfaces. *Phys. Rev. E* **2019**, *99*, 033116. [CrossRef] [PubMed]
41. Verzicco, R.; Camussi, R. Numerical experiments on strongly turbulent thermal convection in a slender cylindrical cell. *J. Fluid Mech.* **2003**, *477*, 19–49. [CrossRef]
42. Kerr, R.M. Rayleigh number scaling in numerical convection. *J. Fluid Mech.* **1996**, *310*, 139–179. [CrossRef]

43. Wu, X.Z.; Libchaber, A. Scaling relations in thermal turbulence: The aspect-ratio dependence. *Phys. Rev. A* **1992**, *45*, 842. [CrossRef] [PubMed]
44. Kimmel, S.J.; Domaradzki, J.A. Large eddy simulations of Rayleigh–Bénard convection using subgrid scale estimation model. *Phys. Fluids* **2000**, *12*, 169–184. [CrossRef]

MDPI

St. Alban-Anlage 66

4052 Basel

Switzerland

Tel. +41 61 683 77 34

Fax +41 61 302 89 18

www.mdpi.com

Fluids Editorial Office

E-mail: fluids@mdpi.com

www.mdpi.com/journal/fluids